POLITICAL FALLOUT

POLITICAL FALLOUT

Nuclear Weapons Testing and
the Making of a Global Environmental Crisis

Toshihiro Higuchi

STANFORD UNIVERSITY PRESS — STANFORD, CALIFORNIA

STANFORD UNIVERSITY PRESS
Stanford, California

© 2020 by the Board of Trustees of the Leland Stanford Junior University. All rights reserved.

No part of this book may be reproduced or transmitted in any form or by any means, electronic or mechanical, including photocopying and recording, or in any information storage or retrieval system without the prior written permission of Stanford University Press.

Printed in the United States of America on acid-free, archival-quality paper

Library of Congress Control Number: 2020930995

Cover design: Amanda Weiss

Text design: Kevin Barrett Kane

Typeset by at Stanford University Press in 10/14 Sabon LT Pro

to TOMIO HIGUCHI,
*who opened my eyes to
the world beyond the seas*

CONTENTS

Figures and Tables ix
Acknowledgments xi
Abbreviations xv

INTRODUCTION
"To Put an End to the Contamination of Man's Environment"   1

CHAPTER 1
A State of Emergency: The Origins of Radiation Protection in
    Nuclear Weapons Testing, 1945–1953   16

CHAPTER 2
"Atomic-Bomb Tuna": The Trans-Pacific Politics of Radiation
    Protection Standards, 1954–1955   41

CHAPTER 3
Epistemic Stalemate: Genetics and the Creation of
    Scientific Committees, 1954–1955   61

CHAPTER 4
Epistemic Divide: The U.S. and British Scientific
    Committees, 1955–1956   88

CHAPTER 5
Epistemic Negotiations: The United Nations Scientific
   Committee, 1956–1958   109

CHAPTER 6
The Local Turn: Community-Based Fallout Surveys in the
   United States and Britain, 1958–1960   136

CHAPTER 7
Fallout: The Making of the Partial Test Ban Treaty, 1961–1963   162

CONCLUSION
"We Can Live in Strength Without Adding to the Hazards
   of Life on This Planet"   190

Notes 201
Glossary 255
Bibliography 259
Index 287

FIGURES AND TABLES

Figure 1. Tests of Nuclear Weapons in the Atmosphere
 and Underground — 3

Figure 2. Locations of Major Atmospheric Nuclear
 Weapons Test Sites in Operation Through 1963 — 18

Figure 3. Areas of Operation for Tuna Catches Registering
 Over 100 cpm, March 16 to August 31, 1954 — 50

Figure 4. Locations of Strontium-90 Surface
 Deposition Sampling, 1956 — 116

Figure 5. "Dr. Spock Is Worried." — 179

Table 1. Results of Tuna Inspection at Seaports, March
 to December 1954 — 49

Table 2. Guidance on Daily Intake (picoCi per day) — 173

ACKNOWLEDGMENTS

I first encountered *hibakusha*, a Japanese word for the atomic bomb victim, during elementary school. Each year, schools across Japan organize trips to Hiroshima and Nagasaki. I did not go there: my school was located too far away to take small children to the cities. My encounter with *hibakusha* was more accidental—and more haunting. One sunny day at the school library, I stumbled upon a large picture book shelved at an obscure corner of the room. I do not recall the title of the book, but it must have been *Hiroshima No Pika*, written and illustrated by a noted Japanese author-illustrator named Toshi Maruki. In it, I saw people smashed into pieces and burned alive; I saw people begging for water, with their skin peeling off of their bodies; I saw piles of dead bodies laid on the ground as far as could be seen. As a kid who grew up in a peaceful suburb of greater Tokyo decades after the atomic bombings, I was stunned speechless. I quietly put the book back on the shelf and left the library, but the lingering images of the *hibakusha* never left me. At the same time, I felt a reassuring sense of distance, both in time and space, from the atomic bomb victims. When I eventually returned to the story of nuclear weapons during graduate school, I was surprised to learn that the radioactive particles from nuclear weapons tests in the atmosphere were still present, albeit in tiny amounts, in all parts of the world, continuously irradiating us from within and outside. I discovered that I, too, was *hibakusha*.

ACKNOWLEDGMENTS

The long journey from my learning of global radioactive fallout to writing this book would have not been possible without the support of my mentors and colleagues at, and the staff of, the History Department and the School of Foreign Service at Georgetown University. David Painter has generously shared his vast knowledge in U.S. foreign relations and provided detailed, line-by-line feedback. His warm encouragement and good humor have always brightened me up even in the most challenging times. John McNeill opened my eye to the environmental dimensions of nuclear affairs and has constantly challenged me to see a big picture—as big as the history of Earth. It has been such a pleasure for me to work with Kathryn Olesko, who introduced me to the exciting history of science and has helped me sharpen up the theoretical focus of my book. I would also like to thank Gregory Afinogenov, Katherine Benton-Cohen, Ananya Chakravarti, Michael David-Fox, Dagomar Degroot, Anthony Eames, and Timothy Newfield for their input and insight that improved the manuscript significantly. Timothy Jorgensen kindly read a draft of the entire manuscript and gave me invaluable advice as an expert of radiation medicine and a science writer. I deeply appreciate his generosity in sharing a masterfully crafted glossary of radiation units, reprinted in this book with his permission. I am also deeply appreciative of the administrative support provided by the able staff, including Amy Chidester, Kathleen Gallagher, Jan Liverance, Sarah Nebbeling, and Djuana Shields.

My project took me to many different academic institutions and conferences, where I met many excellent scholars and made new friends. Sumio Hatano taught me a joy of studying history and encouraged me to pursue my newly found passion outside Japan. Lawrence Wittner has provided helpful advice and warm support for me ever since my first arrival in the United States. David Holloway made me feel at home during a postdoctoral fellowship at the Center for International Security and Cooperation (CISAC), Stanford University. A deep thinker, sincere, and genuinely caring, he has been my role model as a scholar and person. I would like to thank the CISAC faculty and affiliates, especially Lynn Eden, Charles Perrow, and Scott Sagan, for lively conversations during and outside seminars that stimulated my thought greatly. My special thanks go to Gregg Mitman in Madison, who gave generously of his time, feedback, and help in my networking across campus. I am also grateful to the faculty and students of the former Department of History of Science, Medicine, and

Technology and the Center for Culture, History, and Environment (CHE) for enriching my understanding of the history of science and environmental humanities. I am deeply indebted to Hiroshi Nakanishi in Kyoto, who kindly hosted me during my stint at the Hakubi Project. Thanks to the leadership and wisdom of Toshitaka Hori and Koji Tanaka, I could make significant progress with my project. I would like to thank all scholars who read different versions of chapters over many years, especially Kate Brown, Angela Creager, Jacob Darwin Hamblin, Hiroshi Ichikawa, J. Samuel Walker, Audra Wolfe, and Masakatsu Yamazaki. I am grateful to Michael Egan and Rachel Rothschild for opportunities to present my work in front of the engaging audiences at their respective institutions. I also want to extend my special thanks to many friends who have supported me both academically and personally, including Mara Drogan, Robert Forrest, Jonathan Hunt, E. Jerry Jessee, Reo Matsuzaki, Evelyn Krache-Morris, Benoît Pelopidas, Linda Richards, Anand Toprani, and Kristoffer Whitney.

I could not carry out my multinational, multi-archival project without generous funding and research support provided by a number of institutions and individuals. I was fortunate to receive a Mellon/ACLS Dissertation Completion Fellowship; a Mellon/ACLS Early Career Fellowship; and travel grants provided by the Eisenhower Foundation, the John F. Kennedy Library Foundation, the Cosmos Club in Washington D.C., and the Peter and Judith Freeman Fund for the Resident Scholar Program at Oregon State University Libraries Special Collections. For help with research at libraries and archives, I thank Larisa Petrovna Belozerova (S. I. Vavilov Institute of the History of Natural Science and Technology), Christopher Clement (International Commission on Radiological Protection), Malcolm Crick (United Nations Scientific Committee on the Effects of Atomic Radiation), Janice Goldblum (Archives of the National Academies), David A. Schauer (National Council on Radiation Protection and Measurements), and Irina G. Tarakanova (Archive of the Russian Academy of Sciences). I am especially grateful to Paul Simmons for assisting me during my archival research in Moscow. Without his support and expertise, I could have accomplished little during my short-term visits in the unfamiliar settings.

An earlier version of Chapter 2 and part of Chapter 5 appeared in "The Strange Career of Dr. Fish: Yoshio Hiyama, Radioactive Fallout, and Nuclear Fear Management in Japan, 1954–1958," *Historia Scientiarum* 25, no. 1 (2015): 57–77. Chapters 3, 4, and 5 include material from

"Epistemic Frictions: Radioactive Fallout, Health Risk Assessments, and the Eisenhower Administration's Nuclear-Test Ban Policy, 1954–1958," *International Relations of the Asia-Pacific* 18, no. 1 (2018): 99–124. I thank the publishers for permission.

Finally, on a personal note, I would like to thank Paul Fengler, who helped me polish chapters many times over. I am truly blessed to have him as my best friend. I am thankful to my parents, Shiro and Kuniko Higuchi, and my sister, Kaori Suzuki. Their support, love, and understanding from halfway around the world always remind me that there are things other than radioactive fallout that connect us all.

ABBREVIATIONS

AAAS	American Association for the Advancement of Science
ABCC	Atomic Bomb Casualty Commission (U.S.; Japan)
ABERC	Atomic Bomb Effect Research Commission (Japan)
ACXRP	Advisory Committee on X-ray and Radium Protection (U.S.)
ADPC	Atomic Development Problems Committee (U.S., Minnesota)
AERE	Atomic Energy Research Establishment (UK)
ARC	Agricultural Research Council (UK)
ASA	Atomic Scientists' Association (UK)
BEAR	Biological Effects of Atomic Radiation (Committees on, U.S.)
C-14	carbon-14
Ci	curie
CND	Campaign for Nuclear Disarmament (UK)
cpm	counts per minute
CU	Consumers Union (U.S.)
ENCD	Eighteen-Nation Committee on Disarmament
FAS	Federation of American Scientists

FDA	Food and Drug Administration (U.S.)
FRC	Federal Radiation Council (U.S.)
GAC	General Advisory Committee (USAEC)
GSA	Genetics Society of America
Gy	gray (unit)
HASL	Health and Safety Laboratory (U.S.)
I-131	iodine-131
ICRP	International Commission on Radiological Protection
IXRPC	International X-ray and Radium Protection Committee
JCAE	Joint Committee on Atomic Energy (U.S.)
kt	kilotons of TNT
LIPAN	Laboratory of Measuring Instruments of the USSR Academy of Sciences
LNT	linear non-threshold
MAF	Ministry of Agriculture and Forestry (Japan)
MAFF	Ministry of Agriculture, Fisheries and Food (UK)
MHW	Ministry of Health and Welfare (Japan)
milliGy	milligray
milliR	milliroentgen
MPB	maximum permissible body burden
MRC	Medical Research Council (UK)
Mt	megatons of TNT
MWB	Metropolitan Water Board (UK)
NACOR	National Advisory Committee on Radiation (U.S.)
NAS	National Academy of Sciences (U.S.)
NATO	North Atlantic Treaty Organization
NCANWT	National Council for the Abolition of Nuclear Weapon Tests (UK)
NCRP	National Committee on Radiation Protection (U.S.)
picoCi	picocurie
PTBT	Partial Test Ban Treaty
R	roentgen
SANE	National Committee for a Sane Nuclear Policy (U.S.)

SCJ	Science Council of Japan
SI	Système international d'unités (international system of units)
Sr-89	strontium-89
Sr-90	strontium-90
SU	Strontium Unit
UKAEA	UK Atomic Energy Authority
UNSCEAR	United Nations Scientific Committee on the Effects of Atomic Radiation
USAEC	U.S. Atomic Energy Commission
USPHS	U.S. Public Health Service
VNIIEF	All-Union Scientific Research Institute of Experimental Physics (USSR)
WARI	Women's Association for Radiation Information (UK)
WSP	Women Strike for Peace (U.S.)

POLITICAL FALLOUT

Introduction

"TO PUT AN END TO THE CONTAMINATION OF MAN'S ENVIRONMENT"

ON AUGUST 5, 1963, U.S. Secretary of State Dean Rusk, Soviet Foreign Minister Andrei Gromyko, and British Foreign Secretary Alec Douglas-Home gathered in St. Catherine Hall, a great chamber of the Kremlin with marble columns and a vaulted ceiling. With United Nations Secretary General U Thant and seventy U.S., Soviet, and British officials looking on, the three foreign ministers signed a treaty that prohibited the testing of nuclear weapons in the atmosphere, in outer space, and underwater. In the preamble, the three governments proclaimed "as their principal aim the speediest possible achievement of an agreement on general and complete disarmament under strict international control."[1] The Partial Test Ban Treaty (PTBT) came on the heels of one of the most dangerous moments of the Cold War. Just months earlier, in October 1962, the United States and the Soviet Union had come within a hairbreadth of war over the installation of nuclear-armed Soviet missiles on Cuba. The PTBT provided the anxious world with a much-needed respite from the threat of nuclear war. After the signing of this landmark arms control agreement, the dignitaries from the West and the Soviet Union exchanged firm handshakes, light-hearted jokes, and numerous toasts to "peace and friendship." *The New York Times* quoted one diplomat as calling it a "unique day." Another seconded, "Peace—it's wonderful."[2]

The joyful mood at the signing ceremony in the Kremlin, however, belied a deep sense of disappointment. Since Indian Prime Minister Jawaharlal Nehru first called for a nuclear-test ban in 1954, the United States, Britain, and the Soviet Union had unsuccessfully tried to negotiate the prohibition of all tests under international control. In their view, then, the PTBT was partial in every sense of the word. Indeed, it soon became clear that the treaty stopped neither the arms race between the superpowers nor the spread of nuclear weapons to more countries. The United States and the Soviet Union simply went on to test nuclear weapons underground, whereas France and China refused to sign the PTBT and continued to develop their nuclear capabilities. For this reason, historians have typically viewed the PTBT as a significant yet ultimately unsuccessful disarmament initiative. The existing literature on the test-ban negotiations tends to focus on issues related to arms control, including an irresolvable stalemate over the methods of verification, the growing threat of nuclear proliferation, the changing dynamic of alliance politics, and the significant impact of the nuclear disarmament movement.[3]

The Partial Test Ban Treaty, however, was much more than a missed opportunity to stop the nuclear arms race. Both in intention and effect, it was also one of the first multinational treaties concluded during the Cold War that directly addressed a truly global, human-induced environmental issue. From 1945 to 1963, the United States, the Soviet Union, and Britain conducted 435 nuclear weapons tests in the open environment. The total yield of these explosions was approximately 400 megatons of TNT, or over twenty-six thousand Hiroshima-size bombs (Figure 1). The release of this enormous energy set in motion a collateral and dangerous process. The fragments left after the splitting of uranium and plutonium nuclei, the unused portion of the fissile material, and chemical elements capturing neutrons from nuclear reactions all tended to be radioactive. Each radionuclide would spontaneously transmute into a more stable form by emitting high-energy helium nuclei (alpha particles), high-speed electrons (beta particles), and/or penetrating electromagnetic beams (gamma rays). This type of energy, called ionizing radiation, was biologically harmful, injuring or killing organisms by removing electrons from molecules in cells through which it passed.

Part of the radioactive debris blown up by a nuclear explosion would quickly settle ("fall out") to the ground at the test site and not far downwind. The vast majority of the fine dust and gases dispersed widely, however, either thoroughly mixed with the world's atmosphere and oceans or slowly deposited on parts of or even the whole planet.[4] In the preamble of the PTBT, the three governments noted this fact, pledging to "put an end to the contamination of man's environment by radioactive substances."[5] Ultimately, the treaty proved far more effective in its environmental objective than that of arms control. Although France and China continued to conduct atmospheric testing until 1974 and 1980, respectively, the release

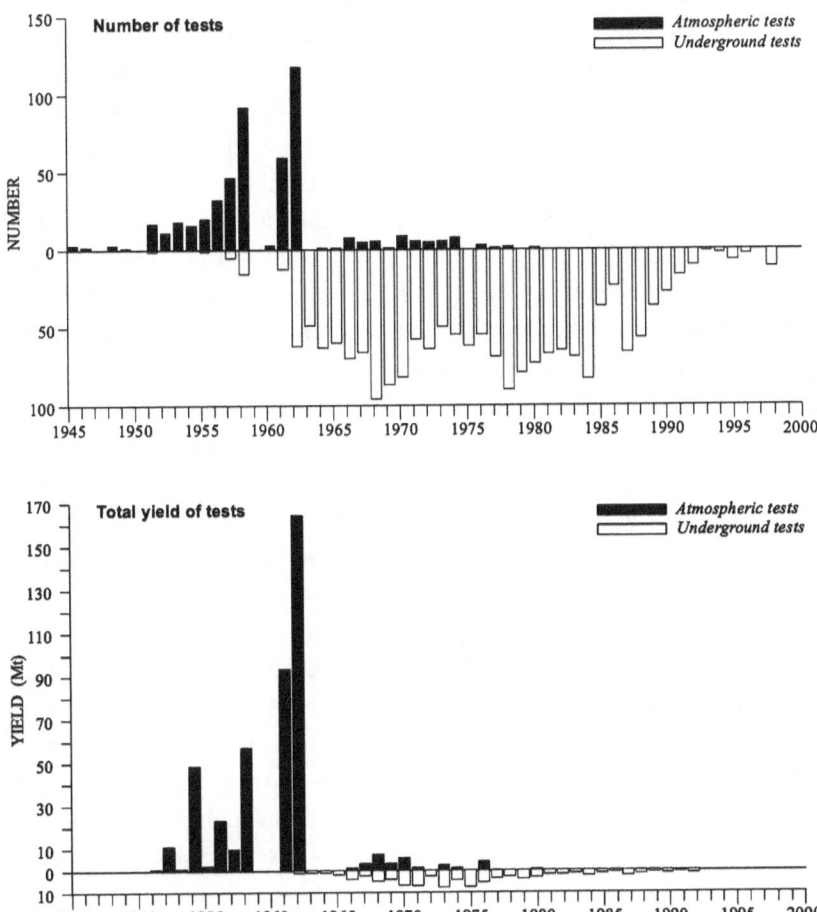

FIGURE 1. Tests of Nuclear Weapons in the Atmosphere and Underground.
Source: *UNSCEAR 2000 Report*, Vol. 1, 160. Reprinted with permission.

of radioactive fallout into the biosphere drastically declined after 1963 (see Figure 1). By concealing the public health and environmental costs of nuclear weapons testing deep underground, the PTBT ironically made the nuclear arms race between the superpowers environmentally "sustainable" for the rest of the Cold War.

Despite its limitations, the PTBT's explicit reference to global radioactive contamination is nevertheless notable for three reasons. First, despite the growing scholarship showing the contrary, the era of unrestricted nuclear testing from 1945 to 1963 is still typically not remembered for environmental consciousness. The PTBT was concluded barely a year after Rachel Carson published *Silent Spring*, an immensely influential book often credited for the rise of modern environmentalism in the United States.[6] Internationally as well, it was not until 1972 that the United Nations hosted its first global environmental summit, in Stockholm.[7] Second, the agreement concerned the nuclear arms race. Unlike most other environmental issues, nuclear weapons were one of the most prominent national security issues that directly contributed to superpower hostility.[8] Finally, the agreement came decades before the issue came to light of the suffering of military personnel and local residents exposed to radiation at much higher doses while present at or near the test sites or downwind than those farther away. Although many of them had long complained about their illnesses from the time of irradiation, it was not until the 1980s and early 1990s that U.S., British, and post-Soviet authorities finally acknowledged at least some of the reported health problems among the atomic veterans and downwinders.[9] If so, then why did the radioactive debris of atmospheric nuclear testing scattered worldwide come to be widely viewed as an unacceptable hazard by as early as 1963? And to what extent, and how, did this shifting understanding of global radioactive contamination affect the course of the test-ban negotiations toward the PTBT?

The question of radioactive fallout and its foreign policy impact is crucial to understanding the entangled relationship between the Cold War and the global environment. The material world in which the international tension between the United States, the Soviet Union, and their respective allies unfolded was neither static nor immobile. Scientists have suggested that Earth has entered a new interval of geological time in which humans have become the dominant influence on the planet's interlocking physical and biological processes. Dutch atmospheric chemist and Nobel laureate Paul Crutzen has popularized the term *Anthropocene* to describe this era of human-driven global environmental change.[10] Although there has been a divergence of views on the origins of the Anthropocene, a common view holds that the human footprint on the planet became unmistakable in the second half of the twentieth century. The exponential growth in food production, fossil fuel use, and other human enterprises during this period drove a wide range of global environmental trends upward, including the atmospheric concentrations of gases of biogeochemical importance, ocean acidification, stratospheric ozone depletion, and biodiversity loss. The Cold War, spanning much of the late twentieth century, coincided with this latest phase of the Anthropocene, called the "Great Acceleration."[11]

The Cold War and the Anthropocene were not merely contemporaneous but also deeply connected through various intersections.[12] Some historians have explained how the global competition between capitalism and communism accelerated the proliferation of large-scale public health and development projects that powerfully reshaped the human environment around the world.[13] Others have pointed out that at times the superpower rivalry also led to greater management of the trans-border environment through international cooperation and, in the case of the Korean peninsula, through the creation of the demilitarized zone.[14]

The Cold War, however, did much more for the Anthropocene than create a favorable geopolitical condition for global economic development and environmental management. The permanent state of military preparedness that characterized the Cold War also had a strong and direct relationship with the enhanced human impact on the planet and the concurrent rise of environmental consciousness. Jacob Darwin Hamblin has demonstrated how a massive research effort by the United States and its NATO allies for environmental warfare against the Soviet Union raised awareness about catastrophic consequences of human-induced environmental change.[15] David Zierler has explained how the destructive use of defoliants by the United States during the Vietnam War sparked a movement to outlaw herbicide warfare.[16]

Of all Cold War military activities, atmospheric nuclear testing stood out for its truly global and long-term environmental impact. Driven by the rising tension between the United States, Britain, and the Soviet Union over the post–World War II settlement, the nuclear arms race led to a staggering increase in the rate of fallout emissions in the atmosphere. If we take the annual yield of atmospheric tests as a proxy measure of radioactivity, it increased three thousand times from 1945 (57 kilotons) to the peak year of 1962 (170 megatons), even with a moratorium on all types of nuclear testing between November 1958 and August 1961.[17] Each explosion scattered a massive amount of radioactive substances over large areas, slightly but steadily raising radiation levels around the world. In this sense, the Cold War radically differed from other conflicts: it was an Anthropocenic war. The fast-changing dynamic of the nuclear arms race not only pushed the whole world to the brink of nuclear war but also simultaneously altered the material conditions of Earth with potentially far-reaching implications for human health and the environment.

Indeed, an increasing number of scientists have recognized the geological utility of radioactive fallout deposited in Earth's crust as a reliable marker to define Anthropocene strata.[18] Nonetheless, many scholars still tend to discount the importance of fallout to understanding the enhanced human impact on the structure and functioning of our planet, noting that the average dose of radiation from the bomb debris scattered worldwide was no more than a fraction of that from natural background radiation.[19] Radioactive fallout, however, offers an important clue for exploring the prefix *Anthropos*, that is, the human part of global environmental change.[20] Indeed, in view of the depth of mutual distrust among the Cold War adversaries, it was far from certain that atmospheric nuclear testing would stop before its public health and environmental consequences became much more significant. While the evolving superpower relationship from confrontation to coexistence ultimately paved the way to the PTBT, this strategic shift alone does not fully account for the decision made by the negotiating parties to include an antipollution objective in the treaty. To understand the global environmental dimension of the landmark arms control agreement, it is crucial to scrutinize the relationship between science and politics under the conditions of the Anthropocene.

Politics of Risk

It might be tempting to attribute the shifting understanding of radioactive fallout and its contributions to the PTBT to the sheer weight of scientific evidence on its unmistakable dangers to humans. On the other hand, some scholars have argued that advocates of nuclear disarmament and the popular culture of nuclear fear exaggerated the quantitatively small risk of harm from fallout out of proportion.[21] Despite their apparent disagreement, both views share the notion that science and politics are clearly separable. The magnitude of global radioactive contamination, however, systematically blurred such a demarcation in three major ways.

First, the severity of fallout hazards was scientifically uncertain. The doses of radiation from fallout dispersed worldwide were so low that it was impossible to directly determine their biological effects to a useful degree of precision. Second, the question was socially contentious. Contamination would affect the world population, including small children and those yet to be born. At the same time, the complex interplay of various natural processes and human activities resulted in the strikingly uneven distribution of fallout across the world. The global reach of fallout and its regional variations inevitably raised a host of questions about social acceptability, especially given the highly contested value of nuclear weapons for world peace. Finally, the question was politically consequential, as any conclusion drawn on contamination was bound to strengthen or undermine the case for the prohibition of nuclear weapons testing on health grounds. The question of fallout hazards thus belonged not to the regular work of scientists within a settled paradigm but rather to what sociologists Silvio Funtowicz and Jerome Ravetz have called "post-normal science," in which "facts are uncertain, values in dispute, stakes high and decisions urgent."[22]

The uncertain, contentious, and consequential issue of global radioactive contamination fused science and politics into an entangled process of knowledge production. Instead of asking whether fallout was harmful in reality, I recast the dangers of fallout as a historical question, the understanding of which evolved over time. How was the problem of fallout constructed? What forces shaped the collection and dissemination of scientific information? Who assessed fallout hazards and determined the "acceptable" risk? And how did these risk assessments affect the policymaking process toward the PTBT? I call this worldwide struggle to determine the biological effects, social acceptability, and policy implications of global radioactive contamination "the politics of risk." Using historical sources from the United States, Britain, Russia, and a few key non-nuclear weapons countries such as Canada and Japan, I examine how the politics of risk, unfolding in the public sphere as well as behind the closed doors of government agencies and scientific institutions, transformed the understanding of fallout from a relatively harmless side effect of the essential military activity to an unacceptable hazard to the whole world.

In charting the politics of risk concerning global radioactive contamination, the concept of "relations of definition" introduced by sociologist Ulrich Beck is particularly useful. Beck has put forth the thesis that industrial society has evolved into "risk society" through the systematic production and distribution of artificial objects that may cause large-scale, long-term, and potentially catastrophic harm. From environmental pollution to genetic engineering, these manufactured risks may overwhelm the existing social norms and institutions grounded in national sovereignty, industrial capitalism, and positivist science.[23] Until and unless materialized, however, risks remain in the realm of possibility. Taking a cue from Karl Marx's concept of "relations of production," Beck has argued that risks are "social constructions and definitions based upon corresponding relations of definition."[24] Beck has described these two types of relations, productive and definitional, as mechanisms of social domination. Just as industrialists derive their power from ownership of means of production, the authority of experts rests on the control of "means of definition," including, among others, "the rules, institutions and capabilities which specify how risks are to be identified in particular contexts."[25] Beck has drawn another similarity between the two types of relations as a focus of social conflicts: whereas class struggles illuminate the relations of production, risk disputes politicize the supposedly natural authority of experts in assessing the possibility of harm.

While following Beck and emphasizing the relations of definition in my analysis of risk politics concerning radioactive fallout, I historicize his framework by placing it in the specific context of radiation protection. Radiation protection was part of the modern regulation of potential harm posed by industrial technologies. As historian Karin Zachmann has demonstrated, insurance companies were among the first to systematically apply probability calculation to ascertain patterns and trends in death, injury, or property loss. This approach to uncertainty as a measurable "risk" spread to many other sectors in the twentieth century as a fundamental principle for managing dangerous machinery, toxic substances, and complex systems such as railroads that were prone to failures and accidents.²⁶ In the case of radiation protection, radiological professionals took the lead in risk management. In the 1920s, as the widespread use of X-rays and radium in industry and medicine resulted in many injuries and deaths among radiation workers, medical and radiological societies in several European countries organized committees to make recommendations for protective measures for their members. To promote and coordinate these attempts at professional self-regulation among nations, the second International Congress of Radiology, held in 1928, created the International X-ray and Radium Protection Committee (IXRPC). This movement crossed the Atlantic when a U.S. representative at the Congress founded the Advisory Committee on X-ray and Radium Protection (ACXRP) a year later.²⁷

Central to the emerging international regime of radiation protection was the concept of "tolerance dose." Proposed by German-born U.S. physicist Arthur Mutscheller in 1924, and formally adopted by the IXRPC and the ACXRP by the mid-1930s, the "tolerance dose" quantified the level of radiation exposure that experts deemed safe for the average worker to receive without bodily damage.²⁸ Although this concept was anchored in the widely accepted hypothesis that there was a threshold in dosage for biological effects of radiation, it also represented the notion of acceptable risk. Mutscheller himself pointed out that the pursuit of absolute safety was impractical, as "the protective shields and apparatus would . . . become entirely too heavy, unwieldy, and costly."²⁹ This utilitarian logic came to the forefront when subsequent scientific research increasingly cast doubt on the existence of a threshold for genetic damage. After World War II, the international and U.S. expert bodies, reconstituted as the International Commission on Radiological Protection (ICRP) and the National Committee on Radiation Protection (NCRP), respectively, replaced "tolerance dose" with "permissible dose" to denote the range of doses in which "the probability of the occurrence of [radiation] injuries must be so low that the risk would be readily acceptable to the average individual."³⁰ Through publications, consultation, and overlapping memberships, the nuclear authorities in the United States, Britain, and the Soviet Union similarly applied this concept of acceptable risk to their weapons work while flexibly adjusting its quantity on the basis of their own assessment of the exposure situations. By the time atmospheric nuclear testing began in earnest, then, the nuclear security states had already agreed on the fundamental rules for the nuclear Anthropocene—radiation exposure up to a certain level associated with the development of nuclear weapons was "permissible" for humanity and the planet.

The dominant position of the nuclear rivals in the relations of definition, however, was fraught with multiple tensions. Historians have demonstrated that the science used to assess radiation hazards spanned a wide range of disciplines in the physical, biological, and environmental sciences. Experts also came from many different countries, each developing a research and regulatory tradition specific to its political, economic, social, and cultural circumstances. Moreover, the institutions tasked with routinely assessing the scientific basis for radiation protection multiplied in number, as radioactive fallout aroused controversy surrounding its potential dangers.[31] While I do not ignore the structural inequality in power and knowledge that systematically promoted a certain definition of radiation hazards in favor of the continuous use of radiation and radioactive material, I argue that the multidisciplinary, multinational, and multi-institutional character of scientific expertise for radiation protection made the relations of definition much more unstable, contingent, and prone to conflicts than Beck has suggested. This became especially clear when the problem of global radioactive contamination cut across multiple political and social divides in the Cold War world. *Political Fallout* thus explores the relations of definition less as a homogeneous and static structure of social dominance than as a complex and dynamic process of risk politics in which many different historical actors competed to shape or challenge understanding of the potential harm posed by radioactive fallout.

More broadly, a historical analysis of the politics of risk concerning radioactive fallout sheds new light on the relationship between science and the Cold War. The earlier historiography has tended to draw a sharp contrast between scientific freedom in the United States and Britain and state control of science in the Soviet Union. No episode seems to better capture this contrast than the so-called Lysenko Affair, a genetics controversy that had a direct bearing on the issue of fallout. The traditional view holds that the "republic of science" in the West stimulated a productive research competition and brought about a genetic revolution, most notably the discovery of the deoxyribonucleic acid (DNA) structure in 1953, whereas the Soviet Union politically suppressed genetics in favor of agronomist Trofim Lysenko's "pseudoscience" of nongenetic inheritance.[32] More recent studies have challenged such a binary view, demonstrating that increasingly militarized states exerted significant influence on science and scientists in the United States, Britain, and the Soviet Union after World War II through a complex system of reward, surveillance, and punishment. The U.S. and British governments became prominent patrons of scientific research in their respective countries, powerfully shaping its development while disciplining scientists who openly dissented from the Cold War consensus.[33] The Soviet government also mobilized scientists in its military and ideological confrontation with the West by a mix of material incentives, ideological appeal, and structured coercion.[34] Cold War science also had a significant international dimension, as the United States, Britain, and the Soviet Union simultaneously promoted and controlled scientific cooperation and dissemination of technical information within and beyond their respective blocs as part of the global geopolitical and ideological struggle.[35]

The Cold War, however, brought about more than historical convergence in scientific research and patronage between different political economic regimes. By driving the testing of nuclear weapons in the atmosphere that scattered radioactive debris worldwide, it also brought the parallel yet diverse practices of Cold War science around the globe into greater synchronization. The scientific basis of radiation protection had been transnational long before the Cold War, as experts from different industrial countries had institutionalized a procedure of standardization among nations through nongovernmental organizations. The worldwide dispersion of fallout also prompted the United States, Britain, and the Soviet Union to expand environmental radiation monitoring beyond the test-site regions for both public health and national security purposes. Even fallout studies conducted on the local and national scales had a global dimension, as their outcomes inevitably added to debates over the safety of atmospheric testing worldwide. Most significantly, the politics of risk led to the greater institutional diversity of radiation protection. Scientists from different countries, including Soviet geneticists, worked through and around these scientific bodies to advance their political, professional, and personal agendas. *Political Fallout* thus looks beyond and beneath nations and blocs, examining the production of knowledge about fallout in the United States, Britain, the Soviet Union, and key non-nuclear weapons countries as an integrative process unfolding in the entangled context of the Cold War and the Anthropocene.

Central to this book is the thesis that the Cold War not only contributed to the escalation of atmospheric nuclear testing but also ultimately helped to mitigate its radiological consequences by driving the politics of risk in an unexpected and unintended direction. The power of the nuclear security states in knowledge production and its limits played a key role in bringing about this paradoxical outcome. Through their dominant position in the relations of definition, the United States, Britain, and the Soviet Union invariably applied the notion of acceptable risk to the whole world, arguing that potential harm of fallout was universally permissible on grounds that it was too small to be statistically detectable and, even if existent in theory, would be still negligible compared to other natural and artificial risks in modern life. This seemingly factual statement, however, entailed controversial judgments about scientific proof and social acceptability. Paraphrasing the aphorism, absence of evidence about harm due to the lack of statistical power was not evidence of absence of harm. Nor would all people accept the risk of harm from fallout simply because it was quantitatively smaller than other hazards.[36] Nonetheless, those politically committed to the continuation of nuclear weapons testing consistently presented their disputable definitions of risk as objective statements of facts.

The political need to reassure the whole world about the safety of tests, however, constantly forced the nuclear security states to overstretch the limits of their knowledge, expose the hidden value assumptions in their claims, and lay their definitions of risk open to contention. Castle Bravo, the largest U.S. nuclear test conducted in 1954 at Bikini Atoll, marked a turning point in this epistemic blowback. The U.S. government's routine safety assurances unwittingly led Japanese authorities to check food and drink with stringent standards, only to discover widespread contamination and fuel a nationwide panic. The flat denial of any harmful effects of fallout by U.S. officials also upset geneticists around the world, spurring them to warn the public about genetic damage and its potentially large impact. While the Soviet Union switched to supporting the prohibition of nuclear testing but struggled to make an effective health argument due to Lysenko's infamous ban on genetic research, U.S. and British officials tried to reassert their definitional power by commissioning studies from the U.S. National Academy of Sciences (NAS), the British Medical Research

Council (MRC), and the United Nations Scientific Committee on the Effects of Atomic Radiation (UNSCEAR).

Contrary to their expectations, the institutional change to the relations of definition further eroded the official Anglo-American position on global radioactive contamination. Working from different scientific, regulatory, and political concerns, the appointed experts at each committee clashed over the uncertainty and acceptability of fallout hazards in light of their worldwide impact and striking regional variations. The credibility of the safety claims also suffered a serious blow within the United States and Britain when concerned scientists and citizens successfully broke government control of information by carrying out fallout surveys in their communities.

Scientific probes on the dangers of fallout across multiple geographical scales in turn reshaped understanding of who was at risk in the nuclear Anthropocene. The nuclear security states based their safety claims on an estimated average and its variance around the world. Risk subjects, however, were much more than a statistical construct rendered legible for planetary risk management by the Cold War rivals. The risk of genetic damage and the contamination of milk put children and those yet to be born at the center of concern, driving many women to speak out against fallout as housewives and mothers. The discovery of unusually contaminated rice in Japan vindicated the feeling of Asian nations emerging from colonial domination that they were the victims of the nuclear-armed Western countries. Together with the ethical doctrine of humanism advocated by the transnational movement of intellectuals, rich discourses of race and gender in the early Cold War world transformed the risk subject from a faceless, statistically constructed average person to countless "innocent bystanders," present and future, caught in the crossfire of the superpower rivalry. This socio-epistemic construction of victimhood led to the redefinition of global radioactive contamination from a harmless phenomenon to an intolerable threat to humanity.

The politics of risk influenced the test-ban debate in a number of subtle yet important ways. Some politicians actively exploited the shifting understanding of fallout to promote their Cold War agendas. Others reluctantly made some gestures toward growing public concern about contamination. The policy impact of risk politics, however, ran much deeper than these superficial changes in political behavior. The emerging idea of fallout as an unacceptable hazard became an inner conviction for some of the top political leaders and scientific advisers on both sides of the Iron Curtain. Even some of the most prominent fallout deniers selectively appropriated the rhetoric of antipollution arguments and spearheaded technological projects to reduce or contain fallout emissions as a solution to contamination without stopping nuclear testing. These multiple and conflicting responses to the politics of risk ultimately gave a distinctly environmental character to the PTBT.

Overview of the Book

Political Fallout traces the politics of risk as unfolding in different stages toward the PTBT. Chapter 1 examines the origins of radiation protection in the testing of nuclear weapons through 1953. The perceived political urgency to develop nuclear weapons led the United States, Britain, and the Soviet Union to define the dangers of radioactive fallout to the general public in a strikingly similar manner as limited to the risk of acute illness in the test-site region. The same logic led the Cold War nuclear rivals to start secretly monitoring the global dispersion of fallout as a useful tracer for national security purposes, further reinforcing the idea of widespread radioactive contamination as a harmless phenomenon.

Chapters 2 and 3 explain how U.S. efforts to contain the repercussions of the 1954 Castle Bravo test backfired, triggering the first round of the fallout controversy. Chapter 2 focuses on Japan, a key U.S. ally in Asia, where the rigorous inspection of tuna backed by the U.S. government and the tuna industry led to the discovery of more contaminated fish and worsened a consumer panic. It examines the process in which Washington and Tokyo struggled to defuse the crisis by introducing the concept of permissible dose. Chapter 3 discusses the entry of geneticists into the definitional struggle on both sides of the Iron Curtain. It shows that a stalemate in public debates over genetic damage forced the U.S. and British governments to cede part of their definitional power to the NAS, the MRC, and the UNSCEAR.

Chapters 4 and 5 follow the politics of risk in the restructured relations of definition. Chapter 4 examines the NAS and MRC reports issued in 1956 and reveals a complex matrix of scientific and regulatory agendas, which led the two committees to reach a similarly reassuring conclusion regarding genetic damage but to disagree over the cancer risk of strontium-90, a fission element that tended to accumulate in bones and teeth.

The disclosed disagreement between two of the most authoritative scientific institutions in the United States and Britain firmly embedded the fallout controversy into the test-ban debate. Chapter 5 turns its attention to the UNSCEAR, an intergovernmental panel in which the Japanese and Soviet delegations led the charge against Anglo-American epistemic leadership. It shows how each group of scientists pushed for the recognition of fallout's worldwide impact and its regional variations to advance their professional standing against their scientific rivals while assisting their governments' advocacy of a test ban. It also discusses the key role that the transnational movement of intellectuals played in rejecting the notion of acceptable risk by questioning the morality of harming innocent bystanders. The chapter concludes by examining how the transformed understanding of fallout hazards incorporated into the 1958 UNSCEAR report affected a series of policy decisions made by the United States, Britain, and the Soviet Union to suspend all tests while test-ban negotiations were in progress.

Chapters 6 and 7 carry the story through to the PTBT by further exploring the powerful and multifaceted policy impact of the emerging idea of fallout as an unacceptable hazard. Chapter 6 examines a local turn of risk politics in the United States and Britain during the moratorium period. It illustrates the role of progressive politics at the subnational level in catalyzing a number of community-based fallout surveys, especially those in Minnesota and Wales. It also discusses the significant effects of these local studies on the Anglo-American decision in 1959 to formally introduce the proposal of an atmospheric ban. Chapter 7 deals with the final phase of the test-ban negotiations after the Soviet Union restarted atmospheric testing in September 1961. While the Soviet action, soon reciprocated by the United States, seemed to suggest their lack of concern about fallout, the continuous and robust influence of the redefined fallout hazards changed the course of the test-ban negotiations and helped to produce the PTBT. The Conclusion reflects on the implications of risk politics over fallout for the subsequent development of the Cold War, the nuclear age, and the Anthropocene.

Chapter 1

A STATE OF EMERGENCY

The Origins of Radiation Protection in
Nuclear Weapons Testing, 1945-1953

IN FALL OF 1945, product inspectors at Eastman Kodak's manufacturing plant in Rochester, New York, discovered small spots on sheets of X-ray film in several cartons stored in the building. At first glance, it looked like just another case of a well-known problem. After all, natural sources of radiation, such as granite rock and radium water, occasionally caused damage to radiographic material. Much to the surprise of Kodak specialists, however, the element enmeshed in the fiber of the containers was cerium-141, a fission product that had previously been believed not to exist naturally. As it turned out, the spots were a direct result of the world's first atomic bomb test, "Trinity," conducted in New Mexico. Radioactive dust created by this secret test was blown by the wind eastward and washed down by rain to the ground when the plume passed over the Midwest. The settled fallout then entered the Wabash River, where a paper mill in Vincennes, southwestern Indiana, unknowingly used the contaminated water to produce the strawboard that stored Kodak's X-ray film.[1] The fogged radiographs, discovered some thousands of miles away from the Trinity site, captured a historic moment in the Anthropocene: a single human-induced event—a nuclear explosion—almost simultaneously left its enduring environmental footprints on parts of or the whole world.

During the next several years, as the world underwent a series of international crises that led to the beginning of the Cold War, the number and yields of atmospheric nuclear tests drastically increased (Figure 2). During the period between 1945 and 1953, the United States conducted forty-three such tests with a total yield of approximately 12 megatons (Mt), the Soviet Union did eight with a yield of 540 kilotons (kt), and Britain did three with a yield of 35 kt.[2] After each explosion, there were occasional reports of transient yet abnormal increases in the levels of radioactivity in areas far from the test sites. The three nuclear powers, however, invariably dismissed the potential health effects of this widespread contamination. U.S. officials repeatedly claimed that radioactivity found in great distances from the test site "definitely is not dangerous to humans, animals, or plants."[3] In announcing the success of Britain's first nuclear test, conducted in the Monte Bello Islands off the northwestern coast of Australia, Prime Minister Winston Churchill stated, "Conditions were favourable and care was taken to wait for southerly winds so as to avoid the possibility of any significant concentration of radio-active particles spreading over the Australian mainland."[4] The Soviet government rarely discussed its nuclear tests in public, but when it did so, no mention of radiation safety appeared in its official statements.[5]

Given their vested national security interest in nuclear weapons, it was no surprise that the United States, Britain, and the Soviet Union similarly denied, dismissed, or ignored the potential dangers of radioactive fallout to humans. As shown in the quotes cited above, however, their denials always took the form of a factual statement, not a political opinion. How, then, did they "know" that the nuclear debris of their weapons tests posed no danger whatsoever to members of the public? To answer this question, this chapter traces the origins of radiation protection for the testing of nuclear weapons, asking how each country developed the institutions, guidelines, and monitors to detect, evaluate, and manage radioactive fallout. Although the specific configurations of the radiation safety programs varied among the three nuclear rivals, their organizing principles were strikingly similar. Fearful of a severe accident caused by a radioactive plume, and pressed to develop nuclear weapons without delay, the U.S., British, and Soviet governments alike defined the exposure situation as an "emergency," concentrating their resources on coping with the well-known dangers of instantaneous exposure to large amounts of radiation. The same logic also led them to start secretly tracking the global dispersion of fallout as a supposedly harmless and useful tracer for national security purposes,

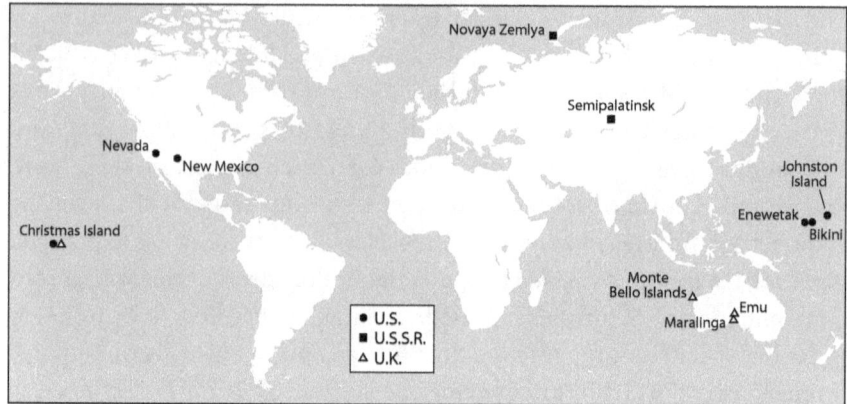

FIGURE 2. Locations of Major Atmospheric Nuclear Weapons Test Sites in Operation Through 1963.

namely to uncover foreign nuclear tests and, for the United States, also to refine estimates of the worldwide effects of nuclear warfare that it was prepared to wage against the Soviet Union. This compartmentalized production of knowledge systematically sustained and deepened ignorance about the widespread and long-term effects of radioactive debris beyond the test-site regions.

United States

In April 1945, with Trinity slated to go off three months later, the Manhattan Project asked physical chemists Joseph O. Hirschfelder and John L. Magee to compile a list of potential hazards that could affect members of the public living near the New Mexico test site. One morning, Hirschfelder recalled, he suddenly realized that something was missing from the checklist. He ran to Magee's office, shouting, "What about the radioactivity?"[6] The two scientists sifted through all known examples in desperate search of any clue about the formation and dispersal of fine particles, whether it was industrial smoke, airplane trails, and even chemical warfare.[7] The scale of a nuclear explosion, however, defied a simple extrapolation from these past experiences. On May 7, a dry run for the Trinity test was staged, with a hundred tons of conventional dynamite with fission products detonated atop a tower. The huge explosion sent the radioactive materials into the sky so high that few were found near ground zero, with the remainder supposedly falling farther away. In reporting this observation, Hirschfelder and Magee pointed out "a definite danger of dust containing active material and fission products falling on towns near Trinity and necessitating their evacuation."[8]

The belated realization of the radiological dangers of the test device to the public came at a critical moment for the U.S. atomic bomb program. Prodded by Albert Einstein's famous letter to President Franklin D. Roosevelt in August 1939, the U.S. government had embarked upon atomic research as a hedge against the German nuclear ambition. The British also organized a bomb project, code-named "Tube Alloys," but the grueling struggle with Germany eventually forced them to suspend their efforts in 1943 and join the Manhattan Project as a junior partner. With the Germans defeated a day after the mock test at the Trinity site, Roosevelt's successor, Harry S. Truman, turned his eye to Japan as the target. At the same time, the president and his aides were keenly aware of the potential political side effect of the atomic bomb, hoping that the demonstration of its power would make the Soviets more reasonable regarding the postwar settlement. As War Secretary Henry L. Stimson later observed, "The bomb as a merely probable weapon had seemed a weak reed on which to rely, but the bomb as a colossal reality was very different."[9] A successful bomb test was thus deemed crucial for a diplomatic success—and a more secure aftermath of the war.

At first glance, the location of the test site, situated in a remote corner of the Great Basin in New Mexico, seemed to have already solved the problem of radioactive fallout. Mindful of safety as well as security, Kenneth T. Bainbridge, the Trinity test manager, chose a place with few people around and remote from major cities, and yet not too far from Los Alamos, where the bomb was assembled. Isolation, however, was an illusion. The Great Basin was dotted with small mining and grazing communities. While all settlements within ten miles of the point of detonation were vacated prior to the test, maps from the early 1950s showed sixty-three ranches, three camps, and fourteen small towns present within a thirty-mile radius of the Trinity site.[10] When Stafford Warren, chief medical adviser for the Manhattan Project, inspected the test site from the airplane, he sighted human traces all over the mesas. "It was quite clear that there were a lot of people living in the desert that weren't on the map," he recalled.[11]

As the assumption of isolation turned out to be false, the radiation safety officers at the Trinity site made two related decisions that set the pattern for U.S. off-site fallout policy for years to come. One was to take no protective action at levels short of those known to be clearly injurious. At the time, the Manhattan Project observed the daily limit of 0.1 roentgen (R) recommended by the Advisory Committee on X-ray and Radium Protection (ACXRP), a nongovernmental standards organization in the United States, for normal operation within the Trinity site. Shortly before the test, however, it established another benchmark, a projected exposure of 75 R in two weeks for external gamma radiation, or approximately fifty times above the daily limit, as a trigger for mass evacuation. The underlying assumption was that civilians outside the test site would be at risk of radiation injury only when a radioactive plume was blown directly toward them due to some unexpected change in weather. Louis H. Hempelmann and James F. Nolan of the Health Group at Los Alamos assured Bainbridge that even the highest exposure level predicted in such a scenario, 68 R in two weeks, "would probably not even cause radiation sickness," adding, "Fatalities probably would not result unless ten or more times this dose were delivered."[12]

The decision to protect civilians against severe injury only in turn shaped the other decision regarding off-site fallout monitoring. Belatedly realizing the threat of radioactive debris to local residents outside the test site, Bainbridge hurriedly set up what was then the world's largest radiation surveillance network. Posts were installed in areas up to fifty-five miles from the point of detonation, whereas officers on vehicles were prepared to drive along highways up to 150 miles downwind. Airplanes were also deployed to track a radioactive plume eastward until it dissipated sufficiently.[13] The spatial and temporal scope of monitoring, however, hardly matched with the actual extent of contamination. Ground measurements were confined to the vicinity of the test site and made only in the duration of the test operation. Moreover, in most cases, monitors measured external radiation coming from a passing plume only, leaving out the whole question of internal exposure through the inhalation and ingestion of radioactive materials. The relatively high reference levels for protective actions and the narrow range of radiation monitoring thus combined to systematically foster ignorance about radioactive fallout dispersed beyond the test-site region.

Off-site exposure, as it turned out, was anything but an unlikely accident. When Trinity went off on July 16, 1945, at 5:29:45 a.m., part of the radioactive cloud churned toward populated areas. That morning, a report came from nearby towns along highway 380 showing an exposure rate of 3.3 R per hour. This greatly alarmed the test personnel, as it could, in the worst case, give an integrated gamma dosage up to 90 percent of the two-week benchmark for evacuation. Fortunately, the rate soon declined. Shortly after the test, however, it was discovered that the plume had passed directly over several ranches tucked away in a canyon approximately twenty miles northeast from the Trinity site. A subsequent radiological survey suggested that the residents were likely to have received external gamma radiation exposure up to a total of 47 R in the two weeks after the test.[14] Seeing no signs of illness among them, however, Manhattan Project officials concluded that there would be no measurable impact on their health. Hempelmann later recalled this incident in an interview, saying that "a few people were probably overexposed, but they couldn't prove it and we couldn't prove it. So we just assumed we got away with it."[15]

The lack of concern about the health effects of low-dose radiation, however, did not mean that nuclear officials were oblivious to the issue of widespread radioactive contamination. Indeed, the idea of a national emergency greatly magnified the perceived value of nuclear dust as a tool of intelligence to detect foreign test explosions from afar.[16] Despite the German defeat shortly before the Trinity test, the threat of a hostile future nuclear power—specifically, the Soviet Union—had already loomed large in the minds of many U.S. officials. The idea of long-range detection first occurred to Magee and his colleague Anthony L. Turkevich in the wake of the Trinity test. The two scientists speculated that some of the fine radioactive particles from a nuclear explosion might remain airborne for a long time and travel around the globe in concentrations high enough to be detectable in distant locations. The source of their inspiration was one of the largest volcanic eruptions recorded in modern history. In 1883, Krakatoa in today's Indonesia exploded with such force that its smoke and gas entered the upper atmosphere and circled the globe for many months. Turkevich later explained that "many of us had read the [British] Royal Society report on the Krakatoa volcanic explosion as an indication of the type of long range effects we might expect."[17] At the request of the Manhattan Project, the Army Air Force flew airplanes equipped with Geiger counters along the West Coast, successfully detecting slightly elevated levels of radioactivity about three weeks after the Trinity test. Although the method used was too crude to pinpoint the source of debris, a report concluded that airborne dust sampling showed a great promise to be capable of uncovering nuclear explosions "almost anywhere with proper meteorological conditions."[18]

Given the importance of national security in the characterization of radioactive fallout, it was no surprise that a gap between monitoring for public health and monitoring for nuclear intelligence widened as the relationship between the United States and the Soviet Union rapidly deteriorated. When the United States mounted Operation Sandstone at the mid-Pacific atoll of Enewetak in 1948, a year after Truman announced aid to Greece and Turkey as part of a global struggle against communism, the radiation safety officers checked radioactivity only inside the designated "Danger Area" and at a few nearby atolls. Meanwhile, the U.S. military secretly collected airborne nuclear dust at twenty-six sites in the continental United States and forty-one installations overseas, supplemented by filter-equipped airplanes flying over the vast area bordered by the Arctic Circle to the north, the Equator to the south, Manila to the west, and Tripoli to the east.[19] A radiological component of the Atomic Energy Detection System, managed by the U.S. Air Force and put online just before the Soviet Union conducted its first nuclear test in August 1949, encircled the Soviet airspace from the North Pacific and the North Atlantic to the Mediterranean. The alert level of radioactivity, initially set at 100 counts per minute (cpm) and soon reduced to 50 cpm, was so close to natural background levels that there had been 111 false alarms before the system successfully identified the Soviet test.[20] This stood in stark contrast to the much higher exposure limits used for the protection of the public from fallout.

The glaring disparity in knowledge about radioactive fallout for public health and nuclear intelligence seemed tenable as long as the United States conducted nuclear tests in the remote corner of the Pacific. The myth of isolation, however, collapsed once again when the U.S. Atomic Energy Commission (USAEC), a civilian successor to the Manhattan Project, opened a new testing ground in Nevada in 1951. With small ranching and mining camps lining the borders of the off-limit zone, the commission's radiation safety officers took more caution than at the time of the Trinity test. The minimum benchmark for protective actions was lowered from a projected exposure of 75 R in two weeks to 25 R in thirty days, and the range of ground monitors expanded to a two-hundred-mile radius.[21] The underlying philosophy, however, remained the same. As the Soviet nuclear test abruptly ended the U.S. atomic monopoly, Truman responded by ordering a crash program to develop even more powerful hydrogen bombs. The outbreak of the Korean War in June 1950 made the quest for nuclear superiority all the more urgent. In this deteriorating national security environment, the radiation safety officers at the Nevada Test Site were reluctant to take any protective measures that might unnecessarily hamper their nuclear weapons program. Noting "the somewhat delicate public-relations aspect of the affair," Thomas L. Shipman, head of the Health Division at Los Alamos, argued that "no drastic action to disturb the public should be taken unless it is clearly felt that such action is essential to protect local residents from almost certain damage."[22]

Much to the surprise of nuclear officials, an immediate problem in public relations for Nevada tests came not from mass hysteria but rather from widespread interference with X-ray film and Geiger counters across the nation. Some of the radiological instruments were so sensitive to the influence of radiation that even a slight change to the amounts of radioactive material in the environment could disrupt the normal functioning of these devices. That was indeed what happened to the Kodak film in the aftermath of the Trinity test. This physical characteristic made the long-range detection of foreign nuclear tests possible, but it also posed a threat to a variety of research and industrial activities in large areas under the path of a radioactive cloud, ranging from uranium prospecting to scientific studies involving low-level radiation measurements. One of the most vocal protests came from the photographic industry. As soon as Nevada tests began in January 1951, air filters installed at Kodak's Rochester plant started intercepting radioactive particles. Alarmed by the return of contaminated air over North America, the president of Kodak directly wrote to the USAEC, warning that his company might have to sue the federal government if damage to film and the cost of preventive measures became substantial.[23]

Blindsided by this forceful protest from corporate America, the USAEC found an ideal troubleshooter in Merril Eisenbud, director of the Health and Safety Laboratory (HASL) in New York. Born in 1915, the eldest son of a Russian-Jewish immigrant physician in New York, Eisenbud started his professional career at the Liberty Mutual Insurance Company as an industrial safety engineer. His initial assignment was to inspect insured factories and offer advice for accident prevention. Over time, however, Eisenbud developed a keen interest in occupational diseases. Unlike accidents, such ailments were often a result of chronic exposure to harmful agents. Realizing that "the risks of many occupational diseases depended on how much of a particular substance was inhaled by the worker," Eisenbud concentrated his effort on collecting airborne dust in the workplace.[24] Widely known as a "dustologist," he was called upon during World War II to investigate problems of handling toxic materials in the war industries. The arrival of the nuclear age enhanced the value of his expertise. In 1947, the USAEC recruited Eisenbud for what was to become HASL, an institution that soon earned distinction as "an extremely useful trouble-shooting arm of the commission."[25]

Put in charge of the task of addressing the complaints of the photographic industry about the bomb debris in the air, Eisenbud decided to take a measure similar to what he had done for factories, this time on a much larger scale. He drew up a plan to routinely collect airborne nuclear dust across North America so that the film manufacturers and others handling radiosensitive objects could receive an early warning of an incoming radioactive cloud. To his dismay, however, the radiation safety office at the Nevada Test Site rejected his request for assistance, asserting, "there can be no significant fall-out which could possibly produce a health hazard of any sort" beyond the two-hundred-mile zone under its watch.[26] Eisenbud was thus forced to improvise a continent-wide fallout monitoring web with the help of USAEC facilities, universities, the U.S. Weather Bureau, and the Canadian government. By 1952, the HASL network had managed to cover much of North America with up to some one hundred ground stations, using gummed paper for around-the-clock collection of airborne dust for radioactivity analysis conducted at HASL and its collaborating laboratories.[27] The trouble of the photographic industry, not any concern about public health, prompted the U.S. government to install the nation's first permanent radiological surveillance across North America.

The film manufacturers, however, were by no means the only ones alarmed by the increasing level of fallout contamination. USAEC monitors and civilian scientists also reported slight increases in background radiation levels across North America following nuclear blasts. In February

1951, shortly after the first test explosions occurred in Nevada, traces of radioactivity were found in snow falling on the Eastern United States and Canada.[28] The radioactive cloud from Shot Easy on May 7, 1952, veered north, passing over a few towns before it reached Salt Lake City, where exposure at no more than 10 milliroentgen (milliR) per hour was recorded.[29] A year later, when Shot Harry went off on May 19, 1953, the wind shifted slightly, blowing fallout toward St. George, Utah, where the cumulative exposures reportedly reached as high as 4.2 R.[30] In the meantime, part of the plume from an earlier test ran into a storm system over the Northeast and deposited radioactivity on the ground in areas stretching from Chicago to Quebec City. After checking widespread contamination in his neighborhood, Herbert M. Clark, a Rensselaer Polytechnic Institute chemist in Troy, New York, published a paper showing that the estimated total exposure to gamma radiation in the first ten weeks was 55 milliR, or more than twice the natural background exposure during the same period.[31]

Initially, nuclear officials simply dismissed the radioactivity detected outside the Nevada Test Site. As reports of fallout continued to come in from many communities, however, the USAEC was forced to publicly clarify the grounds for its safety claims. One was the maximum permissible dose. As the Manhattan Project had followed the dose limits recommended by the ACXRP, so did the commission follow recommendations by its postwar successor, the National Committee on Radiation Protection (NCRP), for normal activities within the test site. For greater operational flexibility, it combined the weekly limit of 0.3 R over the entire duration of a test series (initially ten weeks, later extended to thirteen weeks). In public information material published in January 1953, the USAEC explained this occupational standard, noting that "[n]one of the measurements of fall-out radioactivity outside the Nevada Proving Ground has exceeded the recommended maximum of 3 roentgens per ten weeks."[32] The other benchmark was detectability. In the same publication, the commission admitted that there might be no safe dose at which genetic damage would not occur. Citing the estimated dose required to double the natural rate of human mutation, however, it went on to argue that the doubling dose was "1,600 times higher than the lifetime exposure level of 50 milliroentgens noted in communities surrounding the test site as a result of fall-out."[33] This meant that the fallout radioactivity was too small in quantity to cause a detectable increase in mutations.

Not all scientists accepted these new justifications for accepting exposure of the populace to radioactive fallout. In May 1953, physicist Lyle B. Borst of the University of Utah sent a letter of protest to USAEC chairman Gordon Dean. Borst was no stranger to radiation hazards. He had worked for the Manhattan Project during World War II and then served as the first director of the commission's Brookhaven National Laboratory until 1951. While he agreed that the amounts of radioactivity from nuclear tests remained well below the maximum permissible dose, he challenged "the philosophy implied—that as long as radiation levels do not exceed calculated tolerance values, the situation is safe." Noting that fallout affected not only healthy adults found in the atomic workplace but also pregnant women and children who might be more sensitive to the influence of radiation, he urged the chairman to ensure that his staff take every possible measure to prevent and minimize off-site exposures.[34]

Borst, however, found himself alone in dissent against the definition of risk imposed by nuclear officials. Most scientists in the United States who reported fallout through news media and scientific papers apparently agreed with the USAEC. Clark, for instance, compared the estimated ten-week fallout exposure in Troy, 55 milliR, with the "accepted safe level" of 3,000 milliR for the same period without noting fundamental differences in exposure situations.[35] Borst also received a sharp rebuke from nuclear officials. In reply to Borst, John C. Bugher, director of the USAEC Division of Biology and Medicine, wrote, "Atomic tests are conducted in the interests of national welfare, a circumstance which certainly warrants deviation from normal laboratory practices provided levels of radiation are kept well below levels of danger to the health of our personnel and the general population."[36] Norris E. Bradbury, director of the Los Alamos National Laboratory, was even more blunt. In a letter to a vice president of the University of Utah complaining about Borst, Bradbury wrote, "I think you can believe that an atomic war would be fantastically worse—and that perhaps the AEC's efforts at Nevada make the likelihood of such a war considerably less!"[37]

Indeed, the logic of national security further drove bifurcation in the production of knowledge about global fallout. While publicly denying the dangers of fallout to public health, the USAEC also launched a worldwide radiostrontium survey in 1953 as part of a secret military study. As historian Jacob Darwin Hamblin has demonstrated, the United States and its NATO allies actively studied the catastrophic potentials of nuclear weapons during the Cold War in preparation for total war with the Soviet bloc.[38] One scenario was that a full-scale nuclear conflict might leave the whole world contaminated with radioactive debris to the extent that most, if not all, survivors on Earth might eventually be killed by cancer and other delayed effects of radiation. Worried that the rapid growth of the world's nuclear stockpiles might soon materialize the planetary risk of overkill, the USAEC put physicist Nicholas M. Smith Jr. in charge of Project Gabriel in 1949, a theoretical study to determine the biological limits of nuclear warfare. For this purpose, Smith built a model of global fallout transport with a focus on strontium-90 (Sr-90), a highly carcinogenic fission product that would mimic calcium in the human body and build up in bones and teeth. With no direct measurements made of Sr-90 depositions, however, the model was plagued with large uncertainty. In a report in November 1951, Smith concluded that the most probable limit was in the order of 100,000 standard (20 kt) fission bombs, but he quickly added that the true number might be "100 [times] too low—or 10 times too high."[39] Shocked by the enormous margin of error in the estimates, the USAEC General Advisory Committee (GAC) requested an independent check by RAND Corporation, a California-based think tank known for the technique of systems analysis.[40] RAND, in turn, recruited Willard F. Libby, a nuclear chemist and member of the GAC, as a consultant.

Born the son of a farmer in 1908, Libby grew up in a town north of San Francisco, working at his family's fruit farm. Six feet and three inches tall, with broad shoulders and big arms, he quickly earned a nickname, "Wild Bill," which stuck for the rest of his life. At the University of California, Berkeley, he developed his life-long interest in naturally radioactive elements. From early on, Libby showed an unusual knack as a toolmaker for research on emitters of weak radiation. He later claimed that he was the first in the United States to build a Geiger counter, when he was still a senior college student. He went on to fine-tune the detector such that it could measure feeble radiation from rare earths not previously believed to be radioactive. His pursuit of natural radioisotopes led to a series of breakthroughs that brought him fame. In 1946, he successfully elucidated the natural cycle of tritium, which made it possible to date water and track its circulation in the environment. He also conceived and perfected methods of measuring radiocarbon in organic matter to date their origins, a technique for which he received the Nobel Prize in Chemistry in 1960.[41]

As a leading nuclear chemist, Libby was well-suited for the review of Project Gabriel. He was not a typical lab scientist, however. For one, his scientific interest freely crossed various disciplines. From the outset of his research on radiocarbon dating, he was keenly aware of its potential value for a wide range of fields from archeology to geology.[42] The scope of his research was also no less than global. When later asked why he stumbled upon the idea of radiocarbon dating, Libby said that he always saw the whole world as one system: "Here I was talking about the ocean, I mean the entire ocean mass, the entire biosphere, the entire atmosphere, as though it were in my test tube." To prove that the global distribution of radiocarbon was more or less uniform, Libby had his assistant collect pieces of wood from all over the world with the help of various museums.[43] It was his political conservatism, however, that made him stand out among the atomic scientists in the United States. While most of his colleagues at Chicago including Enrico Fermi were opposed to an all-out effort to build the hydrogen bomb, Libby strongly supported the program to stay ahead of the race with the Soviets.[44] As Project Gabriel was designed to ascertain a threshold of dangers below which Washington could justify the actual use of nuclear weapons, Libby's pronuclear view guaranteed that his review would not be a political embarrassment for RAND and the nuclear security state.

Libby embraced the task with enthusiasm. His intensive work culminated in a three-day expert conference held in July 1953. According to Libby, the "common weakness" among the past Gabriel studies was the lack of a "test" for their conclusions.[45] Then an idea came to him: the whole world already had been contaminated with Sr-90 as a result of the past nuclear tests. "Today we are afforded the opportunity of doing a radioactive-tracer chemistry experiment on a world-wide scale," a RAND report, issued shortly after the conference, stated. "The release in the world of almost 10 kilograms of Sr-90 within less than a decade has probably disseminated enough of the contaminant to provide amounts which are probably now detectable in samples of inert and biological materials throughout the world. An analysis of these materials for Sr-90 will provide us with much of the information which is now lacking."[46] When few cared about the global fallout from testing in terms of human health, Libby found it quite useful as a supposedly harmless tool of research to predict the worldwide impacts of a nuclear war.

As a scientist who treated the entire world as one system, Libby staunchly defended the project's ambitious design. He said at the expert conference that "our problems of worldwide circulation are obviously so serious and so important that I think anything less than a worldwide assay or an assay that doesn't have some samples spread all over the world will be unsatisfactory."[47] Science, however, was not the only reason for his global orientation. It also mirrored the twisted logic of his commitment to the U.S. nuclear posture, seeing the United States as a caretaker of the human race by virtue of its capabilities to wipe out a countless number of innocent bystanders. Libby reminded the experts of the original Gabriel question, saying that "it wasn't whether you killed Americans; it is whether you kill people, and I think that we must assay the world, not our own backyard."[48] He even asked rhetorically, "[A]fter all aren't we really concerned about whether the negros [sic] in South Africa are radioactive?"[49] This nuclear paternalism inspired Libby's call for a truly worldwide research on radioactive contamination from nuclear weapons testing.

Libby's study, code-named Project Sunshine, benefited much from the worldwide reach of U.S. diplomatic and military capabilities. The rise of the United States as a global power in the twentieth century led to the opening of diplomatic outposts in many countries.[50] After World War II, military activities also went global thanks to a chain of overseas bases as well as the spread of naval and air power.[51] The members of Project Sunshine tapped into this operational infrastructure abroad. When Lyle E. Alexander of the U.S. Department of Agriculture toured Western Europe, the Middle East, and North Africa in 1954 and again in 1955 to collect a myriad of samples for Project Sunshine, he received much help from local U.S. foreign officers. At each stop, he took a batch of soil from the yards of embassies, consulates, or residences. In France, the embassy restaurant helped in obtaining a one-year old lamb, and the wife of an agricultural attaché purchased some milk from the market.[52] The U.S. military also proved indispensable. In 1956, the Naval Research Laboratory started making periodical measurements of the gross fission product activity in the surface air along the eightieth meridian west, which eventually resulted in a network of twenty-two ground stations stretching from Greenland to Antarctica.[53] To cover the vast oceans, the Navy also had its vessels in the Pacific collect samples of surface seawater during normal transport operations.[54]

Hard power was not the only platform for Project Sunshine. Soft power played an equally important role. Historians have pointed out how Cold War America promoted scientific cooperation, medical assistance, and educational exchanges as part of its efforts to win the hearts and minds of peoples in the decolonizing world.[55] Local collaborators who joined these intellectual programs turned out to be indispensable for the collection of human bones that otherwise were inaccessible to outsiders. For this purpose, the USAEC approached the Rockefeller Foundation, which supported mission hospitals in India, South America, and South Africa.[56] The Armed Forces Institute of Pathology also agreed to procure human bones from Japan through its ties with the survey of the atomic bomb survivors.[57] Personal contacts supplemented these institutional arrangements. J. Laurence Kulp, a geochemist at Columbia University and chief analyst of human bones for Project Sunshine, wrote to an Austrian missionary doctor in South India, asking for a monthly shipping of human bones of different ages for "this urgent humanitarian scientific study."[58] Over time, Kulp managed to collect specimens from many parts of the world. By April 1956, he had established arrangements with twenty-six individuals in nineteen countries outside of the United States.[59]

As extensive as it was, Project Sunshine was nevertheless not rigorous enough to systematically attack the problem of uncertainty concerning the global dispersion of radioactive fallout. As discussed earlier, it began rather as a modest effort to produce a rough sketch of contamination so that one could detect any major flaws in the theoretical model. With no agreed-upon level of scientific rigor, the project quickly lost its focus. As early as September 1953, a USAEC official complained about Libby's tendency to "rush into the sampling without carefully analyzing the significance of different types of information which might be obtained."[60] Project Sunshine, generously funded in the name of national security, became a profitable enterprise for many scientists. Another nuclear official lamented, "The slogan 'This is important for Sunshine' has replaced 'Open Sesame' for people who want funds and authorization to study everything from Northern Lights to penguin feathers, with ocean bottoms in between."[61] The project leaders, however, quickly brushed aside the need for quality control. Eisenbud, who attended a project meeting held in May 1955, asserted that a high degree of accuracy was "useless" in view of numerous imponderables involved in nuclear warfare and also of practical constraints such as costs, facilities, and personnel. "This philosophical evaluation of the purpose and requirements of Sunshine," a reporter of the meeting concluded, "proved, as always, so troublesome that we cheerfully abandoned it in favor of the customary intuition."[62] These holes and gaps in knowledge about global fallout contamination, however, would loom large when the U.S. government later repurposed Project Sunshine as the most important research to convince the world of the safety of nuclear weapons testing.

Soviet Union

Like their U.S. counterparts, Soviet officials and staff scientists initially relied on geographic isolation to manage both security and safety related to nuclear tests. The instructions issued in April 1947 spelled out an ideal site as an uninhabited region with no farmland within 125 miles, remote from international borders yet close to railways and other logistics.[63] The selected Semipalatinsk Test Site, located south of the Irtysh River in present-day northeastern Kazakhstan, seemed to fit the bill. "Along the way there were neither houses nor trees," a visitor noted. "Around was the stony, sandy steppe, covered in feather-grass and wormwood. Even birds here were fairly rare."[64] The Kazakh steppe, however, was anything but a void. Beginning in the last decades of the tsarist period, peasants of Russian, German, Ukrainian, and Caucasian origins moved into the region, while the Kazakh nomads were pressed into collective farms during the early 1930s. Factories also sprang up, a trend that gained momentum during World War II.[65] This demographic mix created a mosaic of communities scattered over the vast steppe. Although the test director removed a thousand families from the restricted zone, both banks of the Irtysh River were still lined with farming villages and nomadic camps. A 1958 survey revealed twenty-two population centers located downwind of the test site, including the two industrial cities of Semipalatinsk and Ust'-Kamenogorsk.[66]

Stalin, however, had little time to waste in his hurried quest to break the U.S. atomic monopoly. The Soviet Union launched a nuclear program in 1942, but the effort remained modest throughout the war. When Truman informed Stalin of the successful atomic bomb test during the Potsdam Conference in July 1945, the Soviet premier feigned ignorance. Privately, however, he concluded that the Americans intended to use the bomb to intimidate him. Shortly after the atomic bombings of Hiroshima and Nagasaki, Stalin ordered Igor V. Kurchatov, scientific director of the Soviet atomic program, to build the atomic bomb as soon as possible. "Hiroshima has shaken the whole world. The balance has been disrupted," Stalin reportedly told Kurchatov. "Give us a bomb—it will turn a great danger away from us."[67] As the postwar tension between the United States and the Soviet Union turned into a full-scale confrontation, the Soviet project proceeded full steam ahead under the brutal but able leadership of Soviet secret police chief Lavrentii Beria.

The radiation safety officers at the Semipalatinsk Test Site were not oblivious to at least some of the challenges posed by a nuclear test explosion. Although the Soviet Union did not formally participate in the nongovernmental standard organization, the International Commission on Radiological Protection (ICRP), until 1956, it basically followed ICRP recommendations to regulate activities involving radiation exposure. The Third Chief Directorate of the USSR Ministry of Health, a unit created in 1947 to supervise radiation protection at nuclear facilities, adopted the daily limit of 0.1 R for routine work relating to nuclear tests. As had the U.S. nuclear authorities, it also established a separate allowance for radiation exposure up to 100 R in case of urgent tasks or unexpected incidents.[68] This preparedness for a radiation emergency, however, initially did not extend to areas lying beyond the proving ground. When the Soviet Union conducted its first nuclear test, RDS-1, in August 1949, off-site radiation surveillance was almost nonexistent. Mobile units equipped with radiation detectors were placed only in a few towns northeast of the site and halfway between Semipalatinsk and Site "M," a base town (today's Kurchatov) located forty miles north of the hypocenter. Airplanes equipped with Geiger counters were ordered to track radioactive clouds, but only in ranges up to a little less than two hundred miles.[69]

The lack of preparations for off-site radiation exposure became instantly clear as soon as the RDS-1 test took place. As reports on unusually high radioactivity came in from distant places, Kurchatov immediately ordered an investigation. Monitors soon discovered a band of contamination on the ground stretching up to 750 miles long and 187.5 miles wide. The most contaminated area was located near the village of Dolon, 75 miles from ground zero, where the total external exposure was estimated to have reached 200 R.[70] Like their U.S. counterparts, however, Soviet nuclear officials were chiefly concerned about the risk of acute exposure to a large dose of radiation resulting in severe injury and death. Avetik I. Burnazian, director of the Third Chief Directorate, recalled that he visited a hospital shortly after the test to inspect those caught directly under the cloud's path, finding no signs of overexposure among them. At Kurchatov's insistence, he formally certified this reassuring observation via telegram sent to Moscow, confirming that the test explosion hurt no one.[71] In forwarding the field report to Stalin, Beria told the premier, "According to the conclusion of our specialists, radioactive fallout settling farther than 400 km does not represent a serious danger for people and animals."[72]

While the Kremlin was hardly bothered by local residents' exposure to radiation from nuclear fallout, it was seriously alarmed by the United States's successful interception of airborne debris from the supposedly secret Soviet bomb test. On September 23, 1949, Truman announced that his administration possessed "evidence" that an atomic explosion had occurred in the Soviet Union.[73] Astonished by this disclosure, and belatedly realizing the intelligence value of radioactive fallout, the Soviet government immediately ordered a group of nuclear scientists to install a long-range detection system. Isaak K. Kikoin, a senior physicist at Kurchatov's Laboratory of Measuring Instruments of the USSR Academy of Sciences (LIPAN), was put in charge of organizing a task group to develop methods of nuclear debris collection. David A. Frank-Kamenetskii of LIPAN and Andrei D. Sakharov of the All-Union Scientific Research Institute of Experimental Physics (VNIIEF) worked on a basic design for the sampling technique, whereas Boris V. Kurchatov, a LIPAN chemist and younger brother of Igor Kurchatov, led a team to investigate radiochemical and radiometric analysis with the help of the Radium Institute in Leningrad (today's St. Petersburg).[74]

As was the case for the United States, the scope of the Soviet nuclear intelligence program was far more extensive than monitoring for radiation safety. The State Meteorological Service and the Agricultural Ministry installed tablets wrapped with cotton fabric to collect radioactive aerosol at 120 stations across the Soviet Union.[75] The Main Intelligence Directorate in the Ministry of Defense also opened four special outposts in the Far East: Ussuriisk, Sakhalin, Kamchatka, and Dalian in China's Liaodong Peninsula, where the Soviet naval base was located (the unit in China was soon transferred to eastern Turkmenistan upon the withdrawal of all Soviet forces from Manchuria in May 1955).[76] In January 1954, Yulii B. Khariton, director of VNIIEF, decided to activate the radiological surveillance net ahead of the forthcoming U.S. nuclear test series in the equatorial Pacific, code-named Operation Castle.[77] In addition to the ground facilities across the Soviet Union, twenty air balloons were launched at six meteorological stations along the western borders, with airplanes flying between Odessa and Leningrad and also in China between Beijing and Guangdong. The collected debris proved immensely valuable, helping Soviet scientists learn some of the key design features of a multimegaton thermonuclear weapon.[78]

The Kremlin's interest in and commitment to nuclear intelligence contrasted sharply with its indifference to radiation safety at the Semipalatinsk Test Site and beyond. Unlike the United States, the Soviet Union did not immediately move nuclear tests out of the continental limits. In the midst of the fierce arms race, the Soviets could ill afford to expend time and resources in setting up new proving grounds. Soviet nuclear officials and staff scientists were thus forced to continuously rely on erratic weather to minimize off-site exposure. During the 1951 test series, the planners drew up detailed forecasts of fallout patterns ahead of each shot to ensure that a radioactive plume for any given explosion would not add more debris to areas already subject to the fallout.[79] The trade-off was a familiar one: choosing relatively low but widespread contamination over the well-known dangers of acute exposure resulting in radiation sickness or even death.

As it turned out, such a modest protective measure proved utterly ineffective as nuclear weapons tested at Semipalatinsk became more powerful. In May 1953, some of the scientists preparing for the trial of RDS-6, the first Soviet fusion-boosted bomb, suddenly realized that the planned explosion would scatter lethal fallout over a vast swath of land. Sakharov, a chief designer of the device to be tested, remembered the flurry of activity that followed. After consulting the Russian translation of a U.S. civil defense manual published in 1950, the staff scientists called for preemptive evacuation of all residents downwind in areas where total exposures from external radiation were likely to exceed 200 R, an exposure just below the level which the manual described as likely to cause injury and possible death if received by the whole body at once.[80] The military leadership agreed, although warning that the rushed evacuation carried its own risk of casualties. Sakharov remembered Marshal Aleksandr Vasilevskii, deputy defense minister in charge of the tests, as reassuring him: "Army maneuvers always result in casualties—twenty or thirty deaths can be considered normal. And *your* tests are far more vital for the country, and its defense." Sakharov, however, remained anxious about the possibility of human suffering due to his invention. "Catching a glimpse of myself in a mirror," he recalled, "I was struck by the change—I looked old and gray."[81]

With the evacuation plans barely approved two weeks before the test, 620 army trucks immediately started whisking 2,253 people, together with 6,635 heads of large livestock and 37,433 small animals, to shelters located between 125 and 156 miles from ground zero.[82] The operation, however, had not come to completion when RDS-6 went off on August 12, 1953, with a force equivalent to 400 kt, a yield twenty times as powerful as the first Soviet bomb test. The radioactive cloud arising from this enormous explosion passed over a small settlement called Abai, with its 191 residents exposed to external gamma ray estimated at between 10 and 40 R as well as unknown additional doses from the inhalation or ingestion of radioactive material.[83] This and other reports on possible overexposure, however, hardly concerned Soviet nuclear officials. Physicians who checked those exposed to the plume found no signs of radiation sickness, and the evacuees were soon ordered to return as early as two weeks after the test, even when the hourly rates of external exposures still ranged between 15 and 37 milliR, about a thousand times the normal rate.[84] Sakharov later wrote that the incident opened his eyes to the dangers of nuclear weapons to humanity.[85] Despite such an unsettling realization, the Soviet weapon scientists, still operating under the notion of emergency, went back to their laboratories to prepare for further nuclear tests.

Great Britain

As the last of the three countries that went nuclear in the first decade of the atomic age, the United Kingdom presumably enjoyed the benefit of general knowledge about radioactive fallout. Its participation in the Manhattan Project brought to many of the key scientists in the postwar British nuclear weapons program first-hand experience with test explosions. Mathematical physicist William Penny, head of the British delegation to the Manhattan Project, attended the Trinity test and the 1946 Pacific test series as an observer. W. G. Marley, chief radiation safety officer at the Atomic Energy Research Establishment (AERE) in Harwell, also witnessed Trinity in his original assignment as an explosive specialist.[86] The Anglo-American atomic cooperation, however, came to a sudden end when the U.S. Congress passed legislation in 1946 that banned the sharing of scientific information related to nuclear weapons with any foreign country. Although Clement Attlee's Labour administration saw no immediate military threat, it became anxious to develop a strong deterrent to the Soviets. Equally important was its desire to assert Britain's status as a Great Power. As Foreign Secretary Ernest Bevin put it, Britain simply "could not afford to acquiesce in an American monopoly of this new development."[87] In January 1947, Attlee made a secret cabinet decision to develop atomic bombs. The Ministry of Supply was put in charge of the nuclear program, with Penny appointed as scientific director.

Like the other two nuclear powers, Britain decided to test weapons in an isolated location to ensure both safety and secrecy. The British first asked the Americans for permission to use their testing facilities in the Marshall Islands, a calculated move in hopes of revitalizing their atomic partnership. The Truman administration was sympathetic, but it ultimately rejected the request for fear of drawing the ire of Congress in the wake of a major British nuclear espionage scandal that culminated in the arrest of Klaus Fuchs in February 1950.[88] The global reach of the British colonies and Commonwealth nations, however, helped British officials find a suitable place for staging bomb tests. An extensive search in Canada and Australia eventually led to the selection of the Monte Bello Islands, located fifty miles off the northwest coast of Australia, for Operation Hurricane. When it was found necessary to test more weapons immediately after the first trial, Penny hurriedly approved plans for Operation Totem, which took place in October 1953 on the huge claypan of Emu in South Australia. In the meantime, a committee was created to find a permanent test site. After examining more than a dozen places around the world—from Shetland to Somaliland, the Seychelles to the Falklands—the committee returned to South Australia and chose what came to be called Maralinga, located between Emu and the transcontinental railway near the southern coast.[89]

For British nuclear officials and staff scientists, Monte Bello, Emu, and Maralinga seemed to fulfill the requirements for an ideal test site. These were all located not only literally half the globe away from Britain, but also far from Australia's urban centers concentrated on the southeast coast. Monte Bello was an uninhabited archipelago of limestone rock, whereas Emu and Maralinga were both covered by rustic red sand sparsely dotted with small oases. This barren landscape, however, masked the historical presence of mobile people inhabiting Australia's colonial frontiers. The northwest coasts once boomed for pearling, with divers brought in from Japan, the Philippines, and other Asian countries. While the rapid exhaustion of oysters, the Great Depression in the 1930s, and the Japanese attacks during World War II ruined this lucrative enterprise, some of the divers stayed in the region and made seasonal visits to the Monte Bello Islands.[90] Emu and Maralinga also served as a homeland for the aboriginal Anangu people. Although a Christian mission settlement bordering Maralinga to the south was closed in 1952, the natives continued to use a north-south trekking route that cut through the weapon test range for hunting dingoes, delivering goods, and visiting relatives.[91]

When the British formally approached Canberra with the request for permission to conduct the weapon test at Monte Bello, Australian Prime Minister Robert Menzies demanded "some categorical and authoritative statement" declaring that "the effects will be innocuous."[92] In doing so, Menzies was not looking for an excuse to turn down the British request. A famous Anglophile, and keenly interested in securing British assistance for Australia's atomic energy program, Menzies was eager to help.[93] Yet his demand for a word of reassurance put the British in a dilemma. Penny feared that a detailed technical explanation of the effects of a nuclear explosion, including the radiological ones, might allow the Australians to disagree with the assessment of risk or even insist on sending their observers to the test.[94] After some debate, the British decided to offer a public reassurance without disclosing its technical basis. In a joint statement issued on February 18, 1952, the British and Australian governments declared that the trial was to be conducted in a manner that "will ensure that there will be no danger whatsoever from radioactivity to the health of people or animals in the Commonwealth."[95]

What this safety claim actually meant was revealed when British officials established radiation safety guidelines for nuclear tests. As had their U.S. and Soviet counterparts, Penny's staff in the nuclear weapons complex of Aldermaston considered off-site exposure as an emergency situation. Their initial approach seemed cautious, choosing an integrated exposure of 2 R for Operation Hurricane as "reasonable for other people as what was probably a once-for-all exposure."[96] When the trials moved to Australia's interiors in 1953, however, the British adopted essentially the same criteria as those used by the Americans for Nevada tests. Prior to Operation Totem, it was decided that firing should be permitted only when the resulting ground contamination of inhabited areas would be kept at either the "zero-risk level," equivalent to a ten-week integrated external gamma exposure of 3 R, or the "slight risk level," corresponding to exposures up to 25 R. According to Penny's staff, the "zero-risk level" represented not an absolutely safe threshold but rather a range of exposures causing "no measurable effect on the body." The "slight risk level" was defined as posing a risk of "some slight temporary sickness to a small number of people who have a low threshold sensitivity to radiation."[97]

These reference levels, however, were meant for planning purposes only. In fact, there was no systematic effort to record actual exposure levels across Australia and beyond at the time of the Hurricane test. The test director simply called off the plan for a detailed radiological survey after airplanes flying along the mainland coasts found no unusual radioactivity in the air. As it turned out, part of the radioactive cloud blown to the sea later returned to shore, crossing the Australian continent from west to east. On October 10, a week after the test, radioactivity in small amounts turned up in rainwater from Melbourne, approximately two thousand miles from the Monte Bello Islands, which prompted a shutdown in the production of photographic materials in that region.[98] Worried about the lack of long-range fallout monitoring, some Australian researchers took that task upon themselves. One such scientist was Harry Messel, head of the School of Physics at the University of Sydney, who set up a monitoring device on the roof of his office building while Operation Totem was underway in Emu. "I do not believe that there will be any danger to the present population," he told the press, "but you should not go around exploding atomic bombs, however far away, without taking measurements."[99]

Unknown to Messel and other concerned scientists in Australia, the British had been secretly monitoring global fallout for nuclear intelligence purposes. On the basis of a *modus vivendi* with the United States signed in January 1948 that restored the sharing of atomic information in non-weapon fields, Britain operated an extensive radiological sampling network in the British Isles and over the North Atlantic. Beginning in 1951, a joint committee of the AERE and the Ministry of Agriculture supervised routine measurements of gross activity in rainwater at Harwell and also at the Welsh town of Milford Haven.[100] In the meantime, the Royal Air Force also ran radiological ground stations at their airfields in Scotland, Northern Ireland, and Gibraltar, with filter-equipped bombers making routine flights between

Northern Ireland and Gibraltar.[101] By contrast, measurements of fallout in Australia for public health purposes remained sporadic and haphazard. When asked at the Australian Parliament ahead of Operation Totem if the government had any plans to start measuring radioactivity in major cities, Australian Minister of Supply Oliver Howard Beale simply repeated the blanket reassurances given by the British, saying that "every precaution has been taken to ensure that the atomic tests that will take place some time in the future will not in any way affect the Australian people."[102]

Neither government officials nor independent scientists, however, realized the looming danger to a vulnerable population: the Aborigines. Australia's indigenous policy, which systematically segregated the Aborigines from Australian citizenry, made it impossible for the test personnel to know their locations and movements.[103] Following the first Totem shot, a "black mist" reportedly rolled into the Yankunytjatjara community of Wallatina, located approximately one hundred miles from Emu, leaving its residents acutely sick. It took over thirty years until the 1985 Royal Commission finally opened the case of this alleged fallout incident and, after sifting through many testimonies and scientific models, concluded that some Aboriginal people were indeed likely to have been exposed to radioactivity from the nuclear test, resulting in temporary illness.[104] In the absence of contemporary exposure records and medical examinations for the indigenous people, however, the British simply maintained their safety claims in the aftermath of Operation Totem and moved on to test more weapons in pursuit of an "independent" nuclear deterrent in the bipolar world.

Conclusion

Central to the self-regulation of global radioactive contamination by the three nuclear powers in the early years of the nuclear age was the shared notion of emergency. Deeply worried about the risk of a severe accident by local fallout, and equally mindful of the pressing need for the development of nuclear weapons amid international tensions, the U.S., British, and Soviet nuclear authorities created a system of fallout protection designed to prevent exposure of members of the public to high-dose radiation delivered in a short time. Seen from this perspective, fallout in small amounts found in areas far away from the test sites did not appear dangerous; on the contrary, it seemed even advantageous, making it possible to detect foreign nuclear tests and also to estimate the worldwide effects of nuclear warfare. Although reports of small increases in radioactivity across North America eventually forced the U.S. government to publicly address the fact of widespread contamination, nuclear officials dismissed its health effects by stressing its smallness in quantity compared with the maximum permissible dose and natural background levels. In this way, the three nuclear rivals jointly established the dominant definition of radioactive contamination outside the test sites as a harmless phenomenon.

The scientific basis of the safety claims made by the United States, Britain, and the Soviet Union, however, had numerous unknowns. The range of fallout monitoring for radiation protection typically was confined to the test-site region and activated only at the time of trials. While the widespread interference of fallout from Nevada with X-ray film and scientific instruments forced the U.S. government to make routine measurements of gross radioactivity in surface air across North America, this HASL network did not follow the movement of the settled radionuclides through the food chain and into the human body. A separate set of knowledge developed for the military had its own limits. Although the long-range detection networks developed by the United States, Britain, and the Soviet Union covered large parts of the world, the radiological data obtained in this way remained strictly confidential. Moreover, intelligence officers ignored the nuclear debris after deposition, as much of the useful information about its source got lost. The United States tried to address some of the shortcomings in the existing fallout surveys through Project Sunshine, but it was meant to be no more than a quick check of the crude model used to study the radiological consequences of nuclear warfare. The fundamental instability and fragility of definitional power held by the three nuclear powers became instantly clear when the Castle Bravo test conducted in 1954 drastically raised public awareness of the ongoing radioactive contamination on the global scale.

Chapter 2

"ATOMIC-BOMB TUNA"

The Trans-Pacific Politics of Radiation Protection Standards, 1954–1955

O N MARCH 16, 1954, at 2:30 a.m., a truck carrying frozen tuna from Yaizu, a major fishing port located about a hundred miles southwest of Tokyo, arrived at the Tsukiji Fish Market in Tokyo. It was one of the countless numbers of trucks that delivered and picked up some thousand tons of marine products auctioned each day in the world's largest wholesale seafood market. Once unloaded, the tuna was laid down on the concrete floor for a rapid-fire auction. Then a strange thing happened. The Tokyo metropolitan office ordered that the Yaizu fish be quarantined immediately, and a group of men stormed into the market, one with a Geiger counter in hand. As the monitor scanned the quarantined tuna, it ticked faster and louder, with its needle pointing to as much as 250 times natural background levels of radiation. That midnight, with hundreds of spectators watching, the entire truckload of tuna was buried in a deep hole dug at the corner of the market.

The radioactive fish discovered at Tsukiji came from a tuna boat named the *Daigo Fukuryū Maru* (Lucky Dragon No. 5). Two weeks earlier, on March 1, 1954, the small wooden ship with twenty-three crew members had been fishing over a hundred miles east of Bikini Atoll in the equatorial Pacific when the United States conducted the Castle Bravo test. This thermonuclear explosion was so enormous—15 Mt, or approximately a thousand times the force of the Hiroshima-type bomb—that one of the *Lucky Dragon* crew later recalled seeing "the sun rising from west." As it turned out, a radioactive plume was blown by the wind eastward toward the inhabited atolls. The U.S. Joint Task Force in charge of Operation Castle immediately responded by evacuating a few hundreds of the test personnel and local residents from the now seriously contaminated islands. Unknown to U.S. officials, however, the *Lucky Dragon* was also caught in this fallout. As soon as the boat returned to its home port of Yaizu, the crew members, all seriously ill, were immediately taken to the hospital. Although all but one slowly recovered from an acute form of radiation sickness, Aikichi Kuboyama died on September 23. His death was instantly enshrined as the third atomic tragedy that befell Japan, following Hiroshima and Nagasaki.[1]

While the suffering of the *Lucky Dragon* crew drew nationwide sympathy and indignation at the United States, it was the contaminated tuna aboard the ship that terrified the whole nation. By the time the *Yomiuri Shimbun* newspaper scooped the story of the *Lucky Dragon*, the ship's catches already had been auctioned on the spot and shipped to fish markets in Tokyo, Osaka, and other major cities. The breaking news on the escape of the "atomic-bomb tuna" triggered a nationwide panic. Although almost all of the fish, including those buried at the Tsukiji Fish Market, were intercepted before reaching consumers, people stopped eating tuna and other suspicious seafood, forcing thousands of fish retailers and sushi restaurants across the nation to close for many days. Desperate to reassure anxious consumers, the Japanese government immediately ordered all tuna boats returning from areas near Bikini to have their catches inspected for radioactivity. The fears of many Japanese were realized as the number of tuna declared unfit for consumption kept increasing, and the areas from which the contaminated fish came gradually expanded to encompass most of the Western Pacific. Compounded by the discovery of unusually high levels of radioactivity in rainwater, which affected a variety of crops grown in Japan, the "atomic-bomb tuna" consumer scare escalated, sparking a meteoric rise of grassroots movements against nuclear weapons, led by concerned housewives.

Although the tuna aboard the *Lucky Dragon* and the radioactive tuna subsequently discovered at many seaports shared a common source of contamination, there was an important difference between them. While the former were exposed to an intensely radioactive plume coming fresh from the nuclear explosion, the latter contained small amounts of bomb debris dispersed over the Pacific and beyond. The latter kind of fallout had been present in many parts of the world since 1945, but the U.S., British, and Soviet nuclear authorities consistently ignored its potential effects on humans. Why then did this supposedly harmless phenomenon of global fallout contamination suddenly appear, for the first time, as a serious threat to public health?

To understand what prompted the redefinition of fallout hazards in Japan, this chapter examines the politics of radiation protection standards used to judge the fitness of food and water for consumption. The context of this standard setting was trans-Pacific, as Japan and the United States shared not only the Pacific Ocean, where commercial fishing and nuclear testing came into direct conflict, but also political ties, economic interests, and scientific networks as Cold War allies. This multilayered connection required that Japan and the United States consult closely on radiation protection standards in order to cope with the political fallout from the *Lucky Dragon* incident. The U.S. nuclear authorities, however, had long committed their radiation safety resources to the problem of local fallout, creating many gaps and holes in knowledge about global fallout. This considerable disparity in knowledge about two types of fallout pushed the bilateral epistemic negotiations in an unexpected direction, as Tokyo and Washington, in an attempt to reassure the tuna industry and consumers on both shores of the Pacific, agreed to use the most stringent criteria for the inspection of tuna. When standards that did not account for the presence of artificial radioactivity were applied to the material world of the nuclear Anthropocene, the result was disastrous for the Pacific allies.

Tuna and the Bomb

In the view of U.S. nuclear officials, the equatorial Pacific seemed to offer an ideal environment for conducting nuclear tests. Continental test sites posed a risk for local communities of accidental exposure to a radioactive plume. By contrast, there was no permanent human presence anywhere near the Pacific Proving Grounds. The closest inhabited atoll, Rongelap, was located a hundred miles east of Bikini, and all residents of Bikini and Enewetak, approximately three hundred in total, had been relocated to different atolls for an indefinite period.[2] An ocean as vast as the Pacific, however, was not a lifeless void but a home for countless fish—and fishers chasing them. The origins of the "atomic-bomb tuna" were rooted in this dual use of the Pacific as an unlimited source of food and a bottomless bin for radioactive waste.

Japanese fishers had roamed in the equatorial Pacific well before the Cold War. During World War I, Japan seized the Caroline, Marshall, and Northern Mariana Islands from Germany and received a League of Nations mandate to govern their new imperial acquisitions. By the mid-1930s, as motorization made deep-sea fishing possible, tuna had become one of the leading commercial products from the South Pacific Mandate. While the skipjack was dried, smoked, and shipped for the Japanese market as an ingredient in broth, the albacore was canned as "white meat" and exported to the United States.[3] The Pacific War, however, ruined this fledgling business, as it turned the South Pacific Mandate into a front line for violent conflict. The United States eventually captured all of these islands during the war and replaced Japan as a United Nations trustee. Meanwhile, the U.S. occupation authorities in Japan imposed the so-called MacArthur Line, restricting Japanese commercial fishing to coastal waters around mainland Japan.

By 1950, however, the United States had gradually reopened the tuna ground around the Trust Territory for Japan. It was part of the U.S. effort to accelerate the economic reconstruction of Japan as an anticommunist keystone in East Asia. The timing was perfect, as fish stocks within the MacArthur Line had been seriously depleted. Yet Japan's West Pacific neighbors strongly resisted the return of Japanese fisheries to the waters of their interest and even forcefully seized Japanese boats and confiscated their catches. Squeezed from all sides, many Japanese fishers headed south to the equatorial Pacific as the only readily accessible deep-sea ground.[4] While a majority of the tuna caught by the Japanese was consumed in their country, a considerable portion was canned or frozen for exportation. Japan's sales of tuna to the United States rapidly increased from 1951 through 1954 at an average annual rate of 12.4 percent.[5] The boom accelerated after the San Francisco Peace Treaty took effect in 1952 and restored to Japan unimpeded access to international waters. Tuna exports to the United States were so successful that they triggered one of the first postwar trade wars between the two countries.[6]

The U.S. government was well aware of the commercial value of tuna when it opened a nuclear test site in the Pacific. At the time of Operation

Crossroads, mounted at Bikini in 1946, the Joint Task Force included the investigation of oceangoing fish in its radiation safety program. The initial results seemed reassuring. While lagoon fish at Bikini became instantly contaminated following the tests, not a single sample of migratory fish caught near the neighboring atolls was found to be radioactive.[7] As radioactive materials diluted in seawater gradually seeped out of the lagoon and selectively reconcentrated through marine food chains, however, some pelagic fish started showing radioactivity from within. When the Joint Task Force sponsored the Bikini Scientific Resurvey in 1947, a radiobiology team led by Lauren R. Donaldson, a University of Washington fish scientist, discovered that the liver of tuna caught from Rongerik, some 120 miles east of Bikini, registered unusually high levels of radioactivity.[8]

Despite this ominous trend, the U.S. nuclear authorities soon stopped systematically checking pelagic fish. For one thing, the test area seemed devoid of commercially important stocks. Records showed that no more than 1 percent of Japan's tuna catches came from the waters near Bikini.[9] More important, the direction of the radiobiology research connected with Pacific tests changed. Innovative and ambitious, Donaldson shifted the focus of his University of Washington team's study from individual species toward an entire coral atoll, using residual radioactivity left by the nuclear tests as a tracer to map the complex webs of energy and mineral transfers among different components of the atoll environment. This ecological approach led to one of the key findings that later helped explain the phenomenon of the "atomic-bomb tuna"—radioactivity in seawater was selectively absorbed by organisms and accumulated in much higher concentrations. As extensive and holistic as Donaldson's vision was, historian E. Jerry Jessee has noted, it still treated the atoll as spatially and ecologically distinct from the surrounding oceans.[10]

The conceptual boundary between atolls and oceans broke down when the United States conducted the Ivy Mike test at Enewetak in November 1952. Like Castle Bravo, this 10-Mt thermonuclear explosion flooded the atmosphere with a massive amount of radioactive material. Although there was no report of accidental exposure, John C. Bugher, director of the Division of Biology and Medicine of the U.S. Atomic Energy Commission (USAEC), suddenly realized the possibility of radioactive contamination over large areas in the Pacific. Well aware of the status of tuna as "an important economic asset especially to Japan and the United States, including the Territory of Hawaii," Bugher decided in March 1953 to establish a permanent scientific station in Enewetak to speed up marine biological research. He also recruited animal physiologist Willis R. Boss from Syracuse University as a special investigator of pelagic fish in the Pacific.[11] Meanwhile, in November 1953, Bugher flew to Seattle for a meeting with Donaldson and his colleagues to discuss a plan to expand fallout studies to the oceans. Noting that only a very small portion of the total radioactivity from the Ivy Mike test was observed within the test area, he asked about the rest, "The question is where is it?"[12] Before U.S. scientists set out to find the answer, however, the first batch of the "atomic-bomb tuna" appeared in Japan.

Atomic-Bomb Tuna

On March 18, 1954, two days after the *Lucky Dragon* arrived in Yaizu, Japan's Ministry of Agriculture and Forestry (MAF) and the Ministry of Health and Welfare (MHW) issued a joint statement, declaring that all fishing boats operating in or passing through the disaster areas were ordered to enter one of five designated ports (Misaki, Tokyo, Yaizu, Shimizu, Shiogama) to undergo radiological inspection.[13] This measure was intended to reassure consumers in Japan about the safety of tuna. As it turned out, the most outspoken stakeholder came from the other side of the Pacific. Fearing that the panic in Japan might spread to the United States, Van Camp Seafood and other major U.S. tuna canners threatened to cancel their orders on frozen tuna from Japan unless these products were officially certified as radioactivity-free. On March 20, Japanese officials, U.S. embassy staff, and representatives of the U.S. fishing industry met in Tokyo, reaching an agreement: the Japanese government would conduct an extra check for tuna exports so that "no fish showing any signs of radio-activity will be shipped" to the United States.[14] To ensure the proper operation of this tuna inspection, the USAEC hurriedly dispatched its trusted radioactive fallout troubleshooter, Merril Eisenbud of the Health and Safety Laboratory (HASL), to Tokyo.

Japan's pledge to conduct the most rigorous inspection, however, raised a baffling question. Given that all foods contained natural radioactivity in varying degrees, how much radioactivity in tuna was too much? In the United States, upon the USAEC's advice, the Food and Drug Administration (FDA) scanned tuna imports from Japan against the rate of 100 counts per minute (cpm) for gamma radiation, set slightly above normal counts at sea level, just to see if the fish showed any unusually high levels of radioactivity. The underlying assumption was that it was highly unlikely to find any contaminated tuna at all. Bugher, for instance, argued that "the probability of significant contamination of fish outside the immediate test area is inconsequential and that in all likelihood we will be unable to detect these [contaminated] waters more than a few hundred miles away from the Marshall Islands."[15] This reassuring claim, however, lacked supporting evidence based on actual measurements. In an internal memo written shortly after the story of the *Lucky Dragon* broke, a USAEC official admitted that "[scientific] tests will take 10 years to establish this fact [of safety] as a scientific certainty."[16] Nonetheless, the U.S. government released a public statement including Bugher's estimates through the U.S. embassy in Tokyo on March 24.[17]

Until the Castle Bravo test, the U.S. nuclear authorities had repeatedly made blanket reassurances to the public whenever reports surfaced of unusual increases in the radioactivity level. This time, however, was different, as such a statement critically shaped the course of U.S.-Japan consultation on the radiation protection standards for tuna inspection. A day after the USAEC made the press release, Eisenbud arrived in Tokyo for a meeting with Japanese officials and experts. Asked for advice on the maximum permissible concentration of radioactivity in tuna, Eisenbud demurred, saying that it was too early to recommend any specific figure without further data. Predicting that "significantly contaminated fish were not likely to be found," he urged the Japanese to adopt the same inspection criteria as the FDA's "for the sake of uniformity."[18] The Japanese government apparently accepted his advice with extra caution by adding beta radiation to measurements. On March 30, the MHW issued a detailed instruction for the inspection of tuna, determining that tuna was to be scanned at 10 cm from surface and to be rejected if the counter registered 100 cpm and above, with beta and gamma combined.[19]

Contrary to expectations, however, inspectors discovered that some of the tuna examined at seaports showed radioactivity above 100 cpm. Even more surprisingly, the volume of contaminated fish increased over time (Table 1). Much alarmed by this development, the MHW revised its inspection guidelines in July, ordering the rescanning of tuna following the removal of viscera where radionuclides were found to concentrate.[20]

This modest change hardly solved the fundamental problem of contamination, as the westward ocean currents continued to carry radioactive debris from the equatorial Pacific toward Asia. By summer, when the tuna fishing season came to waters around Japan, the fish caught there also had begun to show radioactivity in edible flesh well over 100 cpm (Figure 3).[21] The volume of the discarded tuna started increasing again, reaching its peak as late as October, almost six months after the end of the test series at Bikini. As news media continued their sensational coverage of the "atomic-bomb tuna," Japanese consumers remained wary of this once popular fish. A Japanese official reportedly lamented, "We would like to make an ardent appeal that it is alright to eat [tuna] because the inspection is strict. But customers will not listen to us because the fact is that radioactive tuna continues to come in now and then."[22]

While a variety of factors—deep-sea fishery, extensive inspection, and stringent standards—combined to make Japan the first country hit by the "atomic-bomb tuna," contaminated fish also made landfall in several other Pacific countries. In July, the U.S. Foreign Operations Administration mission stationed in Taiwan secretly reported to Washington that some fish caught in southwestern Taiwan showed radioactivity up to 2,000 cpm from the spleen.[23] In Okinawa, then under U.S. control, two tuna boats returning from areas south of the Philippines underwent a radiological check for an undisclosed reason.[24] Even the United States, located in the opposite direction from the ocean currents that carried bomb debris, could not completely escape the threat of contaminated fish coming through trade. At least twice in May and once in July, FDA inspectors intercepted frozen tuna from Japan that showed radioactivity well above natural background levels. The State Department instructed the U.S. Embassy in Tokyo not to inform the Japanese government of this fact for fear that "public revelation in Japan would result [in] publicity here seriously adverse [to] imports [of] Japanese tuna and possibly domestic tuna sales."[25] It was the United States's careful control of information that prevented the consumer panic in Japan from spreading to other parts of the Pacific Rim.

Meanwhile, the Japanese government was forced to find out the true extent of the contamination on its own. On May 15, as soon as Operation Castle ended, a team of scientists aboard the small trainer ship *Shunkotsu Maru* left Japan toward Bikini. This two-month, nine-thousand-nautical-mile expedition was the world's first oceanographic

TABLE 1. Results of Tuna Inspection at Seaports, March to December 1954.

RATIO OF THE FISH IMPOUNDED AT THE FIVE DESIGNATED PORTS (%) SORTED BY DETECTED RADIATION LEVELS

	Number of Boats Affected	Volume of Fish Destroyed (kg)	Greater than 5,000 cpm	Greater than 3,000 cpm	Greater than 1,000 cpm	Greater than 500 cpm	Greater than 100 cpm
March	2	61,506	100				
April	17	34,140				6.2	93.8
May	86	37,019	2.8	2.8	19.4	25	50
June	126	57,894			34.2	31.6	34.2
July	73	18,290	5	15	25	15	40
August	71	72,953	8.3	8.3	19.5	25	38.8
September	79	64,424	2.8	5.6	47.2	25	19.5
October	126	96,135	3	1.5	16.4	26.8	52.3
November	162	30,484		1.3	11.7	24.6	62.4
December	114	24,151	1.2	2.4	14.6	23.2	58.6

Source: Adapted from Miyake, Hiyama, and Kusano, *Bikini suibaku hisai shiryōshū*, 162, 164.

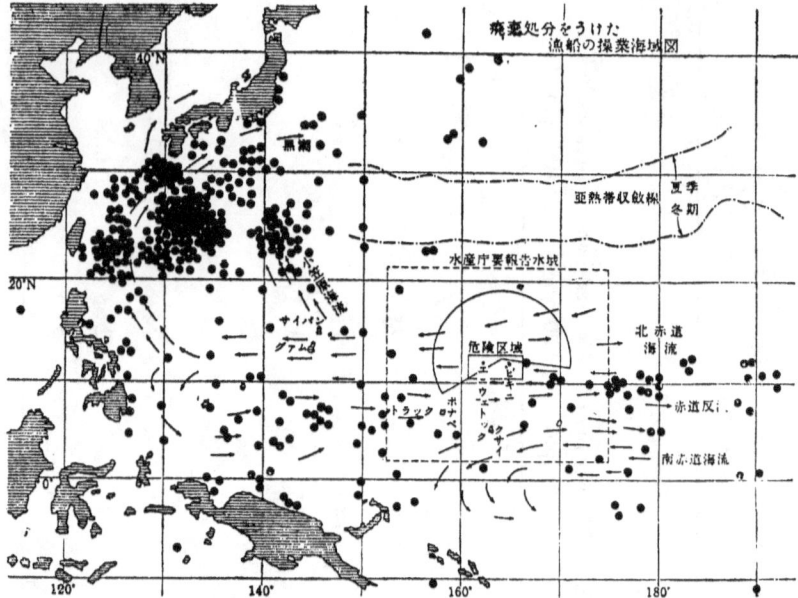

FIGURE 3. Areas of Operation for Tuna Catches Registering over 100 cpm, March 16 to August 31, 1954.

Source: Miyake, Hiyama, and Kusano, *Bikini suibaku hisai shiryōshū*, 103.

expedition to investigate the dissemination of fallout in the Pacific beyond U.S. test sites. Its data clearly contradicted the earlier assumptions made by the USAEC. A sample of seawater, taken two thousand kilometers west-northwest of Bikini, showed gross beta and gamma activities comparable to those at a radioactive sewage pond attached to Oak Ridge National Laboratory.[26] Stunned by this and other findings by the Japanese, the USAEC organized its own expedition in early 1955. Operation Troll essentially confirmed the Japanese findings that contaminated water did not dissipate as quickly as had been expected, but continuously moved on westward ocean currents from Bikini toward the Philippines, Taiwan, and Japan.[27] While the report claimed that at no point contamination neared the maximum permissible concentration, it predicted that radioactivity in seawater would remain "high enough to attract the attention of Japanese scientists," which was of "political as well as scientific significance."[28]

Radioactive Rain

Tuna was not the only object in Japan through which global fallout became perceptible. Another was rainwater. While early reports were ambiguous due to the influence of naturally occurring radioactivity, rainfall starting on May 16, 1954 provided the first unmistakable evidence of artificial radionuclides in rainwater.[29] The contamination of rainwater was particularly disturbing for approximately 750,000 people in Japan who depended on it for drinking. When rainwater collected in Ōshima, an island south of Tokyo, showed increased radioactivity, the MHW urged the islanders to refrain from drinking rainwater at the peak of contamination.[30] As the rainy season began and news media continued to report high counts in rainwater, people grew restless. Gossip about hair loss and other symptoms stemming from acute radiation sickness spread, and the sales of rubber boots, rain coats, and umbrellas soared.[31] People became even more terrified when small amounts of artificial radioactivity also began to appear on the surfaces of rice, vegetables, fruits, and almost everything touched by contaminated rainwater. With traces of radioactive debris found everywhere, the Japanese government decided on June 17 to expand the MHW's expert panel, the Atomic Bomb Effect Research Commission (ABERC), to include environmental scientists.

Unlike the "atomic-bomb tuna," which was discovered by government inspectors, "radioactive rain" emerged through spontaneous radiation monitoring by scientists across Japan. Upon hearing about the fallout from Bikini, researchers with Geiger counters began to routinely analyze precipitation and communicate their findings via news media.[32] According to Yasuo Miyake, director of the Geochemical Laboratory at the Meteorological Research Institute, these scientists were not only curious about the novel phenomenon of radioactive rain, but also bound by a sense of mission to protect people from fallout as well as by indignation at the nuclear arms race.[33] As was the case for tuna, regular rain became "radioactive rain" largely because of confusion regarding reference levels. Assuming that the composition and ratios of fission products in question were identical with those at the time of formation, the USAEC declared as permissible gross beta and gamma activities in drinking water up to the level extrapolated to 5 microCi per liter three days after a nuclear explosion. Many Japanese scientists, however, insisted that one should use a much lower level given by the U.S. National Committee on Radiation Protection

(NCRP) for activities of unknown origin—10^{-4} microCi per liter, or fifty thousand times less than the USAEC's value—until radiochemical analysis determined the composition of radionuclides in rainwater.[34] The application of this stringent standard, corresponding to a few dozen cpm, turned rainwater containing fallout into "radioactive rain." In fact, as Miyake pointed out, almost all rainfall during the monsoon season in 1954 exceeded the NCRP's most conservative permissible concentration.[35]

Initially, measurements in rainwater were higher along Japan's Pacific coast than along its Sea of Japan coast, suggesting that radioactive dust chiefly came from the south, where the United States had been conducting nuclear tests. In September, however, radioactive rain also started falling on the Sea of Japan side. Miyake and other meteorologists suspected that the source of pollution came from the northwest—the Soviet Union.[36] The Soviets vehemently denied this allegation. In February 1955, Radio Moscow broadcast a message to the Japanese, insisting that the Soviet Union, unlike the United States, conducted nuclear tests within its territory so that fallout would affect no one outside its borders.[37] A group of pro-Soviet scientists in Japan even made up false stories, alleging that the United States might have secretly dumped radioactive dust or exploded a small atomic bomb over the Sea of Japan.[38] Miyake, however, successfully matched measurements in Japan with those made in France, both unmistakably pointing to the Soviet Union as the origin.[39] With fallout coming from north and south, Miyake and other scientists warned the people of Japan that they now lived in the "valley of ashes of death."

Crisis

Bombarded by news reports on the contamination of tuna, rainwater, and other foodstuffs, many Japanese started turning their attention to the nuclear tests taking place thousands of miles away. Given the direct economic blow, it was natural that the fishing industry first stood up against further tests. Major tuna ports and markets, including Yaizu, Misaki, and Ishinomaki, as well as the Tsukiji Fish Market in Tokyo, unanimously adopted resolutions calling for the suspension of nuclear testing and for compensations for all losses arising from the tuna scare.[40] The latter demand was especially troubling, as the U.S. government had tried to arrange an *ex gratia* payment to quickly settle the claims related to the *Lucky Dragon* incident without admitting any wrongdoings. Pressed by the fishing industry, however, Tokyo escalated its demands. While the United States made a proposal on May 21, 1954, for making a one-time payment of US$150,000 for direct damage relating to the *Lucky Dragon*, the Japanese counterproposal listed all damages totaling over $6 million, including a large sum of expenses relating to port inspection, the destruction of tuna, and the decline in tuna sales.[41] This substantial discrepancy between the two positions immediately brought the settlement talks to a stalemate.

The fishing industry, however, faced a dilemma. As its spokesman explained, "It might be possible that they do not stop atomic tests if we do not clamor about the effects of radioactivity. But if we complain about it too much, the fishmongers would be in trouble."[42] In the end, the fishing industry switched to stressing the safety of their products through aggressive public relations campaigns. For instance, the Japan Federation of Tuna Fishers hosted a tuna-tasting event, with politicians, scientists, and celebrities eating tuna rolls and other dishes in front of cameras.[43] Sound trucks equipped with loudspeakers traveled across Japan, distributing hundreds of thousands of flyers in Tokyo and other major cities. One such bill read, "Don't be surprised: your body and rice you eat everyday have 20 to 100 [cpm] of radioactivity. We enjoy hot springs because of radioactivity. People are scared of fish without knowing these simple facts. What we should fear is not radioactivity but ignorance."[44]

While the fishing industry tried to walk a fine line regarding dangers of fallout, consumers felt no such reservation. The tuna panic hit those living in cities particularly hard, as fish tended to be consumed in cities much more than in the countryside.[45] This deep connection between cities and the seas made urban consumers unusually sensitive to the influence of fallout contamination. In April 1954, the Japanese Consumers' Co-operative Union (Nihon Seikatsu Kyōdō Kumiai Rengōkai) made an announcement, stating that further tests would "destroy our frugal livelihood by disrupting the nutrition source."[46] Consumers' anger burgeoned into a ban-the-bomb resolution in September, one of the first Japanese-sponsored initiatives in the International Co-operative Alliance.[47] The relay of fallout through the market economy of foods thus triggered an unexpected consumerist turn in the growing antinuclear sentiments in Japan.

The rising tide of consumerism did not occur in a vacuum. Indeed, the core identity of consumers in early postwar Japan was gendered, reflecting the social norm of women as "good wives and wise mothers," whose duty was to manage consumption and health at home for their children and husbands.[48] To these homebound women, the fallout hazards appeared first and foremost as a threat to the peace of their kitchens. A woman in Tokyo, who promptly collected about fifteen hundred ban-the-bomb signatures with dozens of housewives, said, "Of course we aren't saying that we eat tuna all the time. But since this [*Lucky Dragon*] incident, it's been really difficult to fix meals."[49] As fallout trespassed into the women's sphere of concern through foods, "the trouble of fishery people" was instantly transmitted to kitchens as "the trouble of housewives."[50]

Despite their homebound orientation, housewives' concerns hardly remained within the closed circuit of domesticity. The bitter memory of the Pacific War and the image of motherhood as a love-giver and moral redeemer had already attuned Japanese women to the question of peace.[51] Fallout thus provided a spark for women to enter antinuclear activism as "the bearer of the kitchen and that of peace."[52] A women's joint appeal in April 1954, which bore the names of the National Coordinating Council of Regional Women's Associations (Chifuren), the Housewives' Association (Shufuren), and other groups, spoke of the connection between women and peace forged through global fallout: "We the Japanese women are firmly determined not to let our suffering happen again to any other country in the world and not to let the 'ashes of death' fall from the sky worldwide any longer."[53]

The same urgent sense of a mission to rescue the kitchens and restore peace to the world inspired a local movement in Suginami, a residential area of Tokyo with a rich tradition of civic activism. On April 16, during a Women's Week event sponsored by the Suginami Association of Women's Organizations, a fishmonger's wife named Tomiko Sugawara described a threat of contamination to their business and urged others to take action against nuclear weapons. In response to her plea, a book club of housewives organized by Kaoru Yasui, a professor of law at Hōsei University, launched a petition drive demanding a ban on thermonuclear testing.[54] A volunteer reported that "there has been no movement like this to collect [signatures] with such ease..... People were deeply anxious, complaining that we cannot eat fish and vegetables."[55] Within a month, the petition drive collected over two-hundred-and-sixty thousand signatures in the ward of some four hundred thousand citizens.[56] The campaign quickly spread nationwide. Although the movement drew upon a number of concerns, namely the tragic death of Kuboyama and the growing sympathy for the atomic bomb survivors in Hiroshima and Nagasaki, contamination by radioactive fallout remained a major theme. A campaign pamphlet noted the contamination of tuna, rain, rice, drinking water, and vegetables, attributing the movement's success to its "roots in the most immediate demand of citizens' livelihoods and lives."[57] By the end of 1954, the petition had collected nearly twenty million signatures, or one-fourth of Japan's total population. Backed by popular support, the World Conference against Atomic and Hydrogen Bombs was held in August 1955. From this successful event, Japan's first grassroots, permanent ban-the-bomb organization was born, called Japan Council against Atomic and Hydrogen Bombs (Gensuikyō).[58] In this astonishing chain of events, global fallout, made perceptible through the "atomic-bomb tuna" and "radioactive rain," turned Japan into an antinuclear nation.

Unmaking of Atomic-Bomb Tuna

As the fallout scare rocked Japan, Washington became seriously alarmed. Of course, the obvious solution to the problem was to cease nuclear testing. Instead, U.S. officials laid the blame on the radiation protection standards in use for the inspection of tuna in Japan. As early as April 10, U.S. Ambassador to Japan John M. Allison asked Washington whether the USAEC or FDA was "prepared [to] establish [a] health standard for fish products in terms [of] maximum quantitative counts per minute."[59] Perhaps unknown to him, the USAEC had already informed the FDA of revised guidelines for tuna inspection in the United States, which the FDA formally approved on April 20.[60] The guiding principle for this new instruction was the concept of permissible dose. Concluding that highly carcinogenic strontium-90 (Sr-90) was the most dangerous element of all fission products, the USAEC based its fish inspection standards on the maximum permissible concentration for Sr-90 recommended by the NCRP: an hourly rate of gross gamma at 0.1 milliR, or ranging between 300 and 600 cpm when measured at five centimeters from the body.[61]

To persuade the Japanese to revise their inspection standards accordingly, U.S. officials first tried a government channel. On May 6, approximately a month after Allison made his request, the FDA wired the new guidelines to the U.S. embassy. Four days later, however, the State Department decided to withhold the document from formal presentation to the

Japanese. One of the reasons was the apparent respite in the tuna panic. At the time, the Japanese government had just announced the termination of a special check for tuna exports, claiming that the regular port inspection proved capable of intercepting all contaminated fish. Citing a news report on this decision, U.S. foreign officials in Washington believed that the "[i]mmediate urgency of tolerance question [was] now apparently removed." A more important reason, however, was a "possible psychological effect on US tuna consumers of public release of [the] tolerance standard for tuna." Such an announcement, the State Department feared, "[m]ay well set tuna apart from other food products with grave risks to US market for both Japanese and domestic tuna."[62]

In the meantime, Japanese scientists had a heated debate on the question of permissible dose. The Food Sanitation Panel and the Environmental Sanitation Panel of the MHW's expert committee created a joint task force on this subject, but it failed to reach consensus. Ironically, the cause of confusion was none other than the NCRP manuals that the United States relied on, as the protection committee offered different standards for different circumstances. The NCRP handbook listed two versions of the maximum permissible concentration for drinking water: the equivalent of 49 cpm per liter for gross gamma and beta radiation of unknown origin (set extremely low to account for any of the radioisotopes, including transuranic elements) and 392 cpm for Sr-90. While the USAEC used the latter as a basis for its revised references, many Japanese scientists preferred the former due to the lack of information about radionuclides in tuna. Their cautious attitude also stemmed from multiple uncertainties relating to radiological measurements, eating habits, age differences, and the factor of genetic damage for which there might be no threshold at all.[63]

As the formal review process in Japan failed to prompt changes in tuna inspection, U.S. and Japanese officials turned to an informal channel of communication through their scientists. In May, Bugher put forth the idea of holding a bilateral academic conference, and Boss and Donaldson visited Tokyo a month later to sound out a number of senior Japanese scientists on the proposal. A key figure among them was Yoshio Hiyama. An expert on tropical fish in Micronesia, he served as adviser on the matter of tuna inspection for the MHW and Science Council of Japan (SCJ), a representative organization of scientists and consultative body for the government. He had also been in close touch with Donaldson and his colleagues in Seattle through correspondence. Later in his life, Hiyama developed his reputation as an antinuclear intellectual. However, there was a less-known side to him as a staunch supporter of the U.S.-Japan alliance.[64] In a letter soliciting scientific material on the radioactive contamination of fish, Hiyama told Arthur D. Welander, a senior researcher at Donaldson's lab, that "I am earnest to blow out and take off some sort of anti-[A]merican feeling which probably originated by the communists who are utilizing the chance of this unfortunate accident and aiming to let people fear their common food and get in co[n]fusion."[65] His political beliefs, combined with his network of U.S. scientists and insider status in both the Japanese government and scientific community, made Hiyama an ideal broker for the renegotiations of the radiation protection standards between the two countries.

In fact, behind the scenes, Hiyama already had been advocating the upward revision of the inspection standards. As early as April, he had written to Welander complaining about the 100 cpm standard, warning that "if we have to contenue [sic] the examination with so strict level of safety, the discovery, which may occur possibly, of unsuitable fish may excite the public." Hiyama even sought out Donaldson's view on an idea of significantly raising the permissible level before the Japanese government made a formal move. In doing so, he was self-conscious about his role as an epistemic diplomat. "Being a scientist, I do not want to go in[to] the matter related with diplomacy or polotics [sic], but I feel I must tell the right fact about the fish and sea related with this unfortunate accident."[66] It was thus no surprise that Hiyama immediately supported the U.S. proposal for a scientific conference, even suggesting that it should be held immediately to "correct exaggerated stories and reactions still appearing [in] Japanese press."[67] On June 10, Hiyama and three other members of the SCJ held an executive meeting, agreeing that the SCJ should sponsor the conference in Tokyo to discuss the permissible dose, decontamination, and radiological measurements.[68]

The U.S.-Japan Radiobiological Conference began on November 15. During the closed, four-day meeting, the U.S. delegation, led by Paul B. Pearson of the USAEC Division of Biology and Medicine, tried to persuade the Japanese scientists to revise the inspection standard from 100 to 500 cpm on the basis of the concept of permissible dose. Their effort, however, failed. The Japanese biophysicists and geneticists repeated the same arguments raised in the MHW's expert panel, pointing out that the permissible dose shown in the NCRP handbook lacked firm evidence and also overlooked genetic damage.[69] This time, however, the Japanese government acted swiftly. Immediately after the U.S. experts gave a presentation on the inspection standards, the MHW leaked to news media a plan to relax the Japanese inspection standards. Masahiro Kusumoto, director of the Bureau of Environmental Hygiene, was quoted as saying, "We do not change the standards just because the Americans say that radioactivity below 500 counts is harmless, but given various experimental data, we will decide on the significant easing of the rejection criteria within next week."[70] As many scientists complained, the leak was timed so that it gave the misimpression that the conference participants had discussed and approved the change.[71]

At this critical juncture, Hiyama took a curious position. At a public roundtable held on November 29, Hiyama said that he "protested Mr. Kusumoto" for making comments without prior consultation. At the same time, he defended the need to establish the permissible dose not only to solve the tuna problem but also as an essential groundwork for peaceful uses of atomic energy that Japan had just begun to undertake. In fact, in the very month that the *Lucky Dragon* incident happened, conservative politician Yasuhiro Nakasone submitted in the Diet a budget proposal for building an atomic reactor. "As we will undoubtedly enter the atomic age," Hiyama said, "we must study this problem [radiation hazards] seriously to prepare for the time when we will build nuclear reactors and power plants." Having no standard for the future, he said, "will only confuse the public."[72] For him, the dream of atomic power, which was then heavily publicized in Japan to counter the negative image of nuclear weapons, justified the introduction of the permissible dose.[73]

As a fish scientist, Hiyama had no real influence over the issue of the permissible dose. To bring the tuna problem to an end, Hiyama and his associates took steps to introduce the permissible dose through the back door. By then, studies had shown that a main source of radioactivity in tuna meat was not Sr-90 but Zinc-65, an induced radionuclide supposedly coming from metal encasing the thermonuclear bomb.[74] On the basis of these findings, Hiyama drafted a paper on December 10 showing that, even if one ate tuna every day, the amount of Sr-90 absorbed would remain far below the maximum permissible concentration recommended by the NCRP. On

December 22, the MHW called a special meeting of the Food Sanitation Panel of its expert board. The panel members, including Hiyama, reportedly discussed a variety of data, concluding that "tuna meat is safe." Their public statement echoed the logic of permissible dose, presenting radiation exposures up to a certain level as biologically and socially acceptable: "While it is ideal to have no radioactive contamination, we reached a tentative conclusion based on the internationally-recognized maximum permissible concentrations."[75] Upon receiving the report from them, the MHW immediately announced its plan to terminate the inspection of tuna. Although a number of biophysicists and geneticists lodged strong protests, the health ministry called off the inspections at the end of 1954. A few days later, on January 3, 1955, the U.S. and Japanese governments reached an agreement on the settlement of the claims related to the *Lucky Dragon* incident, with a lump sum payment of US$2 million made as an *ex gratia* without setting a legal precedent for future liability.[76] With no more inspections carried out, the "atomic-bomb tuna" disappeared from Japan, at least for a time.

Conclusion

For the Japanese, the *Lucky Dragon* incident was not only about a ship and its crew. All of a sudden, the Japanese woke up to the nuclear Anthropocene in which, in the strictest sense, few objects and places on the planet were free from artificial radionuclides and their potential effects. This chapter has demonstrated that radiation protection standards used to check food and drink drastically changed the ways in which faint fallout in the global environment appeared to the Japanese. What drove the standard-setting process throughout was a series of epistemic negotiations between Washington and Tokyo. The efforts to rescue their Cold War alliance from the political fallout from the *Lucky Dragon* incident, however, consistently suffered from the troubling limits of knowledge about global fallout. Despite the lack of actual measurements, the US-AEC was forced to draw a conclusion that the contamination of tuna with fallout in detectable amounts was highly unlikely. As it turned out, the 100 cpm standard adopted on this optimistic assumption led to regular discoveries of the "atomic-bomb tuna." With little known about radioactivity found in rainwater and other objects, many Japanese scientists also chose the most stringent of all maximum permissible doses recommended by the NCRP to judge their safety. All of this badly shook the U.S.-Japan alliance, stalled the bilateral settlement talks, and sparked the groundswell of antinuclear sentiments in Japan.

This chapter has also examined the subsequent renegotiation of the inspection standards used in Japan. A key epistemic device in this desperate attempt to contain the fallout scare was the concept of permissible dose. Instead of working through official diplomatic channels, however, Washington and Tokyo turned to scientists as their informal agents. Hiyama played a critical role in this track II diplomacy. Of course, these attempts to unmake the "atomic-bomb tuna" by tweaking the numbers failed to undo all political fallout from the consumer panic. The rise of nuclear disarmament campaigns went on to strain U.S.-Japan relations for the rest of the Cold War. Nonetheless, the introduction of permissible dose through the back door, followed by the termination of tuna inspection, helped defuse some of the most immediate and severe threats to the conservative regime in Japan and the U.S.-Japan alliance.

As many Japanese scientists pointed out during the heated debate over the radiation safety of food and water, however, the application of the concept of permissible dose to global fallout raised a myriad of scientific and ethical questions. Did the existing radiation protection guidelines established by the nongovernmental organizations adequately take into account numerous differences among nations, including physiology and diet? How about genetic damage, for which there might be no "safe" threshold in dose? Even if the doses of radiation coming from global fallout were indeed small, was the indiscriminate irradiation of the world's population and its posterity socially "permissible"? While U.S. and Japanese officials and their scientific advisers skirted these fundamental questions, some geneticists and their associates in the United States, Britain, and the Soviet Union decided to bring them forcefully to the public's attention in the aftermath of the radiological disaster at Bikini. The ensuing public controversy opened a new chapter in the global struggle over the definition of risk in the nuclear Anthropocene.

Chapter 3

EPISTEMIC STALEMATE

Genetics and the Creation of Scientific
Committees, 1954-1955

ON JUNE 22, 1954, U.S. geneticist Alfred H. Sturtevant of the California Institute of Technology delivered a keynote address as president of the Pacific Division of the American Association for the Advancement of Science (AAAS). His talk was a response to the remarks of U.S. Atomic Energy Commission (USAEC) chairman Lewis L. Strauss three months earlier. On March 31, at President Dwight D. Eisenhower's news conference held in the wake of the Castle Bravo test, Strauss had read a statement that forcefully denied a series of allegations made about that experiment. Although he spent most of his time discussing the health of U.S. test personnel, Marshallese, and Japanese fishermen caught in the radioactive plume, he also addressed reports on the alleged danger of a fallout of radioactive material in the United States. While admitting that after each test conducted by the United States and the Soviet Union there was "a small increase in natural 'background' radiation in some localities within the continental United States," he nevertheless declared that it was "far below the levels which could be harmful in any way to human beings, animals, or crops."[1] Strauss's statement could have passed as just another blanket reassurance that the USAEC had routinely made regarding global fallout. This time, however, it provoked a public rebuttal by one of the nation's leading geneticists.

Slender, speaking with a drawl, and always sucking on his pipe, Sturtevant was in many ways an unlikely rebel. Born in 1891 the youngest of six children, he grew up on a rural farm in Illinois and later in Alabama. At Columbia University, Sturtevant was admitted to Thomas Hunt Morgan's famed "Fly Room" to work with the common fruit fly *Drosophila melanogaster* to study the chromosomal basis of inheritance. His best-known work was the creation of the first chromosome map in 1913 when he was still a graduate student. He instantly became Morgan's favorite "son," a role which defined much of his career. Aside from his trailblazing work on genetics, Sturtevant was also known for his almost encyclopedic interest. Edward B. Lewis, one of Sturtevant's students at Caltech who later himself played a major role in the fallout debate, observed that his boss "read widely and kept abreast of many topics of current interest, especially politics." Sturtevant subscribed to, among others, the Sunday *New York Times* and the *Manchester Guardian Weekly*, which Lewis recalled he would read "virtually from cover to cover."[2] This combination of genetic knowledge and political awareness made Sturtevant one of the first of the prominent scientists who called Strauss out on his categorical denial of any harm whatsoever from radioactive fallout scattered far outside the test site.

In his address delivered at the AAAS meeting, Sturtevant laid out what is known today as the linear non-threshold (LNT) hypothesis, explaining that the rate of radiation-induced mutations might be directly proportional to the dose received, with no threshold in dose below which no mutations would occur. These mutated genes, he noted, might be cumulative and permanent, almost all of which would be deleterious, ultimately leading to a wide variety of defects in later generations. Sturtevant assured his audience by saying, "The probability of an effect on the germ cells of any one individual may be very low." Nonetheless, he warned, "when many millions of people are being exposed to it, it becomes certain that some of them will be affected." He concluded his speech by denouncing Strauss short of directly naming him: "I regret that an official in a position of such responsibility should have stated that there is no biological hazard from low doses of high-energy irradiation."[3]

In delivering this forceful speech, Sturtevant introduced a whole new perspective to the nuclear Anthropocene. Until then, the U.S., British, and Soviet nuclear authorities had applied a narrow space and time horizon to the problem of fallout, focusing on the dangers of acute exposure in the immediate test area resulting in severe damage to the exposed individuals. Sturtevant, by contrast, elevated the level of risk analysis to the whole world and the distant future. As he pointed out in the speech, the estimated doses from fallout settled all over the world were so low that, according to the LNT hypothesis, the chance of genetic damage to the offspring of any single individual was correspondingly small. Yet if this small risk was applied to the world population, it might ultimately result in large numbers of deaths and defects over many generations. In other words, Sturtevant insisted that one could fully grasp the totality of the global fallout hazard only on its own spatial and temporal scale.

Although the radiological disaster at Bikini provided the immediate context for Sturtevant's critical epistemic intervention, the underlying genetic knowledge predated the nuclear age. In 1927, U.S. geneticist Hermann J. Muller, another student of Morgan, first reported artificial mutations in *Drosophila* exposed to X-rays. The subsequent work done by Muller and others strongly suggested an LNT relationship between radiation dose and mutation frequency. This observation gained a theoretical basis when Russian geneticist Nikolai V. Timofeev-Ressovsky and German physicists Max Delbrück and Karl G. Zimmer published "On the Nature of Gene Mutation and Gene Structure" in 1935. In this seminal paper, the coauthors proposed what came to be known as the target theory, suggesting that ionizing radiation, directly or indirectly through the formation of free radicals in the water of an organism, would strike genes and change their structures.[4] Although further probing into the mechanism of genetic damage by radiation on the molecular level awaited the discovery of the structure of deoxyribonucleic acid (DNA), the target theory was, in the words of British biophysicist Douglas E. Lea, "as certain as a scientific theory ever is in a rapidly developing subject."[5] If so, then what caused a significant delay in the application of genetic knowledge to the assessment and management of risk from global fallout? Why did Sturtevant and many other geneticists decide to speak up about the genetic effects of fallout only in the aftermath of the Castle Bravo test? And how did their sudden activism for the greater recognition of genetic damage shape the subsequent course of the fallout controversy?

To answer these questions, this chapter examines the inclusion and exclusion of genetic knowledge in the global struggles over the definition of risks in the Cold War. This epistemic dynamic involved all three nuclear countries, but it played out differently among them. In the United States and Britain, the radiation protection standards bodies recognized the possibility of no safe threshold in genetic damage from early on. However, multiple attempts at establishing the genetically "permissible" dose, made first within the Advisory Committee on X-ray and Radium Protection (ACXRP) prior to World War II and then between the National Committee on Radiation Protection (NCRP) and the British Medical Research Council (MRC) after the war, repeatedly failed due to persistent concerns among some U.S. members about the negative impact of such a standard on the industrial and medical use of radiation as well as on national security. Meanwhile, in the Soviet Union, genetic knowledge went underground when plant physiologist Trofim Lysenko secured a government ban on the study and teaching of genetics within the country as antimaterialist, useless, and foreign. The radiological disaster at Bikini thus offered a timely opportunity for both the Western and Soviet geneticists to bring the importance of genetic damage—and their knowledge—to the attention of government officials and a wider public. With no institutional support, however, these scientists were forced to act as individuals, struggling with disagreement from within and resistance from without. The resulting stalemate in public debates ultimately led to a major change in the relations of definition, with its institutional locus shifting from the U.S., British, and Soviet nuclear authorities to scientific panels newly created by the U.S. National Academy of Sciences (NAS), the MRC, and the United Nations.

Exclusion of Genetic Knowledge in the United States and Britain

Muller's discovery of radiation mutagenesis in 1927 drove many geneticists to warn about the dangers of excessive radiation exposure for humanity. Muller himself led this campaign. As a scientist, he saw it his duty to educate the wider public about genetic damage by radiation. At the same time, he was a humanist, having an immense faith in human progress through scientific and technological development. To strike a balance between the risks and benefits of radiation, Muller repeatedly urged physicians to reduce their use of X-rays and radium to a necessary minimum. In his address to a group of physicians in 1928, for instance, Muller described a variety of measures to be taken for the protection of physicians and patients from unnecessary radiation. His admonitions, however, offended some physicians as intrusive on their clinical privileges. H. Bentley Glass, one of Muller's students who attended his talk, later remembered that several in the audience stormed out of the lecture room.[6]

Despite such hostile reactions, Muller's work compelled the expert bodies for radiation protection to take his plea for stricter dose control seriously. Beginning in 1933, there had been discussions within the ACXRP about the possible implications of genetic damage in the determination of what constituted a tolerance dose. It took seven years, however, before the experts formally addressed the problem. At a meeting of the ACXRP in December 1940, radiologist Robert R. Newell proposed that the daily limit of 0.1 R be reduced by a factor of ten, with a maximum cap of 200 R in ten years.[7] This proposal, however, drew strong protest from within the committee. While there was no disagreement over the possibility of genetic damage at any size dose, some members were deeply worried about its practical implications. In a letter to his colleagues dated June 16, 1941, radiotherapist Gioacchino Failla complained that the new requirement was so strict that it would create "serious complications in the handling of radon for therapeutic purposes." While acknowledging that "the smaller the dose the less the genetic damage," he insisted that "the possible damage from .1 roentgen [R] per day is so slight that one can just as well stop at this point." He then pushed the burden of proof to the advocates for the change, asking "what evidence members of the Committee have in support of the new tolerance dose."[8] This epistemic maneuver proved successful. In reply, Newell agreed that "we still know very little about genetic injuries in man," suggesting that the committee should revisit the issue when more was known about mammalian genetic damage.[9] At a meeting held in September 1941, it was agreed that "at the present, the Committee lacked genetic evidence sufficient to warrant lowering the tolerance dose and recommended that the matter be investigated further with the help of genetic experts."[10]

The arrival of the nuclear age after World War II, however, forced the radiation protection experts to reopen the case of genetic damage. Keenly aware of hereditary risks involved in atomic energy use, U.S. and British authorities took steps to ensure that geneticists would take part in the discussions on the tolerance dose. In 1947, at the request of the USAEC, the NCRP recruited Muller and Curt Stern as consultants.[11] In Britain, the Tolerance Doses Panel of the MRC decided to seek advice from David G. Catcheside.[12] Meanwhile, U.S. and British officials actively supported radiation genetics research. The USAEC underwrote the genetic study of the atomic bomb survivors in Hiroshima and Nagasaki through the Atomic Bomb Casualty Commission and also established the mouse genetics program at Oak Ridge with William and Liane Russell as project leaders.[13] On a smaller scale, the MRC funded T. C. Carter's radiation work with mice at Edinburgh's Institute of Animal Genetics and later at the Atomic Energy Research Establishment in Harwell.[14]

Initially, however, neither the NCRP nor the MRC saw an immediate need to revise the tolerance dose, as both continued to consider genetic damage only in occupational context. In his report submitted to the MRC, Catcheside concluded that the nuclear workforce represented only 1 percent of Britain's population. This meant, he argued, that even if all workers were exposed to radiation to the maximum limit of the time (0.1 R per day) each day, the overall rise of the mutation rates in the British population would be no more than 5 percent.[15] When he visited Washington in March 1948 as a member of the MRC mission, Catcheside discussed this conclusion with Failla, reaching agreement that, as long as the number of exposed individuals remained small, genetic damage would "probably not be a serious problem."[16]

The consensus across the Atlantic, however, soon broke down when British officials reconsidered their demographic assumption related to radiation exposure. What prompted this reappraisal was the construction of nuclear facilities in Harwell, as the discharge of radioactive wastewater into the River Thames threatened to contaminate a source of drinking water for over ten million people in the Greater London area. With nearly a quarter of Britain's total population at risk, the Ministry of Supply struggled to gain consent from a wide variety of government agencies concerned—the Ministry of Works, the Ministry of Health, the Thames Conservancy, the Metropolitan Water Board (MWB), London County Council—on the maximum permissible concentration of radioactivity in drinking water.[17] British nuclear officials thus turned to the MRC as an impartial arbiter. At an MRC meeting held in August 1948, Klaus Fuchs, a German-born physicist at Harwell soon to be exposed as a Soviet atomic spy, asked the council for help, saying that "some independent authoritative body was required to convince Public authorities, for example, the M.W.B. about various proposals. At the moment, there were no other bodies than the M.R.C. Protection Sub-Committee and Tolerance Doses Panel."[18]

In response, the MRC asked Catcheside to update his assessments of genetic risks involved in atomic work. In 1950, he submitted a new report based on the revised assumption that all fifty million people in Britain would be at risk of exposure to radiation from radioactive wastewater. This shift in focus from atomic workers to the general population led Catcheside to conclude that, to maintain the previous limit of a 5 percent rise in mutation rates, the total average exposure of the gonad during the reproductive period should be kept under 2.5 R, or more than a hundred times lower than the sum of the maximum permissible weekly exposure of 0.3 R for occupational purposes over the same period. Shocked by this analysis, the MRC decided to make a proposal to the Americans that the maximum permissible concentrations of genetically harmful radionuclides in drinking water for public use should be established at one one-hundredth of the occupational levels.[19]

When radiation protection experts from the United States, Britain, and Canada gathered in Harwell in August 1950 for the second session of the tripartite conference to strike out common standards, the Americans strongly opposed the British proposal on two grounds. First, the NCRP relayed grave concern among U.S. officials that the genetic argument would jeopardize their civil defense plans. Since one could logically argue that, on the basis of the LNT hypothesis, the only "permissible" dose was zero, U.S. officials feared that such an idea might trigger a mass panic in the event of a nuclear attack.[20] Second, the NCRP expressed concern over the implications of a "double standard," one for workers and the other for members of the public, for insurance and labor relations.[21] Viewing the British proposal as a threat to the expansion of atomic energy uses, the Americans tried to prevent it from affecting them by stressing the difference between the two countries. Noting that "no large populations in the United States were likely to be exposed via water from contaminated rivers," Shields Warren, a member of the NCRP and director of the USAEC Division of Biology and Medicine until 1952, argued that "[n]o need was seen therefore for the formal adoption of a specific figure of permissible concentration of activity."[22] After some discussion, it was agreed that a decision on the issue of genetic damage should be postponed until the third session of the conference.

Although its delay tactics worked for the moment, the NCRP realized that it was no longer possible to avoid the question of genetic damage. In early 1953, shortly before the tripartite conference was reconvened in Harriman, New York, Failla asked the committee's genetics consultants, Muller and Stern, for advice. A year earlier, during a meeting of the International Commission on Radiological Protection (ICRP) in Stockholm, Muller pointed out that the doubling dose, a dose required to double the rate of spontaneous mutation in humans, was likely to approximate 80 R. On the basis of this estimate, he argued that the human race could tolerate up to a 25 percent increase, equivalent to a cumulative exposure of 20 R during the thirty-year reproductive period, in exchange for the benefits of atomic energy.[23] Failla was ready to accept Muller's proposal as reasonable, but Stern argued that it was too high, suggesting 3 R instead.[24] Although the two geneticists eventually agreed on 5 R as a compromise, this significantly lower figure shocked Failla. When experts met in Harriman, he loudly wondered how one could explain to workers "why they could have doses of up to, say, 500 [R] which is 100 times the proposed 5 [R] per capita." At his insistence, it was agreed that a panel of experts from the three countries should be created for preparing a report on the matter of genetic damage.[25]

With the negotiation of the genetically permissible dose repeatedly delayed, some geneticists in the United States became restless. In a letter to Sturtevant congratulating his AAAS address, Muller vented his frustration at the NCRP. He recalled that, in Stockholm, the British and Scandinavian members of the ICRP were quite receptive to his plea for taking the genetic consequences of unnecessary irradiation seriously. The U.S. group, by contrast, adamantly refused to include guidelines regarding genetic damage in its latest recommendations. Muller wrote that he had even been thinking of asking Stern that "we both resign, rather than to appear to endorse the committee's verdict."[26] Such growing discontent among U.S. geneticists set the stage for their public confrontation with the USAEC over the genetic effects of global fallout. Meanwhile, their Soviet peers also found their knowledge sidelined in their country, but on a more fundamental level.

Exclusion of Genetic Knowledge in the Soviet Union

Contrary to the popular Western image of the Soviet Union as a mysterious, secluded country, cultural and academic exchanges flourished between the Soviet Union and the West during the 1920s and 1930s.[27] Geneticists took active part in these transnational interactions across the political and ideological divide. Muller was one of the first of the Western geneticists who established direct contact with his Soviet peers. In 1922, he visited Petrograd (today's St. Petersburg) and Moscow, where he met Nikolai K. Kol'tsov and Sergei S. Chetverikov from the Institute of Experimental Biology and left with them some of his mutant *Drosophila* stock. Fascinated by this new method of studying life, Kol'tsov and his colleagues started training the first generation of experimental geneticists in the Soviet Union.[28] One of them was Timofeev-Ressovsky, who moved to Berlin in 1925 and went on to play a key role in establishing a model explaining the LNT relationship between radiation dose and mutation rate. Another rising star was Nikolai P. Dubinin, who joined the Kol'tsov Institute in 1932 as head of the Genetics Section. By invitation of Nikolai I. Vavilov, a prominent Soviet botanist and president of the Lenin All-Union Academy of Agricultural Sciences, Muller also returned to the Soviet Union in 1933 and ran a *Drosophila* lab at the Institute of Genetics until his departure in 1937.[29]

Beginning in the mid-1930s, however, a group of Soviet researchers led by Lysenko launched an aggressive campaign against the entire field of genetics. His dogmatic rejection of genetics later earned Lysenko notoriety as a pseudoscientist, but Lysenko initially emerged as a successful plant physiologist. He earned his public fame in the late 1920s for his study of vernalization, a process in which exposing the seeds of winter cereals to the cold and moist environment could accelerate their flowering process and thus shorten a crop cycle. His research had much practical value in Russia, where long, severe, and unpredictable winters often killed winter wheat seedlings. One of Lysenko's key patrons was none other than Vavilov, who recruited this young and ambitious plant breeder to his Academy. Lysenko, however, refused to be content with his subordinate role. A proponent of the Lamarckian hypothesis that an organism could acquire new traits in response to changes in the environment and pass them on to offspring, Lysenko promoted himself as the true heir of Ivan V. Michurin, a Russian plant breeder and champion of the idea of the inheritance of acquired characters. In his speeches and publications, he repeatedly denounced genetics as idealistic, impractical, and foreign.[30]

Lysenko's attack on genetics made a deadly turn when it combined with the waves of the political purge during the Stalin era. The most infamous case was the arrest in 1940 of Vavilov, who perished in jail three years later. By the end of World War II, however, an increasing number of scientists had begun to challenge Lysenko. One of the most vocal critics was Dubinin. A dedicated communist embracing the ideal of science as a path to the truth, he thoroughly despised Lysenko as a pseudoscientist. In his review article on Soviet genetics published in 1947 for the Western audience, Dubinin simply ignored the work of Lysenko and his associates and reasserted the importance of the chromosomes in biology: "[T]he problem of molecular organization of chromosomes and metabolism between the chromosomes and the cytoplasms is the central one in theoretical genetics, as well as for the solution of the entire question relating to the organization of living matter and the origin of life."[31] Science, however, was not the only motive for Dubinin to stand up against Lysenko. Just as self-confident and ambitious as his rival, Dubinin always sought opportunities to promote himself in academia. In 1946, over Lysenko's strong objection, Dubinin was elected to the Soviet Academy of Sciences as a corresponding member. He was also put in charge of organizing the Institute of Cytology and Genetics, posing a direct threat to Lysenko's control of the Lenin Academy and the Institute of Genetics.[32]

Threatened by the growing opposition led by Dubinin, Lysenko immediately struck back. The time seemed ripe, as the beginning of the Cold War prompted Stalin and his lieutenants to tighten control of Soviet intellectuals. In the fall of 1947, Andrei A. Zhdanov, a Politburo member and chief of the Communist Information Bureau (Cominform), launched a cultural campaign to stomp out Western influence from the socialist camp. Zhdanov, however, withheld his formal support for Lysenko for fear that it would not only undermine scientific activities but also boost the position of his political rivals and patrons of the controversial agronomist. Frustrated, Lysenko worked around Zhdanov and eventually secured Stalin's personal approval of a formal ban on the teaching and research of genetics. He triumphantly disclosed this decision at a special session of the Lenin Academy held in August 1948.[33] Not surprisingly, Lysenko saved his worst venom for Dubinin. During the session, Lysenko relentlessly attacked Dubinin, in his absence, as one "regarded by our Morganists as the most eminent among them."[34] Dubinin was soon stripped of his post at the Kol'tsov Institute, with his *Drosophila* lab shut down for an indefinite period.

It is important to note that Lysenko's control of Soviet biological science was far from complete even after the 1948 session. Historian Ethan Pollock has explained that Lysenko's ideological arguments were increasingly out of step with the party's willingness to acknowledge the objectivity of science at least in some fields.[35] Nonetheless, Lysenko managed to hold his dominance in Soviet biology even after Stalin's death in 1953. As historian Michael Gordin has demonstrated, the eventual demise of Lysenko's doctrine did not happen overnight, but instead in a long-drawn-out process of the disestablishment of a Lysenkoite institution.[36] The purge of his opponents from the academia had allowed Lysenko and his allies to secure key posts in the Soviet scientific leadership. In August 1948, the Soviet Academy of Sciences dismissed Leon A. Orbeli, an eminent Soviet physiologist who was also subject to a Lysenkoite attack in his discipline, from his duties as Secretary of the Department of Biological Sciences and appointed Aleksandr I. Oparin, a renowned biochemist and supporter of Lysenko, as his successor.[37] In addition to institutional control, Lysenko also found a new patron in Nikita Khrushchev, an ambitious leader who eventually emerged victorious in the post-Stalin power struggle. Although Khrushchev had little interest in scientific debates and remained inconsistent in his support for Lysenko, he eagerly accepted the agronomist's promise to assist his agricultural policy through the quick and drastic improvement of crops and livestock.[38] Moreover, Lysenko's campaign against genetics went global in the quest for political and scientific legitimacy in the Cold War world. As the Soviet propaganda machine aggressively promoted Lysenko's work on both sides of the Iron Curtain, biologists abroad reacted differently, triggering heated debates around the world.[39]

The rise of Lysenko to prominence in the Soviet life sciences had a uniquely devastating impact on radiation genetics. As noted earlier, Muller and others studied chromosomes as the material basis of heredity. The Lysenkoites, however, rejected this doctrine, insisting that the whole organism was involved in inheritance. "Heredity," Lysenko wrote in 1946, "is the concentrate, as it were, of the environmental conditions assimilated by plant organisms in a series of preceding generations."[40] Such an assertion ran directly counter to both the theoretical and experimental basis upon which to evaluate genetic damage caused by radiation. Moreover, Lysenko's criticisms of genetics extended to its research methods. For him, scientific truth was inseparable from its practical value, and experiments with *Drosophila* seemed to fail the critical test of utility. In a lecture delivered in 1941, Lysenko sarcastically asked, "What problems can receive a practical solution with the Drosophila as the object of research?"[41] In the same vein, he also dismissed the use of radiation as a tool of genetic research. In a speech delivered before a session of the Lenin Academy held in August 1948, Lysenko said that the Michurinists "do not deny the action of the so-called mutagenic substances," such as X-rays and colchicine. "But we insist that such action, which penetrates into the organism not in the course of its development, not through the process of assimilation and dissimilation, can only rarely and only *fortuitously* lead to results useful for agriculture." In this sense, he said, the use of mutagens was "not the road of systematic selection, not the road of progressive science."[42]

The genetic knowledge that Lysenko dismissed as useless for agricultural improvements, however, was essential for radiation protection related to nuclear activities. Placed under the Ministry of Internal Affairs, the Soviet nuclear weapons program developed a secret research complex separate from civilian scientific institutions. This parallel organization offered a protective cover for some geneticists. The fate of Timofeev-Ressovsky was a case in point. After World War II, he was arrested in Berlin and sent to a labor camp. However, Beria's right-hand man, Avraamii P. Zavenyagin, realized the value of Timofeev-Ressovsky's research for the atomic bomb project and secured his release from the Gulag. Assigned to Laboratory B in Sungul', one of the secret nuclear research institutes tucked in the deep forests of the Ural Mountains, Timofeev-Ressovsky managed to continue his study of mutations with *Drosophila*.[43] The influence of Lysenkoism, however, eventually reached this sanctuary for geneticists. Laboratory B was forced to terminate its genetic research when Ivan E. Glushchenko, a follower of Lysenko at the Institute of Genetics, discovered Timofeev-Ressovsky's work.[44] Moreover, as a prisoner scientist handling state secrets, he was cut off from the wider scientific community, and his isolation continued for some time even after he was transferred to the Ural branch of the USSR Academy of Sciences following the closure of Laboratory B in 1955.[45]

When the Castle Bravo test shocked the whole world, then, genetic knowledge was conspicuously missing from the Soviet discourse of fallout hazards. Seeing an opportunity for Cold War propaganda, the Soviet government loudly condemned the Americans for radiation injuries among Japanese fishermen and Marshall Islanders caught in the radioactive plume. In stark contrast, it spoke little of global fallout and its potential genetic risk to humanity.[46] That warning had to come from geneticists in the United States who decided to call out Strauss for his blatant disregard for hereditary damage caused by nuclear tests.

Dissent in the United States

That Sturtevant chose a public address as a platform to sound an alarm about the genetic effects of global fallout created a serious problem for the USAEC. As historian J. Christopher Jolly has noted, the commission was willing to discuss radiation damage among experts, but it steadfastly rejected any public statement that might cast even the slightest doubt on the safety of its nuclear program.[47] When Sturtevant sent an advance copy of his speech to Strauss, John C. Bugher, director of the USAEC Division of Biology and Medicine, immediately wrote a reply on behalf of the chairman. Bugher did not dispute the principles of genetic damage as laid out by Sturtevant, but he strongly disagreed with the conclusion that nuclear tests would ultimately produce "numerous defective individuals." Instead of focusing on the absolute increase in genetic damage as a result of fallout, Bugher urged Sturtevant to put it in two comparative perspectives. One was the background level of risk. Bugher noted that the doses from fallout were much smaller than those from natural background radiation and its large variations from region to region. The other perspective was the alleged benefits of atomic energy for the nation's welfare and security. "Fundamentally, our problems of human adjustment to a world in which nuclear energy is widely utilized are serious enough without exaggerating the significance of vanishingly small probabilities," he said. "In dealing with a subject as sensitive as the atomic weapons program, I feel that the issues involved are of such fundamental importance that sweeping generalizations which are not derived from good observations should receive a prompt and effective correction."[48]

Despite Bugher's strong objection, Sturtevant proceeded with his speech as planned. The press immediately picked up Sturtevant's admonitions as a decisive rebuttal of Strauss's denial of fallout hazards. In a series of nationally syndicated newspaper columns, Joseph and Stewart Alsop picked up Sturtevant's claim, accusing the USAEC of hiding the truth of genetic damage from the public.[49] Nuclear officials, however, continued to speak of fallout only in comparative terms in order to reassure the public without denying the possibility of genetic damage. In a science paper published in July 1955, USAEC commissioner Willard F. Libby explained that the average annual fallout exposure rate in the United States as of January 1, 1955, was 1 milliR, or 2 percent of the lowest normal rate of natural background radiation.[50] He later told Strauss that their scientists "have never deprecated fallout," arguing that "we have pointed to the fact that it was *small* as compared to normal dosages and *therefore* the prophets of doom had better check their reasoning."[51]

The repeated reference to the natural background radiation by U.S. nuclear officials, however, only convinced its critics that the government was simply dodging the issue. Sturtevant decided to speak up once again, but this time, he decided to translate his allegation of genetic damage into a quantitative estimate. As historian Peter Thorsheim has pointed out in his work on the London Fog of 1952, "body count" was a powerful tool to cast the otherwise elusive environmental problem as a public health disaster.[52] In January 1955, Sturtevant published the result of his calculations. On the assumption that the background levels of radiation, 3 to 4.5 R per generation, would yield from 3 to 4.5 mutations per ten thousand germ cells, he said that the estimated average exposure from fallout, 0.1 R per generation, would give about 1 deleterious mutation per hundred thousand germ cells per generation. This meant that continuous irradiation due to fallout at the present rate would produce about 78 mutated germ cells every year in the United States and about 1,800 worldwide. Sturtevant admitted that the estimated excess risk of genetic damage was so small compared to those naturally born with mutated genes that it was statistically impossible to detect. Nonetheless, he argued, "the fact remains that there *is* a hazard, and that it may become a significant one in terms of large populations."[53]

While frequent protests by geneticists kept the public controversy alive, these scientists were far from united in opposition to nuclear officials. To begin with, not all geneticists agreed that a radiation-induced increase in genetic mutation would be biologically harmful. A majority, including Sturtevant, espoused the classical view that an organism was in its best genetic state when it carried two copies of the same gene (homozygote). The spread of recessive genes in the gene pool thus would always be deleterious because a pair of dominant and recessive genes (heterozygote) was not optimal and ultimately weeded out in the form of "genetic deaths." By contrast, a group led by Ukraine-born U.S. geneticist Theodosius G. Dobzhansky put forth the balance view that, in a diverse and dynamic environment, an organism with heterozygotes would often be fitter than one with homozygotes.[54] Even if the classical view was correct, there remained the problem of quantification. The possible range of the radiation-induced genetic deaths diverged by a factor of one thousand due to different estimates on four key variables: the average mutation rate per locus, the number of loci in the human genome, the number of mutant genes in an individual required to trigger a genetic death, and the factor by which the mutation rate was increased under a certain irradiation condition.[55]

The variation in opinions among the geneticists became even wider when it came to the social implications of the fallout danger. Sturtevant was deeply concerned about fallout because it would affect people around the world and their descendants without consent or benefit. In his January 1955 paper, Sturtevant pointed out a profound ethical problem with fallout: "[W]e are all of us submitted, willy-nilly, to fall-out, and while it may be argued that some of this is for our ultimate advantage, it must be recognized that we get fall-out from Russian bombs as well, and that the rest of the world gets it from Russian and American bombs alike."[56] In this judgment, however, Sturtevant found himself in a minority. In his letter to Earl L. Green, a geneticist and member of the USAEC staff, Muller strongly complained about *U.S. News and World Report* articles that cited the commission as a source and downplayed genetic damage. But he quickly added that he was "in full agreement" that the genetic risk from the past nuclear tests had indeed been small. He noted that people apparently accepted the far more serious risk of driving a car, for example, "in order to secure other benefits, such as those of rapid transportation." A fierce critic of Lysenko, he also viewed the Soviet regime as a threat to world peace, strongly supporting the U.S. nuclear weapons program for "strengthening the defense of democratic civilizations."[57] In this way, the Cold War consensus and the unfaltering faith in social progress, both typical of the 1950s United States, brought Muller and other geneticists into agreement with nuclear officials that the genetic risk from nuclear tests was biologically existent yet socially "acceptable."

What ultimately prevented geneticists from expressing a collective view regarding the fallout danger, however, was the lack of an institutional mechanism for them to settle differences among themselves. As noted earlier, both the NCRP and the ICRP put the matter of the genetically permissible dose on hold pending the creation of a panel of experts. Only one geneticist, Curt Stern, served on the USAEC Advisory Committee for Biology and Medicine.[58] The professional organizations of geneticists were also unwilling to take the lead. As historian Audra J. Wolfe has demonstrated, the Genetics Society of America (GSA) was rather slow and reluctant to condemn Lysenkoism, as the political heat of the Cold War made many members wary of being involved in this polarizing issue.[59] Muller clearly remembered this lackluster response from the genetics community when he wrote a reply to the inquiry of University of Colorado radiologist Raymond R. Lanier about the *U.S. News and World Report* articles. He promised to bring the letter to the attention of a number of persons, including an editor of the *Journal of Heredity* and a member of the GSA Committee on Public Education and Scientific Freedom, adding, "Past experience indicates, however, that not much is to be expected from committees of geneticists when it comes to their expressing themselves with regard to matters of general welfare."[60]

With no consensus-building mechanism in place, the geneticists failed to speak with one voice. Indicative of the limits of individual dissent was an opinion piece published by Stern in December 1954, "One Scientist Speaks Up," as a direct answer to missionary doctor Albert Schweitzer's appeal to the scientists to enlighten people around the world on the dangers of thermonuclear tests. Stern argued that the question must be addressed on two different levels. In an absolute sense, even very low doses of radiation from fallout would produce genetic changes resulting in "future misery and death." Relatively speaking, however, the total irradiation from fallout would cause "only a vanishingly small fraction of the individual illnesses" that naturally occurred in each generation. While agreeing that "[e]ach individual case of harm or death due to fall-out, in our time or in the distant future, places a burden on our conscience," he nevertheless insisted that "these cases must be weighted in a balance." Pointing out that "we do not propose to abolish automobiles, x-ray machines, and chemical industry" solely on grounds of the damage caused by them, he concluded that "[t]he question of whether or not nuclear tests are of ultimate benefit to mankind is partly one of political opinion and partly one of scientific prophecy." And for the political aspect, Stern declared, "the scientist's views cannot claim greater attention than those of other citizens."[61] Disavowing any special wisdom or authority in the issue of fallout, Stern and many other geneticists chose silence.

Dissent in the Soviet Union

If geneticists in the United States struggled to bring the genetic risk from fallout to the public's attention, their Soviet peers waged a different kind of struggle. Beginning in January 1954, with repression and censorship gradually relaxed after Stalin's death, Dubinin repeatedly petitioned the Soviet government, the Communist Party, and the Academy of Sciences to create a research institute for experimental genetics. While he justified his case in many different ways, he put a special emphasis on the growing demand for radiation safety related to atomic energy uses. It was part of his conscious effort to change the context in which the Soviet political leadership understood the utility of genetic knowledge. Until then, improvement in agricultural production had been one of the most important measures of success for Soviet biology. This cult of productivity put experimental geneticists like Dubinin at a disadvantage in relation to the Lysenkoites, as breeding using radiation and other mutagens tended to produce random results and required a long time to produce new and useful traits.[62] The knowledge of genetic mutations, however, was essential for radiation protection. Astutely sensing the new opportunity opened by the nuclear age, Dubinin offered an epistemic service to the socialist state in a way that the Lysenkoites could not possibly do.

Initially, much of Dubinin's attention went to considerable progress in peaceful uses of atomic energy. In response to Eisenhower's "Atoms for Peace" speech in December 1953, the Soviet government accelerated its civilian nuclear power program to showcase the regime's superiority, including the Obninsk Atomic Electric Station, opened in June 1954 as the world's first grid-connected nuclear power plant.[63] In a letter to Oparin, Dubinin pointed out that "[t]he release of atomic energy will exert an enormous influence not only on technology, but also on biology," arguing that "it is necessary to keep in mind the influence of radiation on the hereditary properties of organisms."[64]

When the Academy's biology section finally held a hearing on Dubinin's laboratory proposal in April 1955, however, the geneticist introduced the subject of fallout from nuclear tests. By then, as discussed later in the chapter, the fallout controversy had become widely known outside the United States. Moreover, in response to India's call for a halt to thermonuclear tests following the radiological disaster at Bikini, the Soviet government expressed its willingness to consider a test ban as part of comprehensive disarmament. Dubinin immediately noticed a potential linkage between these two issues, seeking to present the propaganda value of genetic knowledge for the Soviet peace offensive. During the hearing, he pointed out that "[t]he testing of a thermonuclear [bomb] gives rise to not only the ionization of the atmosphere, but also the introduction into the atmosphere, soil and water of radioactive isotopes, many of which remain radioactive for a very long time." After discussing the genetic effects of these radioactive materials, he said that "[c]urrently, in the struggle against the testing of a thermonuclear weapon, this argument is one of the very important ones."[65]

Dubinin's effort to underline the usefulness of genetic knowledge for the Cold War propaganda, however, did not lead to immediate success. Instead of creating an independent institute for genetics, the Academy's biology department directed Dubinin to set up a small research unit within the Institute of Biophysics with biochemist Aleksandr M. Kuzin as director. In his memoirs, Dubinin blamed Oparin for this failure, arguing that he continuously blocked Dubinin's proposal until he stepped down from his post at the end of 1955.[66] The backstage sabotage, however, was not the only factor working against Dubinin. The Lysenkoites also exploited large uncertainty in genetic knowledge to cast doubt on its usefulness in radiation protection. In a review of Dubinin's research proposal using *Drosophila*, members of the Institute of Genetics, a bastion of the Lysenko school, argued that Dubinin failed to address the problem of extrapolation from fruit flies to humans.[67] Nikolai I. Nuzhdin, a protégé of Lysenko, repeated the same argument during the hearing of April 1955. "If we put *Drosophila* near an atomic power station right now," he said, "nothing will come out, Nikolai Petrovich."[68] In the United States and Britain, as noted earlier, the problem of extrapolation opened the door to government support of mammalian and human genetics. In the Soviet Union, however, the Lysenkoites used the same problem as an excuse to resist the comeback of genetics.

The delay in the reconstruction of Soviet genetics became clear when scientific exchanges across the Iron Curtain accelerated in the atomic energy field. In a paper presented at an international conference held in Moscow in July 1955, Nuzhdin and his collaborators compared the number of offspring in litters between irradiated mice and controls over several generations, demonstrating that decreases in the fertility of the exposed mammals were heritable. Their paper, however, was based solely on statistical observations, including no analysis of chromosomal aberrations and other markers of genetic damage.[69] Also conspicuously missing from the Soviet scientific literature were the genetic effects of human exposure to radiation. In the Soviet Union, human genetics had been long suppressed as a fascist science of eugenics.[70] When Fedor G. Krotkov, a military hygienist and vice president of the Soviet Academy of Medical Sciences, attended the International Conference on the Peaceful Uses of Atomic Energy held in Geneva a month after the Moscow meeting, he was shocked to see that Western scientists had been actively working on the problem of human genes in the nuclear age. "We would probably be ashamed of our nation," he said during a debriefing session in Moscow, "when it becomes clear that we, due to some tacit collusion, consider these issues as if nonexistent for our country."[71]

As Lysenkoites utterly failed to offer an answer to the pressing problem of radiation protection, outside support started pouring in for the struggling geneticists. In October 1955, 297 Soviet scientists, including many nonbiologists, signed what came to be known as the "Letter of 300," urging the Central Committee of the Communist Party to end the monopolistic control of biological sciences by Lysenko and his associates. In this remarkable letter, the co-signers bluntly pointed out an embarrassing Cold War gap in genetic knowledge exposed at the Geneva conference. While the U.S. delegation demonstrated its leadership in the field of radiation genetics, the letter read, "not a single Soviet report was presented" on this crucial subject.[72] Although Party officials chose to ignore the petition as an unauthorized collective action, the growing support for genetics within the Soviet scientific community, as discussed later in the book, strengthened the position of Soviet scientists during the epistemic negotiations of the global-fallout risk assessment through the United Nations. The initiative for creating that international forum, however, had to come from elsewhere.

Creation of Scientific Committees

On March 6, 1955, the Federation of American Scientists (FAS), a U.S. advocacy group founded by Manhattan Project scientists, issued a public appeal. Noting that "bomb tests now have such scope that their effects cannot be restricted within national boundaries," the FAS proposed the establishment of a United Nations commission to study the problem of radioactive fallout, determine a danger threshold, and recommend steps to be taken to keep contamination below that threshold. What inspired this proposal was a stalemate in the public debate over the genetic risk from global fallout. "The claim has been made that a genetic danger to the race exists, and assurances to the contrary by our Atomic Energy Commission seem to have been insufficient to convince others," the appeal read. The FAS made a key intervention in this impasse by proposing a transfer of responsibilities for the health risk assessment of global fallout from the USAEC and other national atomic agencies to an independent and international scientific committee. "We must leave this problem to the experts in the field," the FAS stated. "However, it is clear that this has now become a problem with international implications, and should no longer be determined unilaterally by one country."[73]

The idea of an international fallout study came out of frustration among activist scientists with the lack of progress in nuclear disarmament. Shocked by the atomic destruction of Hiroshima and Nagasaki, and anxious to prevent a dangerous arms race leading to a global nuclear catastrophe, many Manhattan Project scientists and their supporters called for the international control of atomic energy. The terrifying demonstration of the H-bomb's power at Bikini also prompted many intellectuals, most notably British philosopher Bertrand Russell, to go public with a demand for the abolition of war.[74] Unlike many activists preoccupied with the specter of a nuclear holocaust, physicist and FAS president M. Stanley Livingston decided to focus on the problem of fallout from nuclear tests in an attempt to bring world opinion to bear on the nuclear powers to take immediate steps toward arms control. In the public appeal, the FAS expected that the scientific report of a proposed commission would lead the United Nations to establish "some system of control of bomb tests." Moreover, as a danger threshold established by the commission would "almost certainly be exceeded in the event of a major atomic war," the FAS hoped that all nations of the world would be forced to take "a more realistic attitude toward the consequences of atomic war and the necessity for arms limitation and control."[75]

The proposal made by the FAS instantly drew attention from many countries, but it did so particularly in Britain. Until then, British officials and scientists had mostly remained on the sidelines in the fallout controversy. However, the FAS issued its appeal when the political consensus on nuclear weapons in Britain began to break down. Since Attlee's time, the Labour leadership had been in agreement with the Conservatives on the need of Britain's independent nuclear deterrent. When Prime Minister Winston Churchill announced a decision to develop thermonuclear weapons in February 1955, however, Labour Whip Aneurin Bevan defected from the party line and led the opposition to the thermonuclear program.[76] To contain the rebellion from within the party, James Callaghan, a Labour MP and trusted aide of party leader Hugh Gaitskell, decided to introduce a resolution supporting a UN fallout study and do "no more till we have reports."[77] For this purpose, the Labour leadership skillfully deployed gender to the maximum effect. On March 22, Labour member Edith Summerskill introduced the motion to the Parliament, drawing on her identities as a physician and a mother to make the case for a review on the alleged dangers of fallout to children.[78] Although the Conservatives managed to defeat the motion, it was no longer possible for them to ignore the growing call for an independent fallout study by international authorities.

The explicit link in the FAS proposal between the issue of fallout and that of arms control, however, put Washington and London in a serious dilemma. In April 1954, shortly after the disastrous Castle Bravo test, Indian Prime Minister Jawaharlal Nehru delivered a speech calling for a "standstill agreement" on H-bomb tests as a small step to decelerate the nuclear arms race and reduce international tension. Eisenhower and his secretary of state, John Foster Dulles, initially showed their keen interest in the proposal, but defense and nuclear officials strongly opposed any restrictions on the nuclear test program whatsoever unless the Soviets agreed on comprehensive and verifiable disarmament.[79] In the meantime, Churchill was determined to develop the hydrogen bomb for Britain to replace its obsolete atomic deterrent and secure its seat in the exclusive club of nuclear powers.[80] All sides of the debate over Nehru's proposal however, agreed that the issue was a matter of national security, not public health. The test-ban opponents presented large uncertainty about genetic effects of radiation from fallout as an excuse for nonaction, asserting that "[s]o long as continuing technical studies do not reveal solid basis, these fears do not constitute justification for abandoning our tests."[81]

The unknown, however, could also provide grounds for taking precautionary measures. The USAEC reportedly feared that a debate in the United Nations as to the meaning of the "safe" doses of radiation might create an irresistible momentum for the international regulation of nuclear testing.[82] Edwin Plowden, Britain's atomic chief, echoed this view. In his letter to Lord President of the Council Robert Gascoyne-Cecil, he wrote that, during the proposed fallout review, scientists from the West were likely to admit their ignorance, whereas their Soviet counterparts would deliberately stress dangers. Such contrasting attitudes regarding uncertainty, he warned, might "increase rather than allay public misgivings."[83]

To escape the dilemma posed by an international scientific review, the U.S. and British governments each decided to commission a national study to keep the issue of fallout separated from that of a test ban. British Deputy Defense Minister Richard Powell was the first who conceived such an idea.[84] The Cabinet initially considered the Royal Society for this task, but it eventually chose the MRC for fear that members of the Royal Society were too independent to produce a consensus report.[85] On March 29, a week after Summerskill's motion was defeated, Churchill announced that the MRC would create a special panel to evaluate all radiation hazards, including those from radioactive fallout.[86] Meanwhile, the USAEC decided to cooperate with the NAS for a scientific review funded by the Rockefeller Foundation. On April 7, NAS president Detlev W. Bronk announced the planned study, pledging that his institution would make a "dispassionate and objective effort" to assess the biological effects of radiation.[87]

The proposed national fallout studies by the United States and Britain, however, proved insufficient to persuade other nations. Sweden was particularly insistent on a UN commission. As with the British, the FAS proposal struck the Swedish at the moment of a heated debate over nuclear weapons. In May 1954, two months after the Castle Bravo test, the Swedish Parliament had its first debate on this controversial topic. Five months later, in October, the Swedish supreme commander published a report advocating nuclear armaments to defend the country's time-honored neutrality. Standing at this nuclear crossroads, the left-wing faction of the ruling Social Democrats seized on the global-fallout debate to reassert Sweden's role in promoting nuclear disarmament. Forty-three women's organizations sent a joint open letter to the United Nations, calling for a test ban until an expert panel thoroughly studied biological effects of fallout.[88] In response, Swedish Foreign Minister Östen Undén announced his plan in May 1955 to submit a resolution calling for a fallout study at the session of the UN General Assembly in fall. The British tried to dissuade Undén, but the Swede stood firm on his decision, insisting that the U.S. and British studies would not be enough to reassure world opinion due to an inevitable suspicion of national bias.[89]

A growing pressure abroad for an UN inquiry greatly alarmed many Americans, but it was especially disturbing for Henry Cabot Lodge Jr., U.S. ambassador to the United Nations. In his letter dated May 3, Lodge warned Dulles that the General Assembly would almost certainly discuss the issue of fallout from nuclear tests. He suggested that the United States should take the lead in establishing a UN scientific panel in order to "divert attention from our own tests to those of the United Kingdom and the USSR, and at the same time avoid the pressures that are increasingly building up for a moratorium on tests." Noting that the United States was a chief sponsor of the Geneva Conference on the Peaceful Uses of Atomic Energy scheduled only a few months away, Lodge warned that, without a timely U.S. initiative, the conference might devolve into a debating forum over the dangers of fallout, turning Eisenhower's Atoms for Peace propaganda into a "psychological defeat for the United States."[90]

The scientific panel that Lodge proposed, however, differed substantially from what the FAS had advocated. Instead of an independent investigative body, the ambassador envisioned a worldwide clearinghouse of national studies. When Lodge discussed this watered-down plan with Dulles and Strauss on May 20, he found both still resistant to any initiative whatsoever. Predicting that "[a]ny report by an international body would be considered by a packed jury" against U.S. nuclear policy, Strauss declared that "he would rather accept the onus of opposing anything introduced by Sweden, India or others." Dulles was equally blunt, pointing out that a nongovernmental organization like the International Council of Scientific Unions, which had been also discussing a similar investigation, was "not subject to sufficient control to be entrusted with the job." After some discussion, Lodge managed to secure their approval by reassuring them that the proposed panel was specifically designed to allow the United States to keep all fallout data and analysis under its own control.[91] On June 21, at the Tenth Anniversary Conference of the United Nations in San Francisco, he unveiled what came to be known as the Lodge Plan with much fanfare.

In the subsequent course of consultation with the British and other Western co-sponsors, however, the Lodge Plan underwent some notable changes. One of them pertained to its mandate. The British opposed the idea of creating a mere information clearinghouse, fearing that the circulation of conflicting national reports with no interpretation would rather confuse world opinion. Confident that the studies by the NAS and the MRC would "set the tone of the discussions of the U.N. scientists" and make it difficult for "fellow travelers" to put the subject on the wrong foot, the British pushed for a broader power to be given to the UN panel so that it could produce a coherent and definitive report.[92] The United States eventually accepted this suggestion, and the final draft resolution adopted by the General Assembly included a term of reference allowing the panel to evaluate each submitted report "to determine its usefulness for the purposes of the Committee."[93]

Another major change concerned the panel's membership. The original scheme was that the Secretary-General would nominate scientists upon consultation with a committee of select member states in order to ensure at least some degree of independence of scientists from their governments. Shortly before the General Assembly met, however, the U.S. government redrafted the text of a resolution, now calling upon each state invited to the committee to appoint a scientist as its official representative. While the Americans did not explain a reason for this last-minute change, the Canadians speculated that Washington might have become fearful of a scenario in which, without government control, scientists might take political actions similar to the Russell-Einstein manifesto that had been issued just two months earlier.[94] The result was an intergovernmental panel, with each of its members simultaneously representing the scientific community and his or her government.

As soon as the United States and Britain formally brought a draft resolution to the UN General Assembly in October, however, three challenges threatened to unravel their carefully crafted proposal. First, the Soviet Union seized on the motion to promote its ban-the-bomb propaganda. While declaring its support for the proposed fallout study, the Soviet delegation introduced an amendment allowing the committee to make recommendations as to the protection and treatment of people exposed to radiation. Noting that the debris of nuclear tests was impossible to control, Soviet representative Vasilii V. Kuznetsov argued that the immediate suspension of nuclear tests was the only way to spare the world from radioactive contamination.[95] He was soon joined by the Indonesians and Syrians, who made a joint motion calling for a test moratorium pending the scientific review. Fortunately for U.S. and British officials, the Indian delegation led by Homi J. Bhabha, a prominent Indian nuclear physicist who chaired the Geneva Conference, refused to support the linkage between the fallout study and a test ban.[96] With the Afro-Asian bloc divided, the General Assembly soundly voted down both the Soviet and Indonesian-Syrian amendments.

The second and more serious challenge concerned the committee membership. The original resolution listed eleven countries: Australia, Brazil, Canada, Czechoslovakia, France, India, Japan, Soviet Union, Sweden, United Kingdom, and United States. The U.S. and British delegations, however, faced the snowballing of membership requests. Not surprisingly, the Soviet Union pushed for Romania and also mainland China, a Communist regime and former Korean-War belligerent barred from joining the United Nations and its auxiliaries. The demands that proved more troublesome for the resolution sponsors, however, came from the developing world. Mexico demanded a seat, claiming that its livestock was contaminated with nuclear dust coming from the Nevada Test Site.[97] The Indians strongly supported Mexico and added Egypt under Gamal Nasser, arguing that their participation was crucial to keep the committee nonpolitical and from becoming "an autocratic club dominated by the powers possessing hydrogen bombs."[98] Even some Western countries became restless. Belgium promoted itself as a candidate, citing the contributions of prominent Belgian radiobiologist Zénon Bacq. Although U.S. and British officials tried to resist such a move by stressing "pure scientific merit" as a qualification, the sponsors eventually invited four more countries (Argentina, Belgium, Egypt, and Mexico) to the committee. The result was ominous: as a British observer noted, seven like-minded Western countries (Australia, Belgium,

Britain, Canada, France, Sweden, and the U.S.) now found themselves in a minority on the fifteen-country committee.[99]

The third and most formidable challenge, however, came from India's motion to open the possibility of China's participation in the committee's work through scientific contributions. Fearing that the escalation of the Cold War in Asia might push its neighbor further toward the Soviet Union and make its nonalignment policy untenable, India tried to play a role as a mediator between China and the West.[100] In this spirit, Indian representative Krishna Menon made a motion to allow any country to furnish fallout data to the committee regardless of its UN membership. In explaining his proposal, Menon described fallout as a threat to the whole planet, arguing that the General Assembly were "not concerned with frontiers or political groups" but were an entity "concerning the whole of humanity."[101] His speech electrified the audience, and even British officials admitted that Menon successfully turned the tables and made any restrictions on the collection of data look political. Although the Indian motion was eventually voted down, the margin was so small that "we and the Americans were lucky to defeat them at all."[102] To the relief of U.S. and British officials, the General Assembly proceeded to unanimously approve Resolution 913(X) on December 3, which established the United Nations Scientific Committee on the Effects of Atomic Radiation (UNSCEAR). The remarkable show of support for free exchanges of scientific information, however, suggested that the Cold War logic could no longer go unchallenged when it came to radioactive fallout that affected the whole planet.

Conclusion

The concurrent establishment of three scientific panels to assess biological effects of radiation marked a major turning point in the politics of risk. Until then, the U.S., British, and Soviet governments had monopolized the institutions, rules, and capabilities necessary to evaluate and manage the health risks of radioactive fallout from nuclear tests. Geneticists were the first who successfully challenged this control of definitional power by the nuclear security states. Frustrated with repeated delays in the formal incorporation of genetic considerations into radiation safety standards, the geneticists in the United States saw the government's refusal to admit any potential harm from global fallout following the radiological disaster at Bikini as a final straw. With no institutional base to forge a consensus, however, the geneticists acted as individuals and remained sharply divided over the scientific, ethical, and political aspects of risk analysis. Around the same time, in the Soviet Union, Dubinin seized on the fallout controversy in his lone crusade against Lysenko, but he also faced the limits of individual dissent in the face of the rival's institutional monopoly and large uncertainty surrounding the genetic risks of low-level radiation. This stalemate on both sides of the Iron Curtain was finally broken when the scientists' movement in the United States stepped in with the proposal for a UN fallout study to wrest the means of definition from the nuclear powers while connecting genetic concern to arms control. Although the U.S. and British governments managed to keep the two issues separate by restricting the study's scope and mandate to biological effects of radiation only, their decision to commission studies from the newly created national and international expert bodies carried a danger of exposing large uncertainty and disputable judgments in their safety claim. The stakes were thus high for U.S. and British officials when the NAS and the MRC separately began to scrutinize their definition of fallout hazards.

Chapter 4

EPISTEMIC DIVIDE

The U.S. and British Scientific
Committees, 1955-1956

O N APRIL 7, 1955, the U.S. National Academy of Sciences (NAS) issued an official statement announcing that it had decided to review the alleged dangers of radioactive fallout as part of a comprehensive evaluation of radiological safety in the nuclear age. "Wide differences of opinion regarding the nature and degree of human hazards involved in the use of atomic energy have been revealed by the public utterances of prominent scientists and laymen," the statement read. The Academy pledged, in the words of its president Detlev W. Bronk, to "make a dispassionate and objective effort to clarify the issues which are of grave concern as well as great hope to mankind."[1] Around the same time, the British Medical Research Council (MRC) launched its own study on radiation hazards. In a speech delivered at the House of Commons on March 22, Labour member Ian Winterbottom endorsed such an inquiry. Invoking the image of Britain as a liberal state free from the Cold War dogma that gripped both the United States and the Soviet Union, he asserted that Britain possessed "a body of knowledge and freedom from artificial shackles" that uniquely qualified his country for "an opportunity of putting forward a dispassionate study."[2]

Scientists and politicians in the United States and Britain had a good reason to express their high hopes for the studies undertaken by the NAS and the MRC. Both institutions enjoyed an excellent reputation in their countries as trusted scientific advisory bodies. Founded in 1863 in the midst of the Civil War, the NAS served simultaneously as a private society of the nation's eminent scientists and a public institution tasked with advising the nation on public policy issues involving science.[3] The MRC was established in 1913 to support medical research, but it also provided counsel for the government regarding public health affairs.[4] Consisting of distinguished scholars, run independently from the government, and widely respected by the public, the NAS and MRC seemed to be uniquely qualified to establish objective truth about the dangers of fallout. Firmly convinced that, as a matter of scientific fact, the global dispersion of radioactive fallout posed no undue hazard to anyone in the world, the U.S. and British governments expected that the two expert committees, safely insulated from the political heat, would naturally reach a similarly reassuring conclusion and guide the United Nations Scientific Committee on the Effects of Atomic Radiation (UNSCEAR) toward the same view.

To meet the expectations for a definitive statement, the NAS and the MRC each created a special task force dedicated to the problem of radiation hazards. The MRC moved first, and its two panels, on genetic effects and on individual effects, held their first meetings in May 1955. The NAS postponed the commencement of the Committees on the Biological Effects of Atomic Radiation (BEAR) until fall that year to assemble six committees: genetics, pathology, meteorology, oceanography and fisheries, agriculture and food supplies, and disposal of radioactive wastes. Both institutions took careful steps to ensure that their reports to the public were approved by all members and also in agreement with one another. For this purpose, the BEAR and MRC committees made informal contacts through letters and visits, exchanged drafts, and even agreed that their reports were to be released on the same date: June 12, 1956.[5]

All these coordinating efforts behind the scenes seemed to be successful when the BEAR and MRC reports came out in agreement regarding genetic aspects. While reaffirming the production of genetic damage even at small doses of radiation, the two committees similarly concluded that the relative contributions of fallout to the overall genetic risk had been and would remain very small. Disagreements emerged, however, as to the risk of bone cancer and leukemia from strontium-90 (Sr-90). The two committees agreed that the amounts of radiostrontium in human bones from the past nuclear tests were well below those deemed permissible for the general public, namely one-tenth of the occupational limit. The MRC report, however, went further, declaring one one-hundredth of the occupational limit as a threshold value for considering countermeasures—a level which it warned might eventually be reached if nuclear weapons testing continued unabated. The public discordance between what were billed as the most authoritative statements on the fallout question had an immediate and far-reaching political consequence. As advocates of a nuclear-test ban on both sides of the Atlantic seized on the newly disclosed dangers of radiostrontium to advance their cause, the fallout controversy entered a new phase in which the issue of global fallout became an integral part of the heated test-ban policy debate.

Why did the BEAR and MRC committees think alike about genetic effects of radiation from radioactive fallout but strongly disagree over cancer risks? And how did this epistemic divide across the Atlantic affect U.S. and British nuclear testing policies? To answer these questions requires close examination of the BEAR and MRC reports and how the geneticists and medical researchers in each committee assessed health risks from global fallout.

A key difference between the two groups of scientists was the framing of risk. To compel the standards organizations to take genetic effects of radiation into account, the BEAR geneticists determined what they believed was a low yet practical "genetically permissible dose," a decision soon followed by their British peers. The introduction of permissible dose ironically shifted the focus of concern for the geneticists away from the absolute measure of risk toward a relative one—a measure used by U.S. and British officials to dismiss the health risks from global fallout as negligible. The medical researchers, by contrast, moved in the opposite direction. Skeptical about the existence of a biological threshold in the cancer risk from Sr-90, the MRC experts created a separate and much lower "warning dose" for the general population than the permissible level in order to keep the absolute increase in leukemia and bone cancer small. Their U.S. peers upheld the existing dose limit as sufficiently low compared to the natural baseline. Unlike the geneticists, however, the medical researchers failed to work out the difference across the Atlantic before the public release of their reports. Alarmed by the conflicting definitions of risk offered by the BEAR and MRC reports, an increasing number of government officials, politicians, scientists, and concerned citizens in the United States, Britain, and many other Western countries began to reinterpret the policy implications of the exposed uncertainty as a basis for precautionary action. In this way, the fallout controversy gradually merged with the test-ban debate, spawning a variety of initiatives for an international regulation of nuclear testing.

Genetics

The scientists invited to serve on the BEAR Committee on Genetic Effects of Atomic Radiation were all thrilled by this opportunity. For years, many of them had been pressing unsuccessfully users of radiation and radioactive materials to give due consideration to genetic effects. Ironically, the chief obstacle to such a change lay among the geneticists themselves. While sharing the basic proposition that genetic mutations would occur at any dose level, they disagreed over almost every scientific detail and value judgment involved in evaluating genetic damage. This lack of consensus gave the National Committee on Radiation Protection (NCRP) and the U.S. Atomic Energy Commission (USAEC) a handy excuse to ignore genetic effects. Warren Weaver, an applied mathematician and chair of the genetics committee, aptly summarized such an attitude: "You geneticists have no business to recommend a change in the permissible dose . . . because you don't have any really sound basis for recommending the change."[6] Provided for the first time with an independent platform, the geneticists eagerly set out to forge a unified position to compel the standard-setters and nuclear officials to address the growing threat to genetic health in the nuclear age.

With the new opportunity for the U.S. geneticists came a new dilemma. While some members of the genetics committee disputed the notion that genetic mutations were always harmful under any circumstance, all reconfirmed the linear non-threshold (LNT) hypothesis as a reasonable scientific basis to evaluate genetic damage caused by radiation. Many, however, also believed that this basic principle was not sufficient to change the actual practices of radiation protection. In fact, both the NCRP and the USAEC had already acknowledged, at least tacitly, the likelihood of no safety threshold in genetic damage. What made them reluctant to apply this knowledge to regulatory action was its operational difficulty. Since the LNT hypothesis meant that the only "safe" dose, in the strictest sense of the word, would be absolute zero, the regulators feared that its straightforward application might force them to reduce exposure limits to the point where it would severely interfere with nuclear and other enterprises involving radiation exposure. Weaver, for one, saw some merit in such a concern. "There is the practical problem that goes beyond general principles," he told the panelists. "We have to face the question: 'What are proper safeguards?' It won't be sufficient to say: 'As little as possible.'"[7]

What was proposed as a solution to this dilemma was to determine a genetically "permissible" dose for the general population, which was to be set low enough to minimize genetic risks but also high enough to accommodate expanding uses of radiation and radioactive materials. A leading advocate for this approach was Gioacchino Failla. Although he was not a geneticist, Failla was invited to the genetics committee as an observer. His leadership roles in the NCRP and the International Commission on Radiological Protection (ICRP) made him a key liaison between the old and new expert bodies. Failla had long resisted the establishment of a genetically permissible dose. In the absence of such a guideline, however, Failla realized that the existing standard-setters were losing control of radiation protection, as some engineers and government officials started applying different standards from the genetic point of view. "We must have a figure from the geneticists for the general population," Failla told the committee. "The most important thing to do is to set a value, conservatively but reasonably, so it will not have to be discarded as impracticable."[8]

Not all members, however, readily accepted Failla's logic. Human geneticist James V. Neel raised the strongest objection. As a leader of the Atomic Bomb Casualty Commission genetics study in Japan, he knew first-hand how little scientists understood about genetic damage in humans. Neel repeatedly argued, in letters to Weaver and also during meetings, that he saw no way to choose an exact figure as permissible. He even declared that he was "unprepared to sign any report that states a permissible dose," urging others "not [to] compromise ourselves as scientists."[9] Most of his colleagues, however, strongly disagreed with Neel's argument. William L. Russell, director of the mouse genetics project at Oak Ridge, said that he shared Neel's misgivings but "for practical purposes someone has to set this dose now."[10] A pragmatist by heart, Russell believed that the panel simply could not afford to pass on this unique chance to make a critical intervention in radiation protection before it became too late. "As the result of accepting this responsibility, we will be regarded as the authority,"

Russell told the committee. "This is the opportunity for us to take some control over the situation."[11]

With Neel's reservations politely noted but essentially set aside, the genetics committee proceeded to establish an exposure limit for the general population. When the committee held its meeting on February 5 and 6, 1956, it broke up into two working groups: one for proposing a permissible dose, and the other for estimating the magnitude of the genetic risk entailed by that limit. After some discussions, the first group reported back to the whole committee with the proposal that an average exposure to the gonads per individual during the period from conception to age thirty should not exceed 10 R from all sources other than natural background radiation. A member of the group, James F. Crow, described this figure as one of a compromise between the two figures already proposed to the ICRP and the NCRP. He recalled that Muller had once suggested 20 R, whereas Stern had preferred 5 R. The working group arrived at 10 R as "the geometric mean." Muller then tried to justify this seemingly arbitrary figure. He argued that the value was set well below the most likely estimate of the doubling dose, then standing at 40 R, and also within the experience of some populations in the world, such as those in the Andes and Tibet, who probably received 5 R per generation more than ordinary populations with no apparent genetic problems. Muller also stressed a sense of pragmatism imbued in the proposed limit. Pointing out the estimate showing that in thirty years the U.S. population might already be receiving radiation at above 3 R for medical purposes, he claimed that the 10 R limit would provide ample room for the expansion of both atomic-energy uses and medical radiology. "If we had set 5 [R] as a limit, everyone would have said, 'Impractical—Utopia'," he told the panel. "At 10 [R] many more will strive for protection."[12]

Successfully securing the agreed permissible dose from the geneticists, Failla immediately relayed this information to the standards organizations. During a meeting of the ICRP held in April 1956, Failla reported that "[g]eneticists wanted something less than 10 [R]."[13] Fearing that the BEAR committees might undermine the ICRP's position as a leading authority for radiation protection, he urged the commission to immediately take steps to revise its guidelines before the U.S. experts published their final report. As he wrote in a letter to members, he considered it "very important for the future of the Commission to take the lead in such matters rather than be compelled to accept similar changes by the force of events later on."[14]

The ICRP, however, failed to decide on a specific figure. It was agreed that, as an interim measure, the dose of radiation received by the gonads from all sources additional to the natural background should be limited to "an amount of the order of the natural background in presently inhabited regions of the earth."[15] Failla later explained that this vague statement was intended to allow the commission to withdraw or amend it later once general agreement was reached.[16]

Meanwhile, the MRC Panel of Genetic Effects faced a dilemma similar to that of its U.S. counterpart. During a meeting with members of the panel, Secretary of the MRC Harold P. Himsworth observed that, without a specific, quantitative guide, "the report will be of little value to Parliament and the country, with the consequent risk that the wrong executive decisions may be made, or no decisions at all."[17] For this purpose, the British geneticists initially tried to make estimates of genetic damage expected from different exposure situations without specifying to what extent increases in mutations were deemed permissible. When Himsworth visited Washington in April 1956, however, he was shocked to discover that the U.S. geneticists had already determined 10 R as the genetically permissible dose for the general population. With the final report due only two months away, Himsworth hastily put together a task force to draw up the MRC's own recommendations.

Unlike the BEAR genetics committee, however, the British working group consisted of both geneticists and nongeneticists. Moreover, four of these nongeneticists were directly involved in radiation protection: Ernest Rock Carling, chairman of the ICRP; William V. Mayneord, one of the ICRP's main commissioners; John F. Loutit, director of Harwell's MRC Radiobiology Unit; and W. G. Marley, Harwell's radiation protection chief. Their strong ties with the standards organizations made them wary of usurping the existing standard-setting mechanism. In a letter to a member of the MRC Panel on Individual Effects, Himsworth described the general feeling of the meeting that "we should endorse wherever possible those put forward by bodies like I.C.R.P."[18]

Caught between the old and new expert bodies, the MRC chose the middle of the road. Noting that the existing knowledge was still inadequate to name any specific figure as a limit for the average exposure of the whole population, the MRC report stated that "it is unlikely that any authoritative recommendation will name a figure . . .

which is more than twice that of the general value for natural background radiation."¹⁹ This purposefully convoluted statement essentially brought the British in line with the Americans without declaring so. Like the ICRP, the MRC took natural radiation levels as a baseline, but it doubled the allowance to ensure that the resulting value would be not significantly lower than the 10 R limit proposed by the BEAR genetics committee.²⁰

The successful coordination across the Atlantic in working out a genetically permissible dose radically changed the context in which these experts finally returned to the original question of radioactive fallout. Until then, the geneticists, especially those in the United States, had warned the public about the dangers of fallout as a reminder that no genetic consequence of radiation was too trifling to be dismissed offhand, especially if it was applied to the whole population. With a certain extent of genetic damage now declared as "permissible," however, the focus of the BEAR and MRC geneticists shifted from the existence of the genetic risk from fallout in an absolute sense toward its relative weight among various sources of radiation affecting the populace. Their respective surveys demonstrated that the largest source of exposure was medical X-rays, which accounted for between 22 and 40 percent of the proposed permissible doses for the whole population. The share of external gamma radiation from global fallout, by contrast, was extremely small, no more than 1 percent of the suggested limits even if nuclear testing continued indefinitely at the same rate as the past few years.²¹ "If you have a budget of $10 a year to spend" and it was "50 cents a year for fallout out of our $10," Weaver said in a meeting with his colleagues, "then, I think we can say, 'Well, this still is not a negligible part of our yearly budget. It still deserves scrutiny but we don't need to get all lathered up about it at the moment.'"²²

For some BEAR geneticists, the irony was unmistakable. George W. Beadle from the California Institute of Technology pointed out that such an argument was exactly what the USAEC had been making to dismiss the problem of fallout entirely. He recalled that USAEC commissioner Willard F. Libby had repeatedly argued that the levels of external gamma radiation from fallout were no more than a small fraction of the dose required to double the mutation rate in humans. This, Beadle said, was "a beautiful example of a true, but misleading, statement," as Libby was clearly trying to make the public believe that exposure to low-dose radiation was genetically "safe" as long as it remained below the doubling dose.²³ Beadle was not the only panelist who raised an objection to Weaver's conclusion. Upon reading a draft of the final report, Alfred H. Sturtevant from the California Institute of Technology immediately sent a letter to the chairman, pointing out that even doses less than 1 percent of the genetically permissible dose might ultimately produce some twenty-five hundred or even over a hundred thousand handicapped children in the United States. "To say that we needn't worry about such numbers, merely because other sources of radiation are producing more damage," Sturtevant wrote, "seems... analogous to saying that murder is more serious than robbery, and therefore we'll pay no attention to robbery."[24] At his insistence, the genetics committee approved a last-minute change to the draft report, inserting a paragraph stressing the need to keep the amounts of radioactive fallout released into the global environment from rising to more serious levels.[25] Despite this important qualification, the conclusion remained essentially the same: the genetic effects of fallout, albeit existent in the absolute sense, were nevertheless quantitatively well within the "permissible" limits. Much to the delight of U.S. and British officials, these reassuring statements removed, at least temporarily, the issue of genetic damage from the center of public attention. By then, however, the focus of debate had already shifted to a different type of radiation hazard to humans.

Medical Sciences
When the BEAR Committee on Pathological Effects of Atomic Radiation and the MRC Panel on Individual Effects began their reviews on somatic (nongenetic) effects of radiation, both immediately identified Sr-90, a bone-seeking, carcinogenic beta emitter, as the most dangerous radioactive element contained in the fallout from nuclear tests. Initially, the assessment of the radiostrontium hazard seemed to be fairly straightforward. It was then widely believed that there would be a practical if not absolute threshold in doses for a complex and long-term event such as cancer, and that the maximum permissible concentration of Sr-90 in human bones recommended by the NCRP and the ICRP for occupational purposes, 1 microcurie, represented a truly "safe" dose. On the basis of these assumptions, U.S. and British nuclear officials repeatedly argued that Sr-90 found all over the world as a result of nuclear tests would pose no danger to humans. In a speech delivered in January 1956, for instance, Libby pointed out that the highest amount of radiostrontium

ever measured in the bones of children, a demographic group believed to be the most vulnerable to the influence of radiation, was about 0.001 microcurie, or one one-thousandth of the occupational maximum permissible body burden (MPB). This and other apparently reassuring findings from the global radiostrontium survey conducted by the USAEC led him to conclude that "the worldwide health hazards from the present rate of testing are insignificant."[26] The medical members of the BEAR committees and the MRC scrutinized such a statement from a different set of scientific and regulatory concerns.

When the BEAR pathology committee reviewed the existing literature on ionizing radiation and cancer, it realized that some epidemiological studies cast doubt on the threshold hypothesis. The subcommittee on hematological effects chaired by Eugene P. Cronkite from the USAEC's Brookhaven National Laboratory reported that the whole-body exposure data from U.S. medical radiologists and Japanese atomic bomb survivors both suggested a possible LNT relationship between dose and leukemia incidence. This meant that Sr-90 in amounts well below the occupational MPB might still entail a small excessive risk of leukemia. The subcommittee, however, found the estimated doses in these studies too uncertain to draw any definitive conclusion about the dose-response relationship. For instance, the dosimetry system then used to reconstruct doses received by the Japanese atomic bomb survivors was crude, primarily based on distance from the hypocenter without taking into account shielding and other key mitigating factors.[27] In view of the large uncertainty in the existing epidemiological data, the subcommittee discussed opportunities for a new study, including the existence of newborns and pediatricians in the Seattle area exposed to X-rays during unnecessary fluoroscopic examinations.[28] With no resources to undertake research on its own, however, the BEAR pathology committee relied on past clinical experience instead. While noting that a variety of injurious effects such as leukemia and skin cancer might occur to individuals occasionally exposed to radiation at a relatively low level over a period of years, it declared in the public report that "[a]mong those who have adhered to present permissible dose levels, none of these effects have been detected."[29]

The BEAR medical researchers also addressed the question of bone cancer in such a way as to uphold the existing guidelines. Most members of the subcommittee on internal emitters believed that there would be a threshold, or at least no straight-line function, in the production of solid tumors, and that the occupational MPB for Sr-90 was set well below that limit. Nonetheless, subcommittee chairman Austin M. Brues admitted uncertainty in their conclusion. "One can take the extreme view at one end that any radioactivity is a bad thing and that we should not have it, or one can take the extreme view . . . that the permissible dose of 1 microcurie is based on sound grounds, that it is perfectly all right, and that everybody in the world can have 1 microcurie," he said during a committee meeting held in January 1956. "We are now getting almost into the realm of ethical considerations or metaphysical considerations because we fail to know all the answers."[30] Instead of exploring the possibility of bone cancer induced at doses below the occupational MPB, Brues took the natural radioactivity as a commonsense baseline, claiming that "it would certainly be very silly for anyone to take the position that a small increment from that was a bad thing."[31] After further discussion, the BEAR pathology committee fully endorsed the recommendation made by the NCRP and the ICRP that members of the public would deserve an additional factor of safety, namely one-tenth of the occupational MPB, but no more. Its public report declared that "[t]here seems no reason to hesitate to allow a universal human strontium . . . burden of 1/10 of the permissible, yielding 20 rep [Roentgen equivalent person[32]] in a lifetime, since this dose falls close to the range of values for natural radiation background."[33]

Unlike its U.S. counterpart, the MRC individual effects panel possessed both the reason and resources to challenge the existing guidelines. In fact, some of its members had long been skeptical about the scientific basis for calculating the occupational MPB for Sr-90. The problem of chronic internal exposure to radioactive elements lodged in human bones dated back to the interwar period, when many of the young female radium-painters in the United States contracted bone cancer and other forms of maladies after ingesting radium in tiny amounts over a long period of time.[34] When a panel of experts to determine a safe limit to radium intake was convened in 1941 at the request of the U.S. Navy, a major procurer of radium dials for military equipment, physicist Robley D. Evans measured the body burden of radium in twenty-seven dial painters, showing that those with less than 0.5 microcurie of radium showed no clinical symptoms. The observed threshold, however, was far from conclusive, as bone cancer was such a rare disease—only fifteen out of a hundred thousand women below the age of forty-five would naturally contract bone cancer—that the sample size in Evans's study was too small to ascertain the lower bound of the excess risk from radium.[35] Pressed by the exigencies of the U.S. military to increase defense production, however, the panel was forced to make an informed guess, approving Evans's proposal of 0.1 microcurie as the level at which, Evans said, "we would feel perfectly comfortable if one's own wife or daughter were the subject."[36] This agreed value became a foundation for the extrapolation of the occupational MPB for Sr-90 and other bone-seeking radioactive elements.

In the late 1940s, the problem of radioactive wastewater discharged from Harwell into the River Thames compelled the MRC to take its potential genetic effects of radiation seriously. The same concern led them to reexamine the safety limit for radium. The task fell on Louis H. Gray. An eminent British physicist, Gray was known for his work that defined a unit called rad, later replaced by the SI unit named after him (gray), a measure of the amount of energy deposited in a specified amount of matter by exposure to radiation. After a thorough review of Evans's original data, Gray discovered that, if the LNT hypothesis was as applicable to bone cancer as it was to genetic damage, the risk of the disease at 0.1 microcurie of radium would be not zero but rather 1 to 2 percent. "A 1 % probable incidence of osteogenic sarcoma [bone cancer] may possibly be regarded as an insignificant industrial hazard," he wrote in a memo to the MRC, "but it would seem to be inadmissible as a hazard to which a large section of the population might be exposed."[37] To reduce this theoretical risk that might loom large in the general population, the MRC proposed to lower the safe limit for radium by a factor of one hundred as a basis to calculate the maximum MPB for Sr-90 for members of the public.[38] During the tripartite conferences held from 1949 to 1953, however, the Americans strongly objected to the British proposal, arguing that the suggested value was so low that it would be impossible to enforce without serious interruptions in atomic work.[39] Against this backdrop, the MRC asked two radiologists, William V. Mayneord and Joseph S. Mitchell, to revisit the question. Both confirmed that the occupational MPB for Sr-90 indicated not a safety threshold but rather a degree of risk less than one in ten thousand. This meant that one-tenth of this value, which the BEAR pathology committee declared as universally permissible, might produce about forty tumors per million exposed persons, or a 5 percent increase on the natural incident rate.[40]

The same critical view on the existing guidelines also led the MRC panel to reevaluate the risk of leukemia. As noted earlier, the BEAR committee recognized a possible LNT relationship in the case of leukemia, but it found the estimated doses too uncertain to determine a dose-response relationship. Unlike its U.S. counterpart, the MRC was a research grant provider, capable of commissioning a new study to strengthen the scientific basis of its health risk assessment. At the request of the MRC, radiologist Michael Court-Brown and epidemiologist Richard Doll conducted a large-scale retrospective cohort study with some thirteen thousand X-rayed patients for a type of arthritis called ankylosing spondylitis, suggesting a linear or curvilinear relationship with the number of reported leukemia cases.[41] Although this new data had its own degree of uncertainty and was no more definitive than the previous ones, it made it more difficult for the British experts to set aside the possibility of no safety threshold. In a letter to Himsworth sent in February 1956, radiobiologist Alexander Haddow agreed that the public report should be based on facts, but he insisted that "it should include a certain element of reserve, in view of the uncertainties."[42]

The divergence between the U.S. and British experts did not initially come to the surface. When Himsworth visited Washington in April 1956, he was delighted to see that the reports on the somatic effects appeared to be "on much the same lines."[43] Soon after his trip, however, the British draft was revised substantially to take into account a possible non-threshold model of leukemia and bone cancer. Notably, the impetus for this major change came from within the British nuclear authorities. Shortly before the MRC individual effects panel held one of the final meetings on May 5, W. G. Marley, a member of the panel and Harwell's chief health physicist, met with Harwell's director John Cockcroft. Like the Americans, both endorsed one-tenth of the occupational MPB for Sr-90 as permissible for members of the public. In view of the potential risk to children and other vulnerable people, however, Cockcroft and Marley also suggested another level, equivalent to one-tenth of the population MPB, as undesirable for the entire British population.[44] The MRC panel promptly accepted this proposal and incorporated the "warning dose," one-hundredth of the occupational MPB, into the draft report.

Lowering the reference value was fateful because the amounts of Sr-90 in Britain seemed to increase at a faster pace than the Americans had predicted. On the assumption that radioactive fallout injected into the stratosphere would slowly and uniformly settle to the ground, Libby argued that the peak deposition of Sr-90 in the U.S. soil would be no more than 7 millicurie per square mile (milliCi/mi^2) in 1970. Measurements in Britain, however, showed that the density of radiostrontium in soil had already reached 6 milliCi/mi^2 in 1954 and jumped to 11 milliCi/mi^2 by the end of 1955.[45] Moreover, sheep bones obtained from Wales showed an unusually high ratio of Sr-90 transfer from soil to animals, as plants and animals tended to absorb more strontium to compensate for the shortage of calcium in Welsh soil.[46] All of this pointed to the possibility that the body burden of Sr-90 might reach the "warning dose" in several decades even if the rate of nuclear testing remained constant as in the past few years.[47]

At this critical moment, the mechanism of coordination across the Atlantic broke down. The British experts learned about the genetically permissible dose determined by their U.S. peers in time to formulate their own in step with the Americans. By contrast, most of the major changes to the section of Sr-90 in the draft MRC report were made at the eleventh hour before its public release. Having already finished its deliberations as early as in February, the BEAR pathology committee had no chance to work out disagreement with its British counterpart. The stark contrast in the conclusions of the two public reports regarding the cancer risk of radiostrontium had immediate and serious consequences for Anglo-American epistemic leadership—and also for U.S. and British policymakers, who faced a growing call for the suspension of nuclear tests.

Confluence of the Fallout Controversy and the Test-Ban Debate

By the time the NAS and the MRC published their reports on June 12, 1956, a nuclear test ban had already emerged as a major item on the international political agenda. During the London meetings of the UN Subcommittee on Disarmament in spring of 1956, the Soviet delegation announced its willingness to negotiate a test ban agreement as a separate measure apart from comprehensive disarmament. Around the same time, Adlai Stevenson, the Democratic presidential candidate during the 1956 U.S. presidential election, unveiled a plan to unilaterally suspend thermonuclear bomb tests as part of his campaign platform.[48] These and other advocates of a test ban, however, initially framed the issue solely as a disarmament initiative. The simultaneous release of the BEAR and

MRC reports drastically changed this, as their sharp disagreement over the cancer risk of Sr-90 fueled a call in many Western countries for curbing or banning fallout emissions. In this way, the epistemic divide across the Atlantic set in motion the process in which pressing concerns among politicians, scientists, and ordinary citizens about the escalating nuclear arms race began to merge with growing anxieties about its global environmental consequences.

The first move to address both security and environmental concerns came from the British. When the NAS and MRC released their reports, British Prime Minister Anthony Eden, who had succeeded Winston Churchill in April 1955, was on the verge of announcing plans to conduct the first British thermonuclear test in the Pacific. When his cabinet was briefed on the MRC report, all agreed that "there was nothing in the report which would justify canceling or postponing this particular series of tests."[49] In view of its more thorough and cautious assessment than that of the U.S. report, however, the ministers concluded that "[w]orld scientific opinion and a substantial body of American scientific opinion is likely to agree with the M.R.C."[50] To anticipate the growing pressure for a test ban that would completely halt the British nuclear weapons program, the cabinet agreed to "take the initiative in proposing discussions between the three Powers concerned" for a test limitation.[51] On July 23, Eden told the House of Commons that his administration was willing to enter negotiations with the United States and the Soviet Union toward an international treaty to curb fallout emissions.[52]

As Eden feared, the growing concern about fallout sparked the rise of a nationwide antinuclear campaign in Britain. When Britain's National Peace Council hosted a conference in November to discuss the possibility of concerted action against nuclear tests, activist scientists alerted them to the problem of fallout. Haddow, a member of the Atomic Scientists' Association (ASA) who had served on the MRC individual effects panel, pointed out that a good case could be made for a test ban as a prudent measure so long as scientific evidence about fallout hazards remained inconclusive.[53] The activists accepted the counsel of these scientists and launched an ad hoc campaign called the National Council for the Abolition of Nuclear Weapon Tests (NCANWT). Beginning in February 1957, this umbrella group quickly grew far beyond its pacifist basis, establishing over a hundred local branches across the nation by the end of the year.

Its emphasis on the fallout peril seemed to have a particular gender appeal to women, who constituted about two-thirds of the members.[54] The NCANWT framed the problem of fallout as a moral crisis, claiming that any slight risk of health damage was intolerable for those around the world who did not consent to nuclear tests.[55]

Britain was not the only U.S. ally deeply shaken by the epistemic divide across the Atlantic. Norway, a NATO member, also faced growing troubles with global fallout. In early 1956, snow runoff there showed beta activities in excess of the most stringent maximum permissible concentration recommended by the NCRP and the ICRP. Located along the Arctic Circle, Norway was subject to nuclear dust coming from both U.S. and Soviet tests conducted in the Northern Hemisphere. With a large part of the population depending on melted snow for drinking water, the "radioactive snow" triggered an uproar across the nation.[56] Alarmed by this unexpected development, the USAEC hastily dispatched Charles L. Dunham, director of the Division of Biology and Medicine, to Paris in June for briefing the NATO members, together with British experts, on the BEAR and MRC reports. Although the joint presentation was meant to reassure the anxious European allies, the Norwegians remained unconvinced. Faced with the mounting pressure from its citizens, and encouraged by the abrupt shift of Britain's test-ban policy toward a test limitation, Norway decided to take its own diplomatic initiative. A draft Norwegian resolution, first discussed at the UN General Assembly in November 1956, demanded that all planned nuclear tests should be registered in advance for the UNSCEAR to evaluate if radioactive fallout from proposed experiments would remain within the limits of safety.[57]

In the meantime, Canada, another NATO country close to the Arctic Circle, also grew worried about fallout contamination. Ottawa initially accepted the reassurance made by the BEAR report that the risk of harm from nuclear fallout would remain negligible at the present rate of firing. The Canadians soon realized, however, that neither the BEAR report nor the MRC report could guarantee safety in the future if the tempo of nuclear testing accelerated.[58] This realization came when Canada was in the middle of a search for an idea to break an impasse in the UN disarmament subcommittee in which it took part.[59] At this conjuncture, the Department of External Affairs came to the conclusion by fall of 1956 that "there may well be a case on medical and general health grounds for early action on test explosions." Like many scientists who voiced their concern about global fallout, some Canadian foreign officials understood the importance of assessing its potential hazards in absolute, not relative, terms. For those who opposed "*any* adverse effects on human beings," one such official explained, even the more optimistic estimate would set a point of danger "only a few years in the future."[60] With the lack of information not only on fallout but also on defense needs for the United States, the health, defense, and foreign ministers agreed that Canada should explore a middle ground by calling for some form of test limitation.[61] In January 1957, when Norway formally submitted its UN resolution for an advanced registration program, Canada joined this move as a cosponsor together with Japan.

As the confluence of security and environmental anxieties in the nuclear Anthropocene drove a growing call around the world for an international regulation of nuclear tests, the USAEC, strongly opposed to any restrictions on its nuclear weapons program, tried to counter the momentum by unveiling the "clean bomb" as a technological solution to the problem of radioactive fallout. The clean bomb was a hydrogen bomb with minimum use of fission material to reduce the amount of fission elements of public health concern such as Sr-90. In July 1956, upon his return from the Pacific Proving Grounds to witness the latest nuclear test series, Operation Redwing, USAEC chairman Lewis L. Strauss disclosed the successful test of a clean bomb with much fanfare, stressing its importance "not only from a military point of view but from a humanitarian aspect."[62] Strauss's remarks immediately elicited widespread ire and derision. U.S. physicist Ralph E. Lapp, a former Manhattan Project scientist who became an outspoken critic of U.S. nuclear policy, quipped, "Part of the madness of our time is that adult men can use a word like humanitarian to describe an H-bomb."[63]

For some, however, the clean bomb was much more than a propaganda gimmick. One such person was Edward Teller, a Hungarian-born U.S. nuclear physicist popularly known as "the father of the hydrogen bomb" for his key role in its development. One of the refugee scientists fleeing from Germany under Nazism, and convinced of the evil of communism that swept his home country in the aftermath of World War II, Teller strongly supported every effort to maintain and expand U.S. nuclear superiority over the Soviet Union as an essential condition for peace.[64] He viewed the clean bomb as a major breakthrough in this struggle, believing that it would enable the United States to inflict massive damage to advancing communist forces in the battlefield without causing unnecessary operational disruptions and civilian causalities.[65] As historian Paul Rubinson has demonstrated, however, Teller was also a brilliant political tactician, flexibly changing his arguments to disarm his opponents.[66] In a letter to a USAEC official, he stated that public disclosure of the clean bomb program was worth making to stop the growing momentum for an international regulation of nuclear testing. "I feel strongly that all types of tests must continue," he wrote. "Therefore it is important to publish valid arguments for the test program."[67] While continuously insisting that testing posed no dangers whatsoever, Teller and his allies in the USAEC began to appropriate the rhetoric of antipollution in order to actively campaign against a test ban.

In addition to promising a technological fix, U.S. nuclear officials became more willing to entertain the possibility of a test limitation. When the disarmament talks in London adjourned, Harold A. Stassen, Eisenhower's disarmament adviser, asked the USAEC for a fresh idea to counter the Soviet motion for a test ban. After careful consideration, Libby concluded that a majority of the test ban advocates in the United States and abroad were genuinely worried about the health effects of fallout. "I must say I do not like the thought that my children will have strontium-90 in their bones, as much as I know about the subject, and know how harmless it is," he wrote in a memo.[68] As a solution, he suggested an internationally agreed limit to the emission of fission fallout, namely an annual cap of 5 to 10 Mt fission yields, in order to keep the worldwide levels of Sr-90 contamination below one-tenth of the occupational MPB identified in the BEAR and MRC reports as a threshold of permissibility. With this proposal, Libby aimed to divide the test ban supporters between the health-minded majority and the disarmament-oriented minority, hoping to arrest the growing momentum for a complete test ban.[69]

The quiet search within the Eisenhower administration for a test limitation, however, suddenly came to a halt when the 1956 U.S. presidential election turned the test ban issue into a political hot potato. Having introduced an H-bomb test ban in his campaign platform in April, Stevenson returned to the subject in September with a new emphasis on the danger of Sr-90 in the fallout. In a speech delivered on September 29, he declared that "[t]he testing alone of these super bombs is considered by scientists to be dangerous to man."[70] Much alarmed by this political move, Secretary of State John Foster Dulles abruptly withdrew his support for Stassen's test-ban policy review. Although Eisenhower still directed his disarmament adviser to continue his work, he was forced to allow Republican partisans to viciously attack Stevenson's proposal in public.[71] The president tried to defuse the situation by insisting that Stevenson should be refuted on the basis of "the experts' opinions, completely accurately and factually."[72] For this purpose, Eisenhower's surrogates arranged the release of a joint statement by twelve scientists, who cited the BEAR report as the most authoritative study conclusively showing that fallout at the current levels and in the near future would pose no danger to the public.[73]

Nonetheless, the divide of scientific opinions across the Atlantic and the ensuing public debate had already undermined the credibility of the BEAR report. Stevenson's campaign benefited from this expert disagreement. In his letter to Stevenson, Walter Selove, a physicist and chair of the FAS's Radiation Hazards Committee, argued that the BEAR pathologists put forward "the most optimistic end of the possible range."[74] Laurence H. Snyder, president-elect of the American Association for the Advancement of Science, also sharply criticized an assumption behind the safety claim that the rate of testing would remain constant. "Our tests cause other nations to increase their pace, which in turn increases ours," he said. "It becomes a self-accelerating system."[75] Backed by these dissenting voices of notable scientists, Stevenson dismissed the BEAR report as "already out of date," accusing his Republican opponent of exaggerating "the degree of certainty that is possible on the basis of present scientific knowledge."[76] Then he made the case for a test ban as a precautionary measure under the conditions of uncertainty. A policy memo for his campaign pointed out that "neither side can argue with certainty that continued tests are or are not harmful to the human race," arguing, "With such uncertainty, the argument is on the side of those who wish to stop the large nuclear tests."[77]

Eisenhower's landslide victory in the presidential election put an end to Stevenson's quest for an H-bomb test ban. The successful rollback by test-ban opponents in the United States, however, hardly changed Washington's political isolation in the international arena. The British resumed their pursuit of a test limitation after the presidential election. In January 1957, British deputy foreign minister Allan Noble met with Stassen, discussing the idea of capping the worldwide annual fallout emissions at 15 Mt and dividing the emission credit among the three nuclear powers.[78] Even some

U.S. foreign and military officials were now inclined toward some form of test limitation, either by agreement or self-restraint.[79]

During the Bermuda meeting held in March between Eisenhower and Eden's successor Harold Macmillan, however, Strauss managed to kill the limitation proposal. He turned the policy implication of scientific uncertainty on its head, asserting that it was impossible to determine a fallout emission quota because "no reports on the effects of radiation have fixed a precise danger point." He also repeated the claim that, if testing was continued with due restraint, the health risk of fallout would remain far smaller in quantity than that from X-rays and natural radiation.[80] The British found these arguments unconvincing but nevertheless astutely recognized the political message behind them. As a joint Anglo-French-Israeli invasion of Egypt over the control of the Suez Canal in fall of 1956 had greatly strained the Anglo-American relationship, Macmillan could ill afford to further antagonize Britain's key ally.[81] At Strauss's insistence, Eisenhower and Macmillan issued a joint statement on March 24, simply pledging to conduct nuclear tests "only in such manner as will keep world radiation from rising to more than a small fraction of the levels that might be hazardous."[82] Eisenhower and Macmillan then proceeded to sign a secret note, pledging not to seek a test ban or limitation without mutual consultation in advance.[83]

After successfully forcing the British to shelve their test-limitation proposal, Strauss and his allies moved to defeat a similar initiative made by the Norwegians, Canadians, and Japanese. The USAEC again appropriated the fallout argument, claiming that the progress of the "clean" bomb made it no longer possible to estimate the amount of fission fallout without disclosing weapon designs.[84] The Norwegians eventually agreed to drop the part of its UN resolution calling for a fallout hazard assessment by the UNSCEAR, but not without a warning that world pressure would only continue to build up for a similar regulatory scheme.[85] For now, however, the opportunity was lost for even the most modest form of test limitation. Although the UN Disarmament Subcommittee held a formal hearing on the tripartite resolution in spring of 1957, the idea of the advance test registration was simply left withering during the subsequent talks.

Conclusion

The public release of the BEAR and MRC reports and the ensuing public uproar marked a critical moment in the politics of risk. U.S. and British officials assiduously tried to shift the locus of the global-fallout debate from the public sphere to the NAS and the MRC, hoping that these prestigious scientific bodies would reach a consensus in support of their safety claim. The geneticists and medical researchers in each committee, however, approached the question of risk from a markedly different set of scientific and regulatory concerns. While the former switched a measure of risk from the absolute to the relative term and thus moved closer to the official views of their governments, the latter was split over the possibility of no safe threshold for cancer as well as its public health implications.

The resulting epistemic divide pushed the locus of risk politics back to the public sphere, but with an important twist. As a growing number of experts openly questioned the certainty in the scientific basis of the reassurances offered by the U.S. and British governments, some policymakers and concerned citizens began to interpret the policy implications of uncertainty differently, viewing it not as an excuse for nonaction, but rather as a call for precautionary action. While continuously dismissing the health risks from global fallout as negligible, even some of the staunchest test-ban opponents in the United States also began to appropriate the language of antipollution, using the promise of a technological fix—the "clean bomb"—to contain and defeat a variety of initiatives for an international regulation of nuclear testing. In this important way, the problem of global fallout contamination came to be firmly wedged into the test-ban policy debate.

Although the efforts of U.S. nuclear officials to contain the immediate political fallout from the BEAR and MRC reports were successful, the divide of expert opinions across the Atlantic had a long-term impact on the politics of risk as its locus again shifted from the public sphere to the UNSCEAR. Instead of guiding the international global-fallout risk assessment toward a reassuring conclusion by the sheer weight of a consensus, Anglo-American epistemic leadership suffered a serious blow from within. Moreover, the expert disagreement revealed by the BEAR and MRC reports gave rise to a new way of understanding the policy implications of uncertainty in knowledge about global fallout contamination and its biological effects. Against the backdrop of this shifting definition of risk, the UNSCEAR struggled to produce a consensus report as the most authoritative scientific statement on the risks of global fallout.

Chapter 5

EPISTEMIC NEGOTIATIONS

The United Nations Scientific
Committee, 1956-1958

ON MARCH 14, 1956, scientists from fifteen countries arrived in New York to attend the first meeting of the United Nations Scientific Committee on the Effects of Atomic Radiation (UNSCEAR). The mission of this newly established committee was to assess global levels and effects of ionizing radiation and submit a report to the General Assembly no later than July 1958. Although its work covered all sources of radiation, both natural and artificial, the main focus was on radioactive fallout from explosions of nuclear weapons. In view of the controversy surrounding this topic, UN Secretary-General Dag Hammarskjöld insisted that the UNSCEAR must remain scientific in nature. "We are all aware of this being a scientific committee," he said in his opening speech before the delegates. "This is emphasized by the fact that all the countries in the Committee have sent scientists of world-renown as their representatives." Noting "a lack of knowledge which has caused in many instances an unwarranted reaction to the whole subject," he said that the UNSCEAR would "help move the subject out of the area of emotional sensationalism and place it squarely on the solid footing of scientific knowledge, which will, in turn, change unconsidered fear into sober precaution."[1]

The stark contrast invoked by Hammarskjöld between "emotional sensationalism" and "scientific knowledge," however, collapsed when the UNSCEAR report painted a far more alarming picture of global fallout contamination than U.S. and British officials had presented as a scientific fact. It demonstrated that the influence of global fallout was highly uneven from region to region. For instance, the worldwide fallout rate and deposit of Sr-90 varied with latitude, with areas in the temperate northern hemisphere having deposits of about three times the world average.[2] Moreover, even if weapons tests stopped at the end of 1958, people in Asia taking most of their dietary calcium from rice would receive six times the maximum annual average dose to bone marrow from Sr-90 than those in the Western world from milk.[3] The UNSCEAR report also showed the estimate that the amounts of fallout from tests through 1958 would ultimately cause twenty-five hundred to a hundred thousand excess cases of major genetic defects and, if there was no safe level of radiation for leukemia, also bring about less than twenty-five thousand to a hundred fifty thousand excess cases of leukemia worldwide.[4] Most important, the committee rejected the concept of permissible dose regarding global fallout contamination, pointing out that it involved "new and largely unknown hazards to present and future populations" that were "beyond the control of the exposed persons." Although the UNSCEAR report stopped short of making a call for a nuclear-test ban, it explicitly mentioned "the cessation of contamination of the environment by explosions of nuclear weapons" as one of the "steps designed to minimize irradiation of human populations."[5]

Why did the UNSCEAR reach a conclusion that was radically different from the view of the U.S. and British governments? Did the committee's report, approved in June 1958 and published two months later, play any role in a unilateral moratorium on all nuclear tests by the United States, Britain, and the Soviet Union beginning in November? If so, how? To answer these questions, one must examine the politics of risk unfolding inside the UNSCEAR, with a focus on four major contributors to the final report: the United States, Britain, Japan, and the Soviet Union.

Contrary to Hammarskjöld's claim, the UNSCEAR was not an independent scientific body. Rather, it was established as an intergovernmental panel, whose scientists were selected and appointed by its member states as their representatives. Moreover, it was forbidden from conducting its own investigation and instead tasked with reviewing scientific information submitted by members of the United Nations and also by specialized agencies. This meant that the member states retained a great deal of control over the committee's work through their delegates and research materials. As historian Néstor Herran has demonstrated, the United States and Britain repeatedly tried to "unscare" the world by downplaying fallout hazards while concealing their preexisting fallout monitoring programs for military intelligence.[6] Not surprisingly, the Japanese and Soviet delegations also consistently tried to present scientific information and its uncertainty in such ways as to promote their governments' policies.

The UNSCEAR scientists, however, were not mere pawns of international politics. To the contrary, these experts had their own scientific ambitions, ethical concerns, and personal beliefs to promote a certain understanding of risk even before government officials ever realized its political utility. Moreover, many of the scientists who served on or advised their national delegations to the UNSCEAR took an active part in an emerging transnational movement of intellectuals against nuclear testing, which not only advocated the non-threshold hypothesis as a prudent ground on which to assess the health risks from global fallout under the conditions of uncertainty, but also raised a serious question about the ethics of inflicting even the slightest risk of harm on the innocent bystanders of the nuclear arms race all over the world and in the remote future. Against this broader shift in the definitions of risk in the nuclear Anthropocene, the UNSCEAR scientists from communist and nonaligned countries and even many U.S. and British allies came to view global fallout contamination as an unacceptable hazard. Although the resulting consensus report did not singlehandedly bring about the nuclear-test moratorium, it played a key role as a facilitator by boosting the Soviet initiative for the suspension of nuclear tests while narrowing the range of policy options for the United States and Britain.

Anglo-American Epistemic Leadership

The U.S. and British delegations arrived in New York for the first UNSCEAR meeting from a position of epistemic strength, better armed with fallout data and biological studies than any other members. Since 1953, the U.S. Atomic Energy Commission (USAEC) had been conducting a worldwide radiostrontium survey code-named Project Sunshine. To a lesser extent, Harwell undertook its own fallout monitoring program, regularly procuring samples from the British Isles as well as several overseas colonies and Commonwealth countries. At the time, a review of the biological effects of radiation by the U.S. National Academy of Sciences (NAS) and the British Medical Research Council (MRC) was also underway. In fact, the U.S. and British delegates to the UNSCEAR were all deeply involved in these national studies. The chief U.S. representative, Shields Warren, was a former director of the USAEC's Division of Biology and Medicine and chair of the pathology panel of the NAS's Committees on the Biological Effects of Atomic Radiation (BEAR). His deputies, Merril Eisenbud and Austin M. Brues, supervised Project Sunshine at the USAEC's Health and Safety Laboratory and studies on biological effects of low-dose radiation at Argonne National Laboratory, respectively. William V. Mayneord, a radiologist and member of the MRC Panel on Individual Effects, led the British delegation, and his replacement following the first session, Edward E. Pochin, was also an insider, serving as director of the MRC's Clinical Research Unit at University College London. His alternate, W. G. Marley, provided expertise as head of Harwell's Health Physics Division overseeing British fallout measurements. All these scientists, working in service of the governments and advisory bodies, proved themselves not only scientifically competent but also politically reliable. Provided with direct access to the wealth of scientific data, publications, and resources by their respective governments, the U.S. and British delegations set out to guide the work of the UNSCEAR from within.

From the beginning, the U.S. delegation took full advantage of its information dominance. In preparation for the first meeting of the UNSCEAR, the USAEC decided to declassify Sunshine data involving no sensitive information, authorizing the public release of IBM cards containing basic data pertaining to "several hundred thousand fallout samples which have been collected by the U.S. since the inception of the program."[7] Backed by this mass of data, the U.S. scientists in New York took the initiative to encourage uniform procedures for the collection and measurement of fallout samples. During the first meeting, Warren circulated a sampling guideline called the Chicago Sunshine Method, urging other countries to adopt it to ensure the compatibility of fallout data across the world. The manual explained the sampling and analytic methods used in Project Sunshine in detail, touting their proven record of handling "approximately 250,000 samples collected in the United States and other parts of the world," processing and counting "as many as 600 samples a day."[8] The advantage for the United States of making the Sunshine method a universal standard was clear: to seamlessly integrate the existing and future systems of other countries into its own while minimizing the risk of erroneous reports that might mislead the public about fallout hazards. As a further encouragement to standardization, the U.S. delegation announced that its government was willing to offer technical assistance in the establishment of collection stations, training in methods of analysis, the processing of samples on behalf of other countries, and the exchange of duplicate standard samples for comparing analytical techniques.[9]

While the U.S. delegation described its offer of technical assistance as a gesture for good science and goodwill, not all countries saw it that way. The Soviets, for one, immediately suspected an ulterior motive behind the U.S. proposal. In a report to Moscow, a Soviet diplomat at the United Nations argued that the "assistance" offered by the United States "conceals [its] military and political motives," namely to study the mechanism of radioactive dust transport following a nuclear weapon explosion and also to collect information about Soviet tests and their locations.[10] The Soviets were not alone in resisting the U.S.-led standardization. Other countries that had already undertaken fallout monitoring used different methods suitable for them, and there was strong disagreement as to which procedures would work the best. The UNSCEAR ultimately chose a loose form of coordination through the comparison and evaluation of national methods.[11] Despite this setback, it was clear that the early lead in fallout studies gave the U.S. and British scientists an immense advantage in their effort to guide the work of the UNSCEAR. In a report to London following the first meeting, Mayneord declared that "we and the Americans have been the only delegations who have continually provided exact quantitative data on the subjects under discussion and this has undoubtedly contributed effectively to guiding the work of this Committee." By contrast, the Soviets initially submitted no scientific reports. "Their inability to produce useful data for the Committee," Mayneord noted, "has tended to lower their prestige with the smaller delegations."[12]

Anglo-American epistemic leadership in the UNSCEAR, however, soon faced multiple challenges. Ironically, one of them came from within their countries. The NAS and the MRC simultaneously released their reports in June 1956, which revealed that both scientific bodies were at odds with one another over the levels and effects of Sr-90. This failure in bilateral coordination came when other UNSCEAR members were becoming restless about contamination found in their countries. During the second UNSCEAR meeting, held from October 22 to November 2, 1956, Rolf Sievert from Sweden confided to the British his serious concern about the effects of fallout in Scandinavia located not far from the Soviet borders. Ernest Watkinson from Canada also informed the British of a rapid rise of Sr-90 deposits observed in his country.[13] The U.S. and British delegations barely managed to hold them back from formally raising the issue during the meeting lest it play into Soviet hands.[14] Despite this, it was clear that the balance of opinions in the UNSCEAR started tilting toward the side of caution. In a report to London following the second meeting, Pochin painted a grim picture ahead. Although the committee was barred from making a policy recommendation, it could still do so indirectly by declaring certain rates of release of Sr-90 as biologically hazardous. If so, Pochin said, not only the communist and nonaligned countries but also countries such as Canada and Sweden might demand a test ban on the ground of uncertainties, which in turn might sway more Western members to the position in favor of a test ban.[15]

With the threat of the falling dominoes for a test ban looming, Washington and London drastically expanded their efforts to monitor fallout to shore up their faltering scientific leadership at home and abroad. The focus of the British program shifted to the British Isles to cope with political fallout from the "hot spots" in Wales revealed by the NAS and the MRC reports. In the meantime, the main burden of reassuring the world fell on the United States. On February 4, 1957, USAEC commissioner Willard F. Libby called a meeting of all senior scientists involved in Project Sunshine. "Next to weapons, Sunshine is the most important work in the Atomic Energy Commission," he said in his opening remarks. If the risk from fallout was not properly understood, "weapons testing may be forced to stop—a circumstance which could well be disastrous to the free world." Libby implored them to do "good science" to demonstrate that nuclear testing posed no undue risk to anyone in the world.[16] What he described as an objective fact, however, was inseparable from his political judgment on the test-ban issue. Geochemist Harrison Brown recalled his meeting with Libby after the 1956 presidential election, where the commissioner "leaned back in his chair and said . . . : 'All right, convince me that it's dangerous.' " This episode, Brown said, was a telling example showing that "from the beginning Dr. Libby has been so convinced that H-bomb tests must be continued *a priori* he has taken the attitude that the risks are small."[17] Despite such criticism, Libby was firmly determined to demonstrate that nowhere in the world would contamination reach a dangerous level for the foreseeable future.

With Libby's blessing, Project Sunshine expanded drastically. The USAEC's budget for the sampling and analysis of radioactive fallout more than doubled in the next two years, from US$1.2 million in FY 1957 to $2.6 million in FY 1959. The equivalent of eight hundred scientists and up to a thousand laboratory technicians were engaged in the project at 5 national laboratories, 204 universities, and 19 corporations.[18] U.S. foreign officials, armed forces, and technical experts stationed abroad helped collect airborne dust from a hundred stations in thirty-seven different countries around the clock. Hospital physicians and missionary doctors with ties to the United States regularly supplied specimens of human bone from about thirty stations worldwide. And high-altitude balloons and U-2 spy planes had begun to sample nuclear dust in the stratosphere over the Western Hemisphere.[19] These and many other agents of the U.S. national security state allowed the Sunshine scientists to claim that the entire planet was under their watchful eyes (Figure 4).

The most telling example of this epistemic globalism was the search for "maximum man," a theoretical construct of an individual with the highest amount of Sr-90 in human bone. As J. Laurence Kulp, the project's chief human-bone analyst, testified during a U.S. congressional hearing held in May 1957, the homogenizing effects of modern food markets tended to produce a narrow distribution of Sr-90 doses for people living in cities. "[I]f we were to imagine a primitive culture where the calcium content of the soil was extremely low, and where the people in that area ate food which they grew only on their own half acre," Kulp said, "then these people might be very much higher than anything we have been talking about here."[20] The USAEC immediately followed up on Kulp's statement by sending a scientific expedition to South America, including the Amazon rainforest, where all the conditions described by Kulp were supposed to exist. Amid international outcry—a Soviet newspaper wasted no time to condemn the Americans as "twentieth-century cannibals"[21]—the mission collected food samples from various regions in the continent to reassure the world that the United States was carefully monitoring the well-being

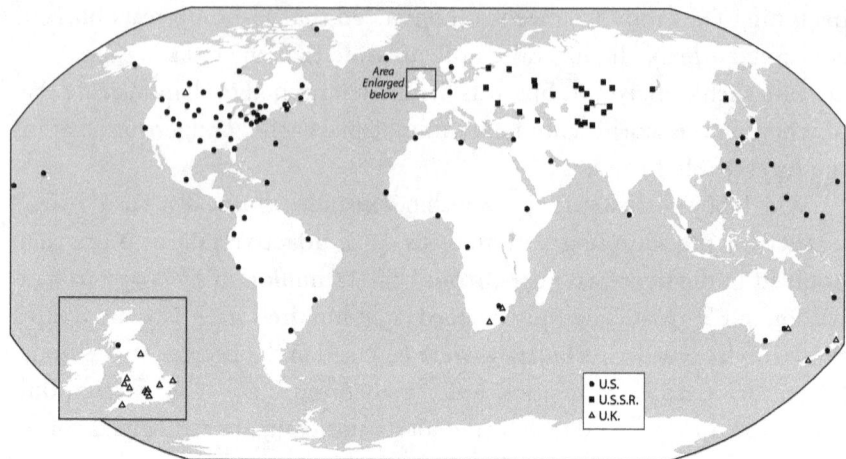

FIGURE 4. Locations of Strontium-90 Surface Deposition Sampling, 1956.

Sources: Bryant, Chamberlain, Morgan, and Spicer, "Radiostrontium in Soil, Grass, Milk and Bone in U.K."; F. J. Bryant, A. C. Chamberlain, and G. S. Spicer, "Strontium 90 Samples from the Commonwealth," December 1957, MAF 291/220, The National Archives of the United Kingdom, Kew; "Data on the Radioactive Scrontium [sic] Fallout on the Territory of the USSR as to the End of 1955," A/AC.82/G/R.163, January 30, 1958, UNSCEAR Library; and "Radioactive Fallout Through September 1955," A/AC.82/G/R.7, August 13, 1956, UNSCEAR Library.

of all humanity, including those believed to be the most vulnerable to global fallout contamination.

As the Sunshine data increased exponentially, the project members grew confident about their knowledge. In a letter dated February 1958, Libby told Kulp, "I think we answered all the questions we had in mind in 1953." What was left, he wrote, was to "ferret out more details."[22] Kulp echoed this sentiment a month later when he boasted Project Sunshine's accomplishments at an international symposium held in Switzerland. In his presentation, Kulp said that the research had successfully identified eight key parameters for estimating the Sr-90 body burden for any individual anywhere in the world:

1. The quantity of fission introduced into the atmosphere with specification of time, altitude, and location.
2. Rate and mechanism of transfer from stratosphere to troposphere.

3. Mean annual rainfall distribution.
4. The local diet, particularly with regard to the calcium-rich components.
5. Place of origin of foodstuffs in diet.
6. Discrimination factor between strontium and calcium from food to bone.
7. Age of the individual.
8. Distribution of common strontium and Sr-90 in present groups of the population.

"With these factors in hand," Kulp said, "it then becomes possible to predict for any given program of nuclear testing how much Sr-90, for example, a 12-year-old boy in Bangkok, New York, Cape Town, or the Amazon jungle would have on the average in 1978 and the distribution curve for 12-year-olds at each locality."[23] In other words, he asserted that the radiological future of the entire planet and its inhabitants was in safe hands.

The glowing promise of total knowledge by the Sunshine scientists, however, masked their keen awareness of large uncertainties in it. In a draft memo dated April 7, 1958, a month after Kulp's triumphant presentation in Switzerland, Hal Hollister, a USAEC official supervising Project Sunshine, argued that the worldwide fallout hazard was "one of many factors that can directly affect national policy in peacetime and wartime" so that its estimate did not need to be more accurate than "that of the other bases of policy."[24] This meant that the United States could tolerate uncertainty in the estimate of the hazard so long as it continued to give priority to "the other bases of policy," namely the compelling need of nuclear weapons testing. However, this pragmatic approach to the unknowns was not necessarily acceptable to all countries at the receiving end of radioactive fallout. As it turned out, Japan, a key U.S. ally in Asia supposedly sharing the same security interest with the Americans, posed one of the most formidable challenges to the U.S. claim of universal knowledge.

Japan and the Management of Nuclear Fear

Of all the non-nuclear UNSCEAR member states, Japan claimed a special status thanks to its possession of numerous data relating to radiation exposure. Each time the Japanese underwent a radiological disaster caused by nuclear weapons—Hiroshima, Nagasaki, and Bikini—not only did the number of victims increase, but also scientific research with them expanded. Historian M. Susan Lindee has traced this coproduction of human suffering and scientific data in Hiroshima and Nagasaki.[25] The Japanese physicians who treated the *Lucky Dragon* crew members added to this wealth of data on medical effects of radiation. Meteorologists and agricultural experts also made extensive measurements of fallout contents in the air, rainwater, and foodstuffs throughout Japan, whereas the oceanographic expedition to Bikini conducted in 1954 and again in 1956 collected numerous radiological samples of seawater and marine organisms in the Pacific. Not surprisingly, the Japanese government tapped these scientists for the UNSCEAR. Its chief delegate, Masao Tsuzuki, led the first Japanese medical survey in Hiroshima shortly after the atomic bombing and also supervised the medical care of the *Lucky Dragon* crew. Yoshio Hiyama, a fish scientist and adviser for the delegation, played a key role in the radiological inspection of tuna in the wake of the *Lucky Dragon* incident. The value of this collective expertise on radiation hazards went beyond the realm of science. As historian James Orr has pointed out, the Japanese forged their postwar nationalism out of pacifism rooted in atomic victimhood.[26] The Japanese scientists thus welcomed the nation's seating in the UNSCEAR as an opportunity not only to publicize their innovative studies internationally but also to represent their nation as the only country having suffered from nuclear weapons three times.

Japan's extensive experience with the radiological consequences of nuclear weapons, however, proved to be a double-edged sword for government officials in Tokyo. In January 1956, shortly before the UNSCEAR held its first meeting, the United States announced its plan to stage a new nuclear test series in the Pacific. Fresh from the consumer panic over the "atomic-bomb tuna," Japanese officials immediately contacted their U.S. counterparts to discuss how to prevent another fallout scare. Noting "the economic implications of establishing too low standards," both agreed that "it would be well to determine practical and economic standards as soon as possible."[27] For this purpose, Tokyo turned to its representatives and advisers for the UNSCEAR. The Ministry of Health and Welfare appointed Hiyama as chair of a task force to determine the "permissible dose" regarding radioactive fallout. In the meantime, by Tokyo's instructions, Tsuzuki privately consulted with U.S. scientists during their visit to New York.

This behind-the-scenes coordination across the Pacific proved highly successful. On May 5, the first day of the U.S. test series, the Japanese health ministry released the report of Hiyama's panel, which followed the 1955 recommendations of the International Commission on Radiological Protection (ICRP) and declared the contamination of drinking water (comparable to food) up to 10^{-7} microcurie (microCi) per milliliter as permissible.[28] Since the ICRP set this value with the assumption of continuous exposure for the span of a year, the Japanese authorities now argued that sporadic exposure to radiation at relatively high doses was "safe" as long as the total doses did not exceed the annual limit. In fact, as Tsuzuki testified before the Japanese Diet, there were no tuna rejected in 1954 that would violate the new standard.[29] In this way, Tokyo and Washington successfully unmade the "atomic-bomb tuna" that had once severely strained their Cold War alliance.

The Japanese scientists recruited for the UNSCEAR, however, by no means merely acted as their government wished. While willing to help contain political fallout from the U.S. nuclear tests then in progress, the delegates were equally adamant about their right to freely publicize scientific studies on atomic radiation. Tensions between Japanese officials and their scientific advisers came to the surface even before the creation of the UNSCEAR. When the Geneva Conference on Peaceful Uses of Atomic Energy was held in August 1955, the Japanese government, upon request from the United States, withdrew several papers on radiation hazards associated with nuclear weapons, including Hiyama's, on grounds that their topics were "unfit for the theme." While Hiyama reportedly told the Ministry of Foreign Affairs that he "found no objection" to this decision, he apparently thought otherwise. In a meeting with foreign officials held prior to the first UNSCEAR meeting, Hiyama loudly complained that the Ministry had not returned the withdrawn papers to their authors, adding a thinly veiled threat that these scholars might refuse to cooperate with the foreign office. He brought up the issue again in a debriefing session upon his return from New York, claiming that the withdrawal of his paper on fish cost Japan "priority to the U.S. and U.K."[30] While Hiyama helped Tokyo and Washington combat the tuna scare, he was determined not to let them stand in his way to beat his Western rivals in the scientific race.

His burning ambition, driven by nationalism, made Hiyama a powerful ally when Tokyo drastically changed its test-ban policy. In early 1957,

Nobusuke Kishi, a powerful politician and architect of the conservative merger that created the Liberal Democratic Party in 1955, became prime minister. Unlike his predecessors, Kishi launched a highly publicized campaign against nuclear weapons testing by any country, including the United States. Behind Tokyo's policy shift lay a hidden political motive. A Japanese official confided to his U.S. counterpart that, by making diplomatic gestures, Kishi was trying to limit the "excess" of antinuclear sentiments, prevent their "abuse" by the leftists, and guide them onto "the right path."[31] To depoliticize the issue of nuclear weapons testing—and to avoid the controversial question of nuclear deterrents on which Japan relied for its security—Japanese foreign officials decided to place a strong emphasis on "humanitarian concerns" about radioactive contamination.[32] Hiyama and his colleagues at the UNSCEAR thus emerged as a centerpiece of Tokyo's effort to redirect nuclear fear in Japan into a politically acceptable direction.

In New York, the Japanese scientists immediately noticed a cultural bias hidden in the scientific basis of the safety claim about radioactive fallout. During the first meeting, Western members made a joint proposal for the international standardization of fallout surveys with a focus on milk and dairy products. This motion was based on the assumption that milk was a principal source of calcium—and hence of Sr-90—around the world. The Japanese delegation, however, raised a strong objection. Stressing the importance of fish in the Japanese diet, it successfully pushed into a resolution a clause calling for the investigation of oceanic and fish contamination.[33] Soon promoted to a deputy delegate, Hiyama was put in charge of an ambitious radiostrontium sampling project in Japan covering various foodstuffs from land and the sea. He even personally appealed to government officials for additional funds, stressing the need to have detailed data for the UNSCEAR.[34]

In May 1957, the Japanese scientists completed a two-volume study of radiological data. Much to their surprise, the most contaminated food was not fish but brown rice. This staple food in Asia was found to contain four to five times as much Sr-90 per gram of calcium as milk. The discovery of "radioactive rice" put the researchers in a bind. The memory was still fresh of the panic over the "atomic-bomb tuna." Not to repeat the same mistake, Hiyama took much care in presenting his findings. The shifting timescale was key to the nuclear fear management. At a press conference attended by many journalists, Hiyama tried to reassure the public that the degree of contamination in rice remained below the maximum permissible levels recommended by the ICRP. While insisting that there would be no immediate risk to health, he suggested that the human body burden of Sr-90 might reach the acceptable limits by the 1970s if nuclear weapons testing continued unabated. "I worry about the future," he said. "There is no choice but to urge wise politicians to ban testing."[35] The concept of permissible dose thus allowed Hiyama to make the case for a nuclear-test ban without triggering another food scare.

The Japanese report to the UNSCEAR caused a sensation among the Asian members. As historian Matthew Jones has shown, many in Asia viewed the U.S. use of the atomic bombs against Japan as racially motivated, demanding the elimination of nuclear weapons as part of their struggle for decolonization.[36] "Radioactive rice," which put the Asians more at risk than the Westerners, appeared to give a scientific basis for this racial interpretation of the nuclear peril. During the fourth UNSCEAR meeting, held in early 1958, the Japanese scientists demanded that the committee should make separate estimates of the body burden of Sr-90 for milk and for rice. To this end, Tsuzuki and his colleagues worked closely with the Indians, also test-ban advocates, "as the same rice eaters" to persuade the "milk-drinking" delegations from the West.[37] To the dismay of U.S. officials, even the British accepted the Japanese request.[38] The UNSCEAR agreed to revise the draft of the final report, showing that Sr-90 in the bones of those living on a rice-based diet might be up to six times that in the bones of those on milk-type diets.[39] By illuminating the striking difference in fallout effects between Asia and the West, the Japanese scientists successfully dismantled the myth of the average risk behind the Anglo-American safety claim. British Prime Minister Harold Macmillan, in commenting on the UNSCEAR report, reportedly feared adverse reaction in Asia.[40] The question still remained, however, as to how to link the shifting understanding of fallout hazards to a breakthrough in the test-ban issue. This was a task that an emerging coalition of Soviet scientists against biologist Trofim Lysenko were willing to take up.

USSR and the Dual Struggle

Since the radiological disaster at Bikini in 1954, the Soviet government had consistently expressed its support for a nuclear-test ban as part of its post-Stalin peace offensive. Its comprehensive disarmament plan, unveiled in May 1955, had already included such an idea among the measures to be taken during the first stage. At the Geneva summit meeting held two months later, however, President Dwight D. Eisenhower countered the Soviet move by proposing a scheme of mutual aerial surveillance to verify compliance with any disarmament agreements.[41] Although the Soviets spurned this "Open Skies" plan as an espionage plot, the idea proved immensely popular worldwide. Suddenly finding itself on the defensive, the Kremlin stepped up its call for a test ban during the disarmament talks held in London from March to May 1956. The time seemed to be ripe for a fresh initiative in this issue. As the Soviets had successfully tested their first multistage thermonuclear weapon in November 1955, a prohibition on further tests would prevent the Americans from widening their nuclear technological lead once again. Shortly before the London talks were adjourned, the Soviet Foreign Ministry authorized its disarmament negotiator, Valerian A. Zorin, to declare his government's willingness to conclude a test-ban agreement as a separate measure apart from comprehensive disarmament.[42] With one of the two Cold War camps now in support, the idea of a test ban had emerged as a real possibility by the time the UNSCEAR began to assess the health risks from global fallout contamination.

The Soviet scientists in New York, however, were grossly unprepared to promote the Kremlin's new Cold War agenda. The U.S. and British scientists continuously outgunned the Soviets in the volume of technical reports submitted to the committee, and the direction of Soviet radiobiological research diverged widely from that in the West. Lysenko and his associates had politically suppressed radiation genetics, a key discipline that made it possible to conceive potential biological hazards from global fallout. Indeed, the Soviet scientists in New York found themselves under pressure to broadcast Lysenko's theories of inheritance. In preparation for the second UNSCEAR meeting, the Soviet government instructed its delegation to stress the importance of the effects of radiation on the whole cell as opposed to its direct damage to chromosomes.[43] The Western members, however, greeted such a contention with cold silence. For them, Lysenkoism represented a perversion of knowledge by totalitarianism, standing as a perfect antithesis to the liberal-democratic ideal of science in the West. Geneticist Hermann J. Muller, who advised the U.S. delegation to the UNSCEAR, once wrote that "some of the excesses of Stalin's dictatorship over science are already proving embarrassing."[44] This Cold War in science thus severely hampered the Soviet challenge to Anglo-American epistemic leadership in the UNSCEAR. In fact, when the committee took up the issue of genetic effects at its third meeting held in March 1957, the Soviets submitted not a single written report on the subject, a stark contrast to the Americans and British, who together contributed five out of the nine publications received by the committee.[45]

Lysenkoism, however, was not the only reason for the initial lack of a Soviet success in the UNSCEAR. The Soviet scientists faced another Cold War divide over the choice of a disease of interest for radiation protection. While the Western researchers tended to focus on less frequent but life-threatening risks of genetic damage and cancer, their Soviet counterparts studied other potential hazards, especially physiological disturbances by radiation. The groundbreaking research of Ivan P. Pavlov, who demonstrated the conditioned reflex, and subsequent government patronage brought Soviet physiology to international attention.[46] The Soviet physiologists were not completely immune to Lysenkoism. In 1950, rivals of Leon A. Orbeli, a leading Soviet physiologist and disciple of Pavlov, attacked him for deviating from the dogma of materialism.[47] Despite this, studies on the physiological effects of radiation continued to thrive. In fact, the dogmatic emphasis on "materialism" might have even facilitated this distinct line of research. As Orbeli himself noted, Soviet physiologists were quick to embrace ionizing radiation, a powerful physical stimulus to the biological systems, as "a new experimental tool for penetrating into the mysteries of physiological processes."[48] The contrast between radiation genetics and physiology never seemed starker when the Soviet Ministry of Health hosted the First All-Union Conference on Medical Radiology in Moscow from January 30 to February 4, 1956. While there was no paper presented regarding genetic mutations by radiation, many participants reported various reactions of the central nervous system and other physiological mechanisms to irradiation.[49]

Armed with a wealth of experimental and clinical data showing a wide range of functional disturbances by radiation, the Soviet delegation to the UNSCEAR, led by Andrei V. Lebedinskii, himself a noted medical physiologist, tried to convince their Western colleagues to include these nongenetic, noncancer effects in the global-fallout risk assessment. Conceivably, this attempt would help Moscow press its Western rivals on a test ban. The research conducted by Mikhail N. Livanov, a leading Soviet electro-physiologist, demonstrated notable changes in higher nervous activity even at exposures as small as the order of 1 R.[50] Western scientists, however, dismissed such findings as insignificant. At that time, the ICRP and the NCRP used the concept of the critical organ, namely bone marrow and the gonads, to assess the effects of whole-body irradiation at low doses.[51] The organ-specific approach tended to set aside potential systemwide effects that were of concern in physiology. This choice of a disease of interest was also related to a more fundamental question: did neurological changes observed at low doses represent a minor and transient "response" or serious and permanent "damage"? Evidence showed that the central nervous system was highly sensitive to radiation even at low doses, and yet it also suggested that visible and enduring neurological damage appeared only at a very high dose.[52] While acknowledging certain scientific value in the Soviet reports, other UNSCEAR members nevertheless refused to put neurological effects at the center of discussion, insisting that their character, duration, and magnitude were too uncertain to alter the existing definition of radiation damage centering on cancer and other irreversible and lethal diseases.[53]

The substance of Soviet contributions to the UNSCEAR, however, began to change when a group of Soviet nuclear physicists and chemists forcefully stepped in to assist the geneticists' struggle against the Lysenkoites. There was a good scientific reason for this emerging interdisciplinary coalition. The discovery of the molecular structure, or double-helix model, of deoxyribonucleic acid (DNA) in 1953 by James Watson and Francis Crick further cracked the disciplinary wall between physics and genetics, opening a new possibility of studying genetic events on the molecular level ruled by the laws of physics and chemistry.[54] Excited by this prospect, nuclear physicist Pyotr L. Kapitsa and his colleagues brought in geneticist Nikolai V. Timofeev-Ressovsky, who had just emerged from his secret *sharashka* work, for a lecture on the genetic code to their physics seminar held in February 1956.[55]

The scientific revolution in genetics was not the only reason for the formation of the anti-Lysenko coalition. It was also part of the resurgence of a cultural phenomenon called the intelligentsia. With roots dating back to the mid-nineteenth century, the intelligentsia was an imagined community of educated elites bound by a common sense of mission to emancipate society from authoritarianism and backwardness.[56] After a period of harsh oppression and partial compromise following the 1917 Bolshevik Revolution, Stalin selectively appropriated the culture of enlightenment, mobilizing the growing rank of educated cadres through cultural and scientific institutions as the "Soviet intelligentsia" in service of the socialist state.[57] This Stalinist project, however, had an unintended consequence, sowing the seed for the revival of the social milieu and ethos of the intelligentsia after Stalin's death.[58] A group of nuclear physicists who contributed to the Soviet atomic project played a key role in this development. These scientists stood to benefit from state patronage, enjoying special privileges, a comfortable life, and an exceptional degree of intellectual autonomy. As the "Thaw" in the Cold War and Nikita Khrushchev's de-Stalinization policy led to the gradual lessening of restrictions on research activities and international contacts, the nuclear weapons scientists strove for the expansion of intellectual freedom beyond the confines of their profession.[59] Patriotic, cosmopolitan, and intolerant to hypocrisy, these socialist heroes stepped in for geneticists in the struggle against Lysenkoism.

In their joint struggle against Lysenkoism, Soviet nuclear physicists and geneticists seized on the issue of fallout hazards to underline the political utility of genetic research in the Cold War. Geneticist Nikolai P. Dubinin had already made a similar argument shortly after the *Lucky Dragon* incident in an attempt to persuade the Academy's leadership to create a new research center for experimental genetics. Although the Lysenkoites tried to sidestep the issue of genetic damage by stressing large uncertainties in it, nuclear physicists and their allies, speaking with authority as experts in nuclear affairs, gave a critical boost to Dubinin's effort. In March 1957, when the Presidium of the USSR Academy of Sciences discussed a plan to establish the Institute of Cytology and Genetics with Dubinin as director, Aleksandr P. Vinogradov, a notable geochemist and member of the Soviet nuclear weapons program, strongly supported the motion. "No ordinary slogan or threat will decide this struggle for peace, the ban of tests, and the halt of the military use of atomic energy," he said. "The only solution—on this front it is genetics, which will put an end to all the discussions and provide an unambiguous solution."[60] Backed by Vinogradov and other nuclear scientists, the Presidium passed the resolution, making a breakthrough in the reconstruction of Soviet biological sciences.

As Lysenko's enemies had predicted, the genetic argument soon reemerged in the test-ban debate after the BEAR and MRC reports had temporarily assuaged public concerns by stressing the relatively small probability of genetic damage. A turning point came in April 1957, when Albert Schweitzer, an Alsatian missionary doctor and world-renowned humanist, delivered a passionate plea for a test ban through a radio address broadcast worldwide. Although he abhorred nuclear weapons, he had until then avoided speaking on what he considered a political issue. A meeting in fall of 1956 with Norman Cousins, a U.S. political journalist and supporter of Adlai Stevenson's presidential campaign, changed his mind. Realizing that global fallout posed a danger to a countless number of innocent bystanders in the Cold War, including those yet to be born, he concluded that it constituted a serious moral crisis.[61] In his public address, titled "A Declaration of Conscience," Schweitzer reflected on the suffering of individuals in the distant future from genetic damage caused by radioactive fallout, declaring that "every increase in the existing danger through further creation of radioactive elements by atom bomb explosions" was "a catastrophe for the human race."[62]

Schweitzer's powerful message triggered a chain reaction of test-ban activism among scientists around the world. One of its most prominent leaders was U.S. chemist and Nobel laureate Linus Pauling. Influenced by his pacifist wife, Ava Helen, and convinced that scientists had a special responsibility for using their knowledge to advance peace and justice, Pauling forcefully spoke out against nuclear weapons. As historian Paul Rubinson has shown, Pauling's brand of antinuclear activism made him not only a target of constant harassment by the U.S. national security state but also a pariah in the scientific community.[63] Inspired by Schweitzer, however, Pauling reasserted the moral obligation of scientists to oppose the testing of nuclear weapons. In a speech delivered at Washington University in St. Louis on May 15, 1957, Pauling cited Schweitzer's philosophy of life, declaring that "no human being should be sacrificed to the project of perfecting nuclear weapons that could kill hundreds of millions of human beings."[64] Heartened by enthusiastic reactions from the audience, Pauling and a handful of hosting faculty members launched a petition campaign, urging scientists to support a test ban as those who possessed "knowledge of the dangers involved."[65] Their effort to link science to ethics proved highly successful. By the time the petition was submitted to the United Nations on January 14, 1958, it showed the names of 9,235 scientists from forty-four countries, including thirty-six Nobel laureates and other prominent researchers.[66]

Another impetus for transnational activism among scientists for a test ban came from an international conference in July 1957 held in Pugwash, a small village in Nova Scotia. Organized as a follow-up to the famed Russell-Einstein Manifesto issued in 1955, this private meeting of leading scientists from both Cold War blocs was convened to discuss the dangers of nuclear weapons, including global fallout from nuclear tests.[67] The Soviet government initially approved the participation of two biochemists known for their support of Lysenko, Aleksandr I. Oparin and Norair M. Sisakian. Shortly before the conference was held, however, the USSR Academy of Sciences replaced them with Aleksandr M. Kuzin, director of the Academy's Institute of Biophysics, which then temporarily hosted Dubinin.[68] Although he was not a geneticist, Kuzin was intellectually prepared for a constructive dialogue with his Western colleagues. After extensive discussion, the conference unanimously agreed that the linear non-threshold hypothesis was a prudent basis to assess the genetic and cancer effects of low-dose radiation.[69] As was the case for Schweitzer and Pauling, however, Pugwash's most significant contribution to the politics of risk was an ethical one. The final statement noted that the effects of fallout were "global, and exerted upon citizens of all countries, regardless of whether they or their governments have approved the holding of tests." In such a situation, it declared, "the usual criteria as to whether a given hazard is justifiable cannot be applied."[70]

Watching the rise of transnational campaigns for a test ban with keen interest, Soviet officials moved to seize this opportunity to step up their pressure on the Western rivals. In a letter to the Central Committee dated July 20, Foreign Minister Andrei Gromyko recounted Schweitzer's radio address, Pauling's petition, and the Pugwash conference, among others. Then he proposed to "use a campaign among foreign scientists," pointing out that "an authoritative opinion of scientists has a large effect upon public opinion."[71] On August 13, with the nod from the Kremlin, a statement signed by 196 Soviet scientists was released, which declared their solidarity with scientists all over the world in the face of the common threat of radioactive fallout from nuclear tests. The signers called for an international conference of all kinds of experts—not only scientists but physicians, philosophers, economists, historians, sociologists, and teachers—to "add their voices to the demands for the prohibition of nuclear weapon tests."[72] In the meantime, 216 members of the USSR Academy of

Sciences signed Pauling's petition, joining 9,000 other scientists in their demand for the immediate suspension of nuclear testing.[73]

With the Kremlin's interests firmly aligned with theirs, a group of anti-Lysenko scientists made their first major intervention in the controversy over the "clean bomb." In summer of 1957, the U.S. delegation to the UN Disarmament Subcommittee repeatedly denied the need for a test ban on health grounds by citing Libby's assertion that the United States was developing a thermonuclear weapon designed to minimize Sr-90 and other harmful fission products. Baffled by this technical claim, Zorin sent a telegram to Moscow with an urgent request for a counterargument. In response, Igor V. Kurchatov and other nuclear scientists at the Scientific-Technical Council of the Ministry of Medium-Machine Building carefully scrutinized Washington's claim.[74] While the "clean bomb" reduced the amount of radioactive elements produced by nuclear fission, its fusion reaction emitted neutrons that turned the nitrogen atoms in the atmosphere into carbon-14 (C-14). As the world's foremost authority on radiocarbon research, Libby knew about such an outcome, but he dismissed its effects on humans by claiming that the estimated annual dose from bomb-produced C-14 would be extremely small. The Soviet scientists, however, decided to calculate the magnitude of genetic damage to the world population from the total doses from both fission and fusion products over their mean lifetimes—for example, approximately eight thousand years for C-14. This exercise, made possible by the linear non-threshold hypothesis derived from radiation genetics that the Lysenkoites had once suppressed, demonstrated that the "clean bomb" was hardly clean in terms of its genetic consequence. In a paper published in December 1957, Soviet physicist Ovsei I. Leipunskii cited the data from Dubinin and others, demonstrating that a 10-Mt fusion explosion would be ultimately responsible for forty-nine thousand cases of genetic damage while a fission version would cause forty-one thousand.[75]

The reconstructed biological knowledge, however, was not the only reason for the growing Soviet epistemic competence in the global-fallout debate. Soviet physicist Andrei D. Sakharov recalled that, in the course of preparing propaganda material for a test ban, he learned about Schweitzer's argument against harming innocent bystanders for the purpose of nuclear weapons.[76] This prompted Sakharov to ask penetrating questions about ethical issues behind the dry figures of probability in risk analysis. In his draft paper, Sakharov estimated that the nuclear tests conducted through 1958 would ultimately lead to a minimum of five hundred thousand excess deaths over a period of the order of the next eight thousand years. With humanism as a moral compass, he refuted each of the relative measures of risk cited to dismiss the effects of global fallout. He objected to a reference to natural radiation, arguing that "added to the existing distress and death of human beings are the death and distress of hundreds of thousands more, and these include people in neutral countries and in future generations." He also took exception to an analogy to automobile accidents and other hazards in modern life, pointing out that the problem with global fallout was "the total impunity of the crime . . . and also the total defenselessness of future generations with respect to our actions." Most important, he rejected a choice between the risk from nuclear warfare and that from weapons testing as false, presenting a test ban as the best step to reduce both dangers at once.[77] All these arguments made Sakharov's paper unusually strong propaganda material. Personally edited by Khrushchev and translated into five languages, the publication was distributed worldwide.[78] Although Sakharov and other Soviet nuclear scientists did not join the UNSCEAR, their propaganda activities provided the Soviet scientists in New York with a trove of materials to be used against the Americans and British.

With robust support from Moscow, the Soviet delegation to the UNSCEAR successfully revamped itself shortly before the committee held a meeting in January-February 1958 to discuss a draft report. Kuzin joined the delegation as a genetic adviser, with relevant data supplied by Dubinin and two geneticists working under Kurchatov. Lebedinskii and Nikolai A. Kraevskii, a pathologist at Lebedinskii's USSR Ministry of Health Institute of Biophysics, supervised the work on cancer and other nongenetic effects. The number of Soviet publications submitted to the UNSCEAR jumped from a few in 1956 to over thirty by 1958.[79] Armed with the enhanced scientific firepower, the Soviet delegation took on the Anglo-American knowledge claims. As Lebedinskii told members of the USSR Academy of Sciences, the UNSCEAR faced "an interesting situation in the history of science," where there was large uncertainty in knowledge about the environmental distribution and biological effects of global fallout.[80] The Soviet strategy was thus to exploit the scientific unknowns to the maximum political effect. When the UNSCEAR reviewed the risk of leukemia from Sr-90, Austin Brues, a U.S. delegate who acted as rapporteur, summed up the discussion as suggesting that the existence of threshold remained an open question. Lebedinskii immediately lodged a protest, arguing that theories suggesting non-threshold effects "must be given high value in this question since medical statistics did not present enough evidence."[81] Both sides eventually reached a compromise, agreeing to use both the threshold and non-threshold hypotheses to make estimates of excess leukemia incidence.

The most successful intervention made by the Soviet delegation, however, related to ethical aspects. When the UNSCEAR discussed the conclusion of the draft report, Lebedinskii introduced an amendment, proposing a sentence recommending a nuclear-test ban. The Indians followed suit, presenting a draft text that included an explicit call for the suspension of all tests. In explaining his motion, Lebedinskii claimed that even the slightest risk of harm from global fallout was ethically indefensible because it would affect all people in the world with no compensating benefit in return. This argument, it turned out, won strong support among many members, even the Swedes and Belgians, who usually sided with the Americans and British.[82] To prevent others from joining the Soviets and Indians, the U.S. delegation endorsed a Belgian amendment, which adopted most of the Soviet text except for a demand for a test ban. After tense exchanges, the Soviets and Indians agreed to have their amendments added to the Belgian text in footnotes. With this compromise reached, the UNSCEAR unanimously adopted the draft report in June 1958, which conclusively refuted the relative measures of risk that underpinned the U.S. and British position. Nonetheless, it stopped short of recommending the cessation of nuclear tests, declaring that the questions of how to control radiation exposures "involve national and international decisions which lie outside the scope of its work."[83] With this statement, the UNSCEAR left the task of developing specific policies to combat radioactive pollution to the policymakers in Washington, London, and Moscow.

Toward the Test Moratorium

On March 31, 1958, with the UNSCEAR report completed except for the conclusion, Gromyko made a surprise announcement that the Soviet government had decided to suspend nuclear testing unilaterally. This bold move reflected the Kremlin's careful political calculations. In October 1957, the Soviet Union had shocked the world by successfully launching Sputnik, the first man-made satellite, into orbit. During the same month, Georgy Zhukov, a Soviet marshal and Khrushchev's last-standing political rival in the Presidium, was dismissed from his post. Khrushchev, a winner in the post-Stalin power struggle, seized on this political momentum to take a new disarmament initiative for peaceful coexistence.[84] The Soviets were thus eager to use the UNSCEAR report as a universal validation for its test-ban proposal. At a press conference held on June 17 following the formal voting on the report, Lebedinskii described the consensus as underlining the need to stop the tests immediately. "That is why the peaceloving Soviet people give their wholehearted support to the highly humane decision by the Soviet Government to stop unilaterally the nuclear weapons tests," he said. "We do still hope that the Governments of the United States and of the United Kingdom will decide likewise."[85]

The Soviet announcement of the moratorium put both the Americans and British on the defensive, but it was particularly troubling for the latter. As a lesser nuclear power, Britain refused to suspend its testing unless the United States agreed to share its nuclear weapon technologies. Moreover, it tried not to alienate France, as this key European ally and economic partner planned to conduct its first atomic bomb test shortly.[86] To escape this dilemma, the British resumed their pursuit of a test limitation. During a meeting with U.S. Secretary of State John Foster Dulles in early May, British Foreign Minister Selwyn Lloyd discussed the idea of a moratorium on high-yield nuclear tests as an interim measure until an international verification system was installed to enforce a total ban. What informed this proposal was the growing sense in Whitehall that the damning conclusion of the UNSCEAR report made it no longer possible to simply ignore the dangers of global fallout. In a personal memo to Lloyd, Macmillan wrote, "It seems to me that we will not be able to stand up against the pressure in the [General] Assembly and public opinion here after the medical report of the United Nations."[87] During a summit meeting in Washington held a month later, the British prime minister mentioned the UNSCEAR report, pointing out that it "contains inexact estimates of the consequences of the tests and some speculation which could be used to our disadvantage in propaganda." Eisenhower agreed, saying that, despite the contradicting opinions of Edward Teller and other scientists, "public opinion is a most important element."[88] Although still skeptical about the dangers of global fallout, both Eisenhower and Macmillan concluded that the UNSCEAR report turned them into a political fact to be reckoned with.

Despite such acknowledgment, the U.S. government initially refused to limit or suspend nuclear testing on health grounds for three reasons. First, it was on the verge of conducting a new test series in the Pacific, and feared that its cancellation might give the Soviets a further technological lead following the Sputnik launch. Second, the U.S. Congress was slow to revise the Atomic Energy Act to permit full-scale Anglo-American nuclear partnership, a key condition for the British consent to a test ban.[89] Finally, a unilateral halt to tests would run counter to the long-standing U.S. policy that any arms control agreement must be rigorously verified and enforced through international inspection. In fact, Eisenhower believed that a test ban could be used to force the Soviets to open up their country to the West. "We want something in return for agreeing to suspend tests and the great price will be to obtain Russian cooperation in an inspection system," the president told Macmillan during the meeting.[90] A comprehensive test ban coupled with an international verification system was thus a more desirable option than a mere limitation designed to reduce fallout emissions. To push the focus of the test-ban debate back to disarmament aspects, Eisenhower sent a letter to Khrushchev dated April 28 with the proposal that experts from the Western and Eastern Blocs should meet to discuss the feasibility of a detection system to enforce a comprehensive ban.[91] The Soviet premier accepted the invitation, and the technical conference was held in Geneva from July to August. Only when the delegates successfully worked out a blueprint for the control system did the U.S. and British governments finally announce their willingness to suspend all tests for one year beginning in November, on condition that the agreed system should be installed by the end of the period.[92]

The apparent U.S. indifference to global fallout, however, belied three major ways in which the planetary radioactive contamination drove Washington and London toward the moratorium. First, the public release of the UNSCEAR report set a time limit for the United States and Britain to react to the Soviet initiative for a test ban. The publication was scheduled to be discussed at the UN General Assembly in fall, where many countries, including India and Japan, were expected to use the occasion to push for the suspension of tests. When Dulles met John A. McCone, who had just succeeded Lewis L. Strauss as USAEC chairman, to discuss the proposal of announcing a moratorium following the conclusion of the Geneva conference, the secretary pointed out that such an announcement "would in any event be imperative for the next General Assembly, particularly having regard to certain aspects of the prospective UN report of experts."[93]

Second, test-ban opponents in the United States revived the idea of limits on nuclear testing as a politically acceptable alternative to a comprehensive ban. In May 1958, two months after the Soviets announced a test moratorium, the USAEC's General Advisory Committee (GAC) recommended that the United States should transfer most tests to underground while seeking an international agreement to limit annual fallout emissions to one megaton fission equivalent.[94] Two months later, as the White House leaned toward a moratorium, Libby and Teller urged McCone to support the GAC proposal. "A most powerful objection to continued testing would be removed if radioactive fallout from future tests could be limited or eliminated," the two scientists stated in an urgent teletype to the chairman.[95] Although Eisenhower and Dulles ignored their last-minute plea, this once again showed the surprising flexibility of fallout hazard deniers, who strategically appropriated the changing norms against fallout to defeat the advocates for a halt to all nuclear tests.

Finally, the U.S. government's public silence on the problem of global fallout actually spoke of its substantial impact on U.S. test-ban policy. This became clear when Macmillan sent a telegram to Washington in his final attempt to dissuade the Americans from announcing a moratorium on all tests. In it, he revived the idea of a threshold ban enforceable by national means of detection, asserting that it would go a long way toward addressing public concerns about fallout without compromising the principle of verifiability. Eisenhower and Dulles, however, rejected this alternative. In a reply to Macmillan, Dulles explained that his government had decided against explaining the planned moratorium in such a way because it might imply that fallout was indeed harmful. He pointed out that if the United States and Britain suspended tests temporarily as a step toward disarmament, they could keep pressure on the Soviets to work toward a safeguarded test-ban agreement. If they stopped tests on health grounds, Dulles stated, "then indeed we might have burned our bridges behind us," which would make it impossible for them to resume testing even if the Soviets refused to implement the agreed verification system.[96] The conspicuous lack of reference to radioactive fallout in the U.S.-British joint announcement released on August 22 was thus tantamount to their tacit acknowledgment that the dangers of global fallout contamination had become a political fact, if not a scientific truth, to be taken into account in the test-ban debate.

Conclusion

Once the locus of risk politics in the nuclear Anthropocene shifted from the United States and Britain to the international level, a series of assumptions made by the U.S. and British nuclear authorities about global radioactive contamination collapsed. Japanese scientists dismantled the myth of the average risk by discovering heavily contaminated rice, which underscored the highly uneven distribution of fallout among people and across regions. This inequality of global-fallout hazards emerged through the decolonizing, Cold War world of the 1950s. Driven by their scientific ambition and pan-Asian nationalism, and yet constrained by their government's security need for the U.S. nuclear umbrella, Japanese scientists effectively leveraged their scientific findings to manage nuclear fear, rallying the "rice-eating" East against the "milk-drinking" West while cutting off the issue of a test ban from the larger problem of nuclear weapons.

Meanwhile, the Soviet scientists who opposed Lysenko debunked the "clean bomb" argument and also questioned the ethics of harming innocent bystanders in the Cold War. Much of the Soviet success in redefining the risks of fallout was due to the transnational movement of intellectuals, which seamlessly linked the scientific hypothesis of no safe threshold in radiation exposure with the ethical principle of humanism. It was this interplay between the reconstruction of biological knowledge in the Soviet Union and the transnational activism that guided the UNSCEAR toward a consensus that deemed global fallout contamination as unacceptable. Although policymakers in the three nuclear powers still considered a test ban primarily as an arms control initiative, the UNSCEAR report changed their political calculations in a number of subtle yet important ways that facilitated the temporary suspension of all nuclear tests.

Despite the decisive shift in the definitions of risk on the international level, many government officials and scientific advisers in the United States and Britain did not consider the question as settled. Although even the staunchest opponents to a test ban now found it politically expedient to pay lip service to the growing concerns about fallout, the U.S. and British governments still maintained that the testing of nuclear weapons posed no undue danger to any individual. Further change had to come from within and below the two Western countries. Deeply frustrated by the blanket reassurances from government officials, activist scientists and concerned citizens in the United States and Britain began to check radioactivity in their own communities. This local turn in the politics of risk drastically changed how U.S. and British political leaders understood the policy implications of global fallout contamination.

Chapter 6

THE LOCAL TURN

Community-Based Fallout Surveys in the
United States and Britain, 1958–1960

O N MARCH 20, 1958, heavy rains started pounding northern California and continued for a few days. The storm came amid the news of multiple Soviet nuclear tests conducted thousands of miles away. This coincidence worried Karl Lindbergh-Holm, a University of California biochemist in Berkeley. He scanned collected rainwater with his Geiger counter, which detected gross radioactivity at unusually high levels, 208 times as much as the maximum permissible concentration recommended by the National Committee on Radiation Protection (NCRP) for radioactive materials of an unknown source in drinking water. He immediately contacted state and local authorities, who rushed to check surface water and vegetables. The results confirmed his findings: some of the samples showed gross beta radioactivity ranging from 10 to 100 picocuries (picoCi) per gram. The federal government, however, refused to act. It issued a brief statement reassuring anxious Californians that there would be no harm even if radioactivity in some foodstuffs temporarily exceeded the NCRP guidelines. Noting that the "permissible dose" assumed a lifetime consumption, government officials argued that no individual would consume a diet consisting of isolated items of significant contamination for a long period of time.[1]

Although Lindbergh-Holm's lone effort to alert citizens to widespread fallout contamination in their surroundings was unsuccessful, it was but one of many similar initiatives in the United States and Britain to conduct radiological surveys in communities. What drove this local turn in the politics of risk was the growing awareness that the impacts of global fallout were not distributed equally among people and across regions. The complex interplay among various natural forces and human activities—wind patterns, rainfall, soil characteristics, diet, and metabolism, to name but a few—rendered some areas, foodstuffs, and individuals on Earth more contaminated than others. Beginning in 1957–58, concerned scientists, activists, and citizens living in some of the "hot spots" in the United States and Britain, in particular Minnesota and Wales, systematically checked radioactivity in their food and drink. At the same time, Consumers Union, the largest consumer advocacy group in the United States, conducted a milk survey in fifty North American cities. The results of these subnational surveys revealed the strikingly uneven radiological burden of national security within the country.

Why did local communities and consumer groups in the United States and Britain decide to monitor fallout on their own? And how did their community-focused actions affect test-ban negotiations during the moratorium period? The answers lay in the monopoly of fallout information by the national security states. The U.S., British, and Soviet nuclear authorities created similar regimes of self-regulation for weapons testing, monitoring radioactive debris, and judging its potential hazard by themselves. Even after the United Nations Scientific Committee highlighted the unequal distribution of global fallout and its potential health effects across space and time and rejected the notion of acceptable risk, U.S. and British officials continued to view their countries as homogeneous, asserting that all citizens would fall within a small variance from the estimated mean radiation dose while receiving the same benefits of nuclear testing to national security. As the nuclear arms race intensified international tensions and radioactive contamination, however, an increasing number of scientists and activists in the United States and Britain called into question both the Cold War consensus and the concept of the average risk. Instead of expressing their dissent in political and moral terms, these concerned citizens chose to take Geiger counters to their communities, measure exposure, and disseminate the collected data as an act of self-defense and also of protest against the government's claim to the exclusive right to determine what was an acceptable risk for the entire nation.

The most notable example of what historical sociologist Kelly Moore has called the "information-provision" model of scientific activism was the Greater St. Louis Citizens' Committee for Nuclear Information. Led by U.S. biologist Barry Commoner, this civic group launched the Baby Tooth Survey in December 1958 that collected milk teeth for the analysis of radiostrontium content.[2] As historian Michael Egan has demonstrated, this grassroots fallout survey gave birth to a new form of environmental activism with an emphasis on technical information and public discussion.[3] The scientists and activists discussed in this chapter shared the same agenda and method with Commoner, but their focus on food and drink had a much more significant impact on the politics of risk because the revealed trends and variations were drastic and directly contradicted the claims made by government officials. Moreover, the outcomes of these diet-focused surveys came out at the critical moment in the test-ban negotiations when opponents in the United States blocked progress by developing the technique of underground testing, the latest "technological fix" to radioactive fallout, which greatly complicated treaty verification. Faced with the diminishing prospect of a comprehensive ban and alarmed by the mounting reports of unusually contaminated food and drink in their countries, political leaders in the United States and Britain came to view an atmospheric ban as a solution to both the inspection and fallout problems.

United States

The hot spots in the United States first came to the public's attention through the fallout studies conducted by the U.S. Atomic Energy Commission (USAEC). Since 1952, the Health and Safety Laboratory (HASL) had been running the National Radiation Surveillance Network across the contiguous United States. In a speech delivered in October 1956, USAEC commissioner Willard F. Libby discussed the latest HASL data, noting that some communities in Illinois and Wisconsin had 50 percent more Sr-90 deposits in soil than those observed elsewhere in the world.[4] U.S. nuclear officials, however, refused to take a more detailed look at the Midwest. In a paper submitted to the U.S. congressional hearing held in May and June 1957, HASL director Merril Eisenbud calculated the future levels of Sr-90 in humans on the basis of data from the Northeast, showing that the maximum concentration from the past tests, 4 to 6 picoCi of Sr-90 per gram of calcium (called Strontium Unit:

SU), would be well below the population maximum permissible concentration of 100 SU recommended by the Committees on the Biological Effects of Atomic Radiation of the National Academy of Sciences in its 1956 report. Although he admitted that the actual values might be higher in other parts of the country, he maintained that, on the basis of a spot check of milk from six domestic and two foreign sources, "a factor of three applied to the estimates for northeastern United States should be adequate . . . to bracket the highest foreseeable value, from tests to date, in any region of the United States."[5]

The radioactive hot spots in the Midwest, however, unexpectedly interacted with the region's rich tradition of political dissent. As the Midwest became integrated into the national market and exposed to its exploitative influence by the end of the nineteenth century, the Populist movement sprang across the region, calling upon the government to curb the power of business and promote the welfare of farmers. Although the Populist movement as a nationwide political institution soon collapsed, the Midwesterners continued to elect many Progressives to political offices on both the local and national levels.[6] Agrarian radicalism also shaped their distinct view of international affairs, turning the Midwest into a bastion of noninterventionism. Although World War II and the beginning of the Cold War diminished the isolationist impulse, a small but vocal group of pacifists in the region continued to support disarmament and other initiatives intended to reduce international tensions.[7] All these regional political traditions were bound to amplify the political impact of global fallout contamination in the Midwest far beyond its physical quantity.

Of all Midwestern states, Minnesota was among the first to create a scientific advisory committee for radiation protection. In 1957, Democratic Governor Orville L. Freeman established the Atomic Development Problems Committee (ADPC) as part of his plans to introduce nuclear power plants in the state. Minnesota's keen interest in atomic energy spoke to a close association between rural electrification and agrarian populism. Before World War II, only a small proportion of U.S. farmers enjoyed affordable access to electricity.[8] Rural electrification quickly became a rallying cry for agrarian populists, and Franklin D. Roosevelt elevated it to a key element of his New Deal policy. Two of his signature agencies, the Tennessee Valley Authority and the Rural Electrification Administration, aimed not only to revitalize rural economy after the Great Depression, but also to broaden public participation in resource uses.⁹ Minnesota looked to nuclear power for what hydroelectric dams had done during the New Deal era. Following the passage of the 1954 Atomic Energy Act that began the privatization of nuclear industries, the Rural Cooperative Power Association at Elk River, located approximately forty miles northwest of Minneapolis, announced its plan to install a small nuclear power reactor. Completed in 1962, the Elk River Station came to be known as "Rural America's first atomic power plant."[10] To put the atom to work for farmers, the state government sought the counsel of experts to study a wide range of problems arising from nuclear-powered rural development.

Initially, radioactive fallout was only one of many sources of radiation that the ADPC studied as a baseline for radiological health in the nuclear age. The results of a preliminary survey, however, caused a great alarm among the investigators: levels of gross radioactivity in rainwater and surface waters exceeded, in a considerable number of instances, a provisional limit recommended by the NCRP for unknown sources of radiation in drinking water.[11] Federal officials in Washington, however, remained unresponsive to their concerns. During a meeting with members of the ADPC in March 1958, Gordon M. Dunning and John H. Harley of the USAEC Division of Biology and Medicine tried to reassure the Minnesotans by discussing fallout data obtained from other Midwestern states. One such piece of information showed that Sr-90 contents in milk from North Dakota and Wisconsin ranged between 5 and 20 SU. Although these values were below those high enough to raise radiostrontium concentrations in humans to the 100 SU limit, Maurice B. Visscher, a University of Minnesota physiologist and head of the ADPC's subcommittee on environmental contamination, found them far from comforting. In a letter to the committee chairman, Visscher quoted his colleague biochemist William O. Caster as saying that milk was by no means the only dietary source of calcium and strontium, and that the total Sr-90 intake might be closer to the limit than might appear from the milk values alone. The USAEC, however, had no detailed radiostrontium data on foodstuffs other than milk. Dunning and Harley responded to the concerns among the Minnesotans by offering an analysis of materials for Sr-90. At Visscher's suggestion, the ADPC accepted this proposal and organized a comprehensive survey of milk and edible plants across the state.[12]

Why did Visscher decide to conduct a fallout survey in Minnesota independently from the USAEC? Part of the reason was ethical. A devoted humanist, Visscher had long been worried about any increase in human suffering as a result of nuclear tests. During the 1956 presidential campaign, he joined a group of scientists who signed an open letter endorsing Adlai Stevenson's pledge to suspend thermonuclear tests, asserting that "[a]ny appreciable increase in the number of [genetic] defectives in the world at large should not be ignored."[13] It was the USAEC's restrictive public information policy, however, that drove Visscher to develop his own fallout data. Although the HASL program was unclassified, the USAEC had a long-standing practice of releasing its information to the public only through preapproved speeches and publications by its officials. Likewise, during the March 1958 meeting, Dunning and Harley asked the Minnesotans not to disclose the data without permission. This prompted Visscher to send a letter to the editor of the *New York Times*. "To withhold [data] from the public means one of two things," he wrote, "either that our Washington administrators do not trust our intelligence, or that they hope to control opinion by monopolizing information."[14]

Shocked by this accusation, Charles L. Dunham, director of the USAEC Division of Biology and Medicine, immediately wrote to Visscher, claiming that the nondisclosure request was made not to suppress the information but rather to ensure its accuracy before publication.[15] To Visscher, however, the exclusive right to decide on when and how to release data opened doors to abuse, allowing the USAEC, with its vested interest in the continuation of nuclear testing, to slip its political opinions into the scientific information. Visscher contended in his editorial letter that "scientists who talk and write about this problem mix up science and public policy." Declaring that scientists had "no special competence in arriving at policy judgment," he argued that "[p]olicy in a democracy must be based on informed public judgment and therefore clear distinctions between fact, guess and policy judgment must be made."[16]

Visscher's call for a separation between science and politics spoke to an emerging "information" model of scientist activism in Cold War America. Symbolized by the infamous security clearance hearing of J. Robert Oppenheimer in 1954, influential scientists who openly opposed U.S. nuclear weapons policy on political and moral grounds found themselves not only harassed by the national security state but also marginalized in the scientific community. As historian Paul Rubinson has noted, Linus Pauling and his supporters refused to separate science from morality in their test-ban petition campaign. Many mainstream scientists, however, accepted the boundary of "objective" science and refrained from expressing political and ethical opinions about nuclear weapons.[17] Visscher tried to turn this professional norm of scientists against the national security state, demanding that it should release all scientific information about fallout and leave value judgments to citizens. The radiological survey in Minnesota thus served as a perfect vehicle to challenge the USAEC's monopoly of fallout data. The stark contrast between Pauling's model of scientist activism and Visscher's became clear when Pauling asked Visscher in early 1958 to join a planned lawsuit seeking judicial restraint of U.S. nuclear tests. Visscher politely declined the request, explaining that his quiet work on fallout through the state government offered the best hope for obtaining "independent evidence" to demonstrate that "pooh-poohing the consequences of bomb testing approaches the quality of dishonesty."[18]

Visscher and his colleagues in Minnesota were not the only Americans seeking to gather fallout information in their communities. In summer of 1958, Consumers Union (CU), the nation's largest consumer advocacy group, decided to conduct an "independent, non-government sponsored check on the validity of government-published data" on Sr-90 contents in fresh milk.[19] Like the Minnesotans, consumer activists had grown frustrated by Washington's slow response to the radioactive contamination of food and drink. The USAEC had been measuring radiostrontium contents in milk obtained from many parts of the world under Project Sunshine, but its domestic investigative arm, the HASL, took samples from no more than several U.S. cities at one time, because its chief objective was not to provide comprehensive geographical coverage, but rather to obtain just enough information to build a reasonable model of Sr-90 transfer in the milk food chain. When the 1956 presidential campaign put a spotlight on the radiostrontium contamination of milk, however, concerned consumers, particularly parents with small children, began to ask for information specific to milk marketed in their neighborhoods. In response, the U.S. Public Health Service (USPHS) agreed to make a periodic check of fresh milk for radioactive contents. This program, however, developed only slowly. It started in mid-1957 with five cities, and one year passed before it finally added five more. As government officials continued to cite administrative difficulty as a reason for the lack of coverage, CU drew up a plan to collect one milk bottle each from fifty cities across North America to demonstrate how easily the federal government could expand its milk surveillance network if it was serious about protecting public health.[20]

CU's attention to radioactive fallout reflected an "information" turn in consumer activism in the United States. Founded in 1936, CU sought to combine the technique of checking consumer products with pro-labor activism. With the beginning of the Cold War, however, the organization toned down its advocacy of industrial workers and refashioned itself as a supplier of unbiased and scientific information for middle-class consumers. This strategic shift helped CU expand its reach nationwide, with 850,000 subscribers to its monthly magazine *Consumer Reports* and over four million potential readers by 1960.[21] Despite its successful adaptation to postwar consumerism, CU did not completely shed its progressive roots. The organization used its well-developed skills of product testing to harness the purchase power of consumers for promoting reform in a wide range of social issues from inflation to health care.[22] It thus came as no surprise that CU seized the radioactive contamination of milk to channel consumers' concerns about food safety into the question of disarmament and peace.

The first step toward forging the linkage between consumer protection and peace activism was to describe fallout not only as a potential hazard to humans but also as a blatant violation of consumer rights. An article published in the September 1958 issue of *Consumer Reports* noted that while the estimated doses from fallout to date were smaller than those from X-rays, there was a legitimate cause for concern because some elements in the fallout such as Sr-90 would build up in the environment and the human body, and also because there existed no reliable expertise, technological control, or consumer choice for this indiscriminate contamination.[23] By offering no "buying advice," the article powerfully reminded the readers that the only way to protect consumers was the suspension of nuclear tests. The fresh milk tests planned by CU were thus intended as a follow-up on this statement, seeking to demonstrate that the government, by conducting nuclear tests without carefully monitoring their radiological consequences, violated the fundamental rights of the consumer.[24]

To determine Sr-90 contents of milk on a broader geographical basis than had been covered by the federal government, CU carefully drew up a plan to be carried out for a one-month period between July and August 1958. The sampling sites, forty-eight in the United States and two in

Canada close to the U.S. border, were selected on the basis of locations, populations, and climatic and soil characteristics. Unlike government agencies, however, the private consumer group lacked resources to conduct such an extensive probe on its own. To compensate for this, it tapped into its network of secret shoppers deployed for regular product testing across North America. A volunteer in each location received detailed instructions as to when to buy milk, which brand, how much, and how to ship the sample. Once a sample arrived at the headquarters' lab in New York, CU closely followed the procedures established by the USAEC to make its tests credible. For this purpose, the organization used the testing service offered by Isotopes, Inc., one of the commercial laboratories hired for Project Sunshine. It also consulted a number of scientists, those working for the government as well as Pauling and other critics, before publication.[25] This strategic combination of consumer activism and external expertise made possible the largest radiological check of fresh milk undertaken by a single private organization.

Britain

In Britain, as in the United States, nuclear authorities first discovered the existence of hot spots. In a paper referenced in the 1956 Medical Research Council (MRC) report, the Atomic Energy Research Establishment (AERE) in Harwell reported that samples of soil, grass, and sheep bones from Wales consistently showed larger amounts of Sr-90 than those from other parts of the country. Particularly worrisome were the bones of upland Welsh sheep. Their radiostrontium contents in 1956 were up to ten times higher than the average for lowland sheep. Although Sr-90 concentrations in the bones of young children in Britain were no higher than 1.3 SU and well below the MRC's population average 10 SU "warning" level, the findings of unusual fallout contamination in Wales were a political bombshell, as Britain was moving toward its first thermonuclear test, scheduled in May 1957.[26] Labour MPs elected from Wales repeatedly pressed the Conservative government of Harold Macmillan in Parliament for further information on the reported contamination and its potential hazard.[27] Although the thermonuclear test took place as planned, it was clear that the problem of hot spots in Wales would only continue to grow more serious as nuclear testing accelerated around the world.

A rural, hilly region with coal mines and heavy industries clustered in the south, Wales had a distinct tradition of left-wing politics that shared many characteristics with Minnesota. Even after the Labour Party left office in 1951, a coalition of miners, industrial workers, and small farmers reliably elected Labour members to the Parliament. Campaigning for peace and disarmament was an integral part of the leftist activism in Wales. Labour MP Aneurin Bevan from South Wales unsuccessfully tried to swing his party to opposition to the thermonuclear weapon program, and the Campaign for Nuclear Disarmament (CND) established in 1958 enjoyed popular support in many parts of the region.[28] This certainly did not mean that people in Wales always viewed themselves as a separate political community. As historian Kenneth Morgan has observed, Welsh nationhood remained an open question in early post–World War II Britain, where the drive for centralization was particularly strong. Viewing politics primarily through the lenses of class, Bevan and most other Welsh Labour MPs were less inclined to promote causes specific to the region.[29] On the occasion of a Welsh Day debate in the House of Commons in 1944, Bevan famously declared that "I do not know the difference between a Welsh sheep, a Westmoreland sheep and a Scottish sheep."[30] Over a decade later, however, the Welsh sheep became more radioactive than those from the rest of the country. The phenomenon of hot spots, then, added a unique, material layer to the question of Welsh identity in the nuclear Anthropocene.

Well aware of the political danger of heavier fallout contamination in Wales and other regions with similar environmental conditions, the British government took action. In summer of 1957, at the suggestion of R. N. Quirk, undersecretary in the Office of the Lord President of the Council, representatives from the UK Atomic Energy Authority (UKAEA), the MRC, and the Ministry of Agriculture, Fisheries and Food (MAFF) held meetings to review the government's fallout monitoring program. E. P. Keely, a senior agricultural official, identified two major issues that hampered the similar U.S. effort. One was public pressure for fallout information specific to one's own community. The USAEC consistently prioritized research over coverage in its radiological survey and rejected requests for expansive sampling. Keely, by contrast, viewed the problem primarily as one of public relations, suggesting that monitoring should be installed "on a sufficiently wide scale—wider in fact than might be necessary from the strictly scientific point of view."[31] The other issue was a conflict of interest. While the USAEC adamantly defended its mandate as sole interpreter of fallout data, Keely warned that, if the British nuclear authorities did the same, "they might be criticised (as the American Atomic Energy Commission were) for being judges in their own case."[32] After further discussion, the Agricultural Research Council (ARC), the MRC, and the UKAEA agreed to divide the fallout monitoring task among themselves, and that the ARC, MRC, and the Development Commission would create a joint committee to coordinate the work of the three agencies. The new program placed a special emphasis on the problem of hot spots. The coordinating committee drew up a plan to collect approximately 120 samples of soil, herbage, vegetables, liquid milk, and animal bones in the 1958–59 period from six heavily contaminated regions: one in Wales, two in England, two in Scotland, and one in Northern Ireland.[33]

Initially, the overhaul in the British fallout monitoring program seemed to achieve its objective. Although some local officials in Wales contacted London to express their concerns about fallout, they merely asked for additional information. These restrained reactions, however, soon changed. In summer of 1958, some upland farmers in Wales approached John H. Fremlin, a nuclear physicist at Birmingham University, with samples of grass from grazing land for radiological analysis. A talented public speaker, Fremlin was one of the University's extramural lecturers. He was also a socialist and active member of the CND, strongly supporting the suspension of nuclear tests as a step toward armed reduction and peaceful coexistence.[34] The combination of scientific expertise, public outreach, and political leftism thus made Fremlin an ideal partner for the disquieted Welsh.

To Fremlin's surprise, the samples from Wales showed gross radioactivity about a hundred times higher than background radiation. He initially considered issuing a serious warning about contamination at a town hall meeting organized by a local CND committee in the small Welsh town of Talgarth. Before delivering the lecture, however, he consulted the secretary of the ARC and decided to moderate the tone of the statement considerably.[35] In part, he seemed to be genuinely reassured that Sr-90, an element of concern, made up only fractions of the measured radioactivity in grass. His restraint also conceivably reflected his effort to observe the boundary between science and activism. His daughter Margaret Fremlin later recalled that some of his colleagues were hostile to Fremlin's leftist politics, so he took extra care to avoid the impression that he was speaking on behalf of his institution.[36] In his Talgarth speech delivered in August

1958, Fremlin reassured the farmers that there was "no danger at all from the tests so far in Welsh sheep eating the grass on the mountains." At the same time, he also warned about future trends, urging local authorities to make regular checks on the amounts of Sr-90 in the milk produced in their communities.[37] A region-specific fallout survey, in short, was his answer to the dilemma between alarmism and indifference.

Fremlin's call for regional fallout studies came when sensational reports of the local press on radioactive contamination were triggering a massive wave of public concerns.[38] Upon learning about unusual Sr-90 buildups in the region, David Cole, editor of the Welsh newspaper *The Western Mail*, launched a private investigation. Despite the newspaper's conservative-leaning editorial policy, Cole was determined to publish information that he believed would expose serious flaws in the government's reassurances. This media campaign caused much alarm inside Whitehall. In October 1958, shortly after *The Western Mail* began to run a serial feature on the problem of radiostrontium in Wales, Minister for Welsh Affairs Henry Brooke told Macmillan that he and his deputy, Lord Brecon, "both sense a great risk that, at the next election, the Government's critics in Wales will exploit these feelings to the detriment of our cause." According to Brooke, the problem was not so much contamination itself as the perceived disregard for the Welsh. He believed that critics would attack the government for having remained silent "as though it did not matter at all that the Strontium 90 deposits in Wales were much heavier than elsewhere."[39]

Brooke's fear soon became reality when a local public health officer from Wales sparked a regionwide rebellion against the government's safety claims about fallout. In January 1959, T. Alun Phillips, a public health officer for Caernarvonshire in northwest Wales, told the rural district council that the levels of gross radioactivity in local drinking water had already reached 30 percent of the maximum value permitted for the River Thames.[40] As a fierce public debate broke out over Phillips's findings, local councils across Wales passed resolutions calling on London to approve funding for radiological monitoring in their districts. To support such an initiative, Phillips and a group of municipal medical officers wrote to the Welsh Association of Local Authorities in August with a proposal for establishing a regional radiological laboratory for analyzing local samples. "We see no evidence that the monitoring of the Atomic Energy Authority takes special account of local problems," the letter stated. "Having regard to the apparent indifference of central policy in this matter it seems that initiative must come from local authorities."[41]

Unlike Americans similarly concerned about hot spots, however, the Welsh were rather slow to conduct a systematic fallout survey in their areas. Not until October 1960, toward the end of the moratorium period, did the Monmouthshire County Council in southeast Wales finally approve a proposal to establish a monitoring service and a testing center for Wales on a pilot scale.[42] Comparison to the United States suggests two possible reasons for this striking trans-Atlantic difference in the grassroots politics of risk. First, the British government's fallout monitoring program, involving various agencies other than the atomic establishment, successfully preempted the allegation of a conflict of interest. Although the Welsh voiced their frustration about the lack of responsiveness from London, this did not grow into a wholesale criticism against information control by the nuclear security state. Second, the worried British citizens lacked an institutional basis to carry out a comprehensive fallout survey on their own. Unlike their U.S. counterparts, local authorities in Wales tended to look to London for assistance and guidance, and no civic group comparable to CU stood to offer an independent check on the government's fallout studies. British officials thus managed to keep the explosive issue of hot spots under control during the moratorium period. In the meantime, the stunning results of the local fallout surveys in Minnesota and other parts of the United States began to arrive, jolting the USAEC's credibility as the only authoritative source of fallout information in the country.

The End of Information Control

The scientists from Minnesota fired the opening salvo in the grassroots revolt against the USAEC's knowledge claim about global fallout contamination. On February 6, 1959, Visscher issued a statement during Governor Freeman's press conference, announcing that the Sr-90 level in the 1958 wheat produced in Minnesota was found to be on the average 155 SU. This value was approximately ten times as high as that in milk from the region reported in fall of that year. Moreover, the range of values in wheat varied widely, with one batch from 1957 crop showing as much as 600 SU. Visscher quickly added that there would be no immediate danger to those who consumed contaminated wheat.

Unlike Japan, where brown rice was a leading calcium (and hence radiostrontium) source, cereal grain products accounted for a relatively small portion, namely between 5 and 10 percent, of the total calcium intake in the average U.S. diet.[43] The Minnesota scientists nevertheless included wheat in their study because it was "an important crop in Minnesota and an important part of the diet."[44] In a letter to Freeman sent on the eve of the press conference, Visscher explained what the discovery of radioactive wheat meant for his state and beyond. He noted that the USAEC always sought to reassure the entire 150 million people in the United States on the basis of the national average. "As to the survival of the nation this is true, but as to the survival of children in some Minnesota towns or on some Minnesota farms, it is simply not true," he said. "Averages of figures with a hundred-fold spread are utterly meaningless when applied to individuals or small groups." The apparently low averages often cited by nuclear officials, then, "conceal rather than reveal facts."[45]

Successfully dismantling the myth of the average risk that underpinned the fallout safety claims made by U.S. nuclear officials, Visscher seized this opportunity to take aim at what he saw was a root cause of the problem: the USAEC's monopoly of fallout data. In the press release, he declared that the role of the scientist was not to provide bland reassurances in the absence of factual information, but instead "to provide a sober interpretation of facts, to allow people to take precautionary and preventive measures before it is too late."[46] A member of Minnesota's liberal political circle, Visscher immediately contacted two Democratic U.S. senators from the state, Hubert Humphrey and Eugene McCarthy, saying, "I have a new job for one of you." Noting that the USAEC had spent millions of dollars for Project Sunshine and yet grossly overlooked the serious contamination of wheat, the physiologist called for a congressional investigation into the commission's failure. More important, he asked the senators to push for a ban on nuclear tests as representatives of those living in the hot spots who were forced to bear biological costs of national security more than those in other parts of the country. Wondering aloud "why the Plains States and our region should be made to take the brunt of so much fall-out burden if we are to continue bomb testing," Visscher urged all politicians from the Upper Midwest to "get together and get stoppage of further contamination of this region."[47]

As radioactive fallout knew no political borders, it was natural that the uproar over the unusual contamination of wheat in Minnesota instantly spread northward to Canada. This NATO country located in the northern high-latitude zone found itself in a double bind between the defense needs of the Western bloc and the disproportionate burden of global fallout. The news on radioactive wheat deepened this dilemma, as pressure for a test ban intensified in the prairie provinces that had elected Canadian Prime Minister John G. Diefenbaker and many of his political allies to the Parliament.[48] Historian Daniel Heidt has shown how Canada's increasing trouble with the nuclear Anthropocene drastically changed the outlook of Howard C. Green, Diefenbaker's Secretary of State for External Affairs. An ardent anticommunist, Green had long supported military cooperation with the United States and other Western allies. His exposure to the 1958 report of the United Nations Scientific Committee on the Effects of Atomic Radiation (UNSCEAR) and the subsequent fallout scare over milk and wheat, however, convinced him that Canada was uniquely vulnerable to the influence of nuclear tests and atomic warfare.[49] Driven by public opinion and his inner conviction, Green introduced a UN resolution in October 1959, calling for greater cooperation in fallout studies among member states and specialized agencies. The impact of radioactive wheat on this Canadian initiative was clear when he briefed the cabinet on his proposal. Noting that the problem of wheat was "of extreme importance to Canada," Green argued that "[i]f Canadian wheat became suspect in world markets because of radiation, irreparable damage might be done."[50] Although the Americans, British, and even the Soviets were reluctant to divert attention away from their ongoing test-ban talks in Geneva, the Canadian resolution unanimously passed on November 2, serving as a powerful reminder that fallout was an unacceptable hazard to the international community.[51]

U.S. nuclear officials tried to downplay the significance of radioactive wheat. In a presidential briefing held on March 6, Libby argued that any effect of unusually contaminated crops would be averaged out over a period of time and also in total diet. While admitting that biological tolerance for Sr-90 "really isn't known," the USAEC commissioner nevertheless told President Dwight D. Eisenhower that the news from Minnesota hardly changed the fact that the risk to human health from global fallout was "very small compared to other hazards of life."[52] This time, however, Libby found it difficult to reassure the president. Instead of attacking the USAEC for its moral or political judgment, Visscher deliberately chose to let the fallout data speak for itself. This evidence-based approach successfully amplified the political effects of his findings. The report of radioactive wheat came when Eisenhower had become seriously worried about the devastating consequences of an all-out nuclear war. In response to Libby's briefing, the president commented on "the difficulty of any assumption [that] there could be a nuclear war," declaring that "the radioactivity level from a massive attack [with nuclear weapons] would be just tremendous compared with what is evident in Minnesota wheat as the result merely of a few tests."[53] Outside the White House, Libby also faced a powerful bloc of Democrats in Congress mobilized by the Minnesotans. Politically checkmated, Libby took on an unusually conciliatory tone. He told members of Congress that he was "very concerned" about the data from Minnesota and supported a special hearing on this subject.[54] Libby even wrote directly to Visscher and made a gesture of taking the blame. "We have been concentrating on human bone, milk, rain, and soil," he stated, "and I am afraid thus somewhat underemphasizing grain and vegetables."[55]

No sooner had the news on radioactive wheat from Minnesota shocked the nation than consumer activists dropped another bombshell. In the March 1959 issue of *Consumer Reports*, CU published a ten-page special report on the measurements of Sr-90 in fresh milk from fifty cities across North America. To some extent, the results were reassuring: the mean value, 8 SU, was in good agreement with the USPHS's estimate. CU's more detailed data, however, revealed a strikingly uneven distribution of Sr-90 in milk, ranging from 2 to nearly 16 SU and possibly even higher.[56] This delivered another blow to the myth of the average risk. As Irvin Michelson, director of CU's Division of Special Projects in charge of the milk study, later told members of Congress, "averages for the country . . . are no substitute for knowing local levels."[57] More important, by demonstrating that even a private organization of limited means could carry out an extensive fallout survey, CU debunked the excuses made by the government agencies for the limited coverage of their work. In a letter to Clinton P. Anderson, Democratic U.S. Senator from New Mexico and chairman of the congressional Joint Committee on Atomic Energy (JCAE), Surgeon General Leroy E. Burney was forced to admit "the generally unsatisfactory nature" of the USPHS program in reassuring the public about the contamination of milk.[58]

Like the Minnesotans, the consumer activists carried out their milk check for Sr-90 as a means, not an end. By demonstrating the blind spots in the government's fallout monitoring program, CU stepped up its pressure for a fundamental change in the existing mechanism of knowledge production dominated by the nuclear authorities. Noting that the problem of fallout in the United States still remained "the province of the AEC," CU stated in its article that "it is hard to see why judgments on matters of public health should have to depend primarily on the reports of the very agency charged with the responsibility of manufacturing nuclear weapons." To detach the function of radiation protection from the polluter, the consumer group suggested that the USPHS should take over the responsibility for radiological surveillance and assessments.[59] Like the Minnesotans, CU seized on its findings of contaminated milk to push for an end to nuclear tests. Describing a reported effort by government scientists to develop techniques to remove Sr-90 from milk as "only palliatives," it declared that it was "the diplomat who holds the key to the solution of the base problem: cessation of nuclear explosions in the atmosphere," adding that it would not only stop further contamination but also mark "a first step toward the prevention of nuclear war."[60]

The last straw that broke the public's trust in the USAEC's safety claims, however, came from within the national security state. A key piece of information in the global-fallout puzzle was the distribution patterns and descending rates of nuclear test debris injected in the stratosphere. This large uncertainty in knowledge about the upper atmosphere surfaced during the U.S. congressional hearing held in May and June 1957. Libby argued that radioactive dust would spread in the stratosphere more or less uniformly and fall to the ground only slowly, namely over up to ten years. Lester Machta of the U.S. Weather Bureau, however, strongly disputed this claim, pointing out the possibility that, depending on latitude of injection, nuclear dust in the stratosphere might be dispersed unevenly and descend at rates much faster than Libby had asserted.[61] To settle this dispute, the Air Force launched the High Altitude Sample Program in November 1957, which deployed U-2 airplanes to make direct measurements of radioactivity in the stratosphere.[62] The results painted a mixed picture. It turned out that there was not as much nuclear dust in the upper atmosphere as had been expected. A clearer sky, however, also meant that the debris was descending to the ground at a much faster pace—up to a few years for low latitudes, and no more than a year in high latitudes—before being fully mixed.[63] Once informed of the existence of these latest data, Machta and his fellow meteorologists asked for their prompt declassification.[64] Given the sensitive nature of the program using the top-secret spy equipment, however, the Department of Defense, the CIA, and the USAEC agreed to disclose the Air Force data to the meteorologists on a strictly secret basis.[65]

Withholding the key fallout information from the public, however, ultimately backfired. In February 1959, Herbert B. Loper, the Pentagon's atomic energy adviser, sent the Air Force data to Anderson. Learning about this, Libby immediately informed the senator that the information was still premature for public release. In a subsequent letter to Anderson, Loper confirmed that the forwarded information contained no military secrets except for a few lines, but he urged the senator to handle it as confidential because "there is not full agreement as to the interpretation of the data that has been obtained so far."[66] This request raised a red flag for Anderson. A fervent critic of the USAEC and the Republican administration, he immediately published a series of his correspondence with Libby and Loper. Presenting them as further proof of the USAEC's cover-up, Anderson demanded a congressional hearing to interrogate the commission for its public information policy.[67]

A flood of criticisms from within the government and without finally forced nuclear officials to end their control of fallout data. On March 24, 1959, shortly after Anderson's disclosure of the letters, USAEC chairman John A. McCone made a public pledge to publish all fallout information without delay. "[S]o long as I am Chairman of the Atomic Energy Commission," he declared in a statement, "I shall not be a party to the suppression or distortion of any information bearing on the safety and health of the American public."[68] Beginning in September 1959, the commission published its fallout monitoring results on a quarterly basis.[69] This made it no longer possible for nuclear officials to guide the interpretation of the radiological data through comments and dismiss their opponents as merely "speculating" with little or no scientific evidence.

Critics of the USAEC, however, demanded more than greater transparency. In their view, the problem had deep roots, namely the commission's conflict of interest in conducting nuclear tests while assessing their health hazards. This growing criticism reflected a broader shift in U.S. political culture. Political scientist Andrew Stark has pointed out that the earlier notion of a "conflict of interest" in the public administration was fairly narrow, limited to financial stakes and judged case by case. After World War II, however, it became a universal principle, extending to any kind of commitment which could potentially affect one's judgments.[70] Against this backdrop, and driven to enlarge its own bureaucratic turf, the USPHS took steps to challenge the USAEC's exclusive mandate to manage radiological health relating to atomic energy use. In February 1958, the Surgeon General established the National Advisory Committee on Radiation (NACOR) as an institutional adviser for the Division of Radiological Health. On March 26, 1959, amid the public uproar over the USAEC's public information policy, the NACOR released its first report with the recommendation that the USPHS should take charge of radiation protection from all sources of radiation, including those under the USAEC's jurisdiction, to ensure both unified control and regulatory independence.[71]

Faced with the revolt from within the federal government, the White House tried to deflect the mounting pressure for the separation of nuclear promotion and regulation by ordering the Bureau of the Budget to review the administrative structure of radiation protection.[72] The bureau agreed with the NACOR that the USAEC's dual function was a source of distrust in its regulatory activities, but it also opposed leaving the task of radiation protection to the USPHS alone, asserting that it would allow public health officials to dictate "the whole future of the use of radiation."[73] After further discussion, the bureau concluded that the president possessed the ultimate legal authority to balance the risks and benefits of the use of radiation and radioactive materials. Accordingly, in August 1959, Eisenhower announced the creation of the Federal Radiation Council (FRC) within the White House. Comprising representatives from the Departments of Commerce; Defense; and Heath, Education, and Welfare as well as the USAEC, the FRC was a coordinating board rather than an independent institution with its own resources and expertise.[74] Despite this severe limitation, the formation of the FRC nevertheless marked the end of the era in which the USAEC could singlehandedly determine what constituted an "acceptable" risk associated with its activities for the whole nation.

Repercussions

The conflicts over control of fallout information came at the critical moment in the diplomatic negotiations for a nuclear-test ban. Eisenhower and his secretary of state, John Foster Dulles, pursued an agreed ban on all nuclear tests to slow down the nuclear arms race, reduce international tensions, and force the Soviets to accept the principle of safeguards in disarmament. To finalize the design of a control system for such a treaty, the conference of experts was reconvened in Geneva in November 1958. Soon thereafter, however, U.S. experts discovered that the originally agreed system might not be as reliable as had been believed. The new seismic data on a series of underground tests conducted at the eleventh hour before the moratorium went into effect indicated that the threshold of detectability might be in the order of 20 kt rather than 5 kt. This meant that it would require a vastly greater number of ground stations or on-site inspections within the Soviet territory to achieve the same level of assurance as before. Predictably, the Soviets charged the Americans with bad faith and refused to accept any basic changes from the originally agreed scheme. As the test-ban talks came to a complete halt by February 1959, James J. Wadsworth, a senior State Department official and chief U.S. negotiator in Geneva, effectively abandoned any hope for an agreement and urged Washington to call up a recess.[75]

Underground testing was the latest technological solution pursued by test-ban opponents in the United States to address concerns about radioactive fallout without stopping weapons testing. As was the case for the "clean bomb," physicist Edward Teller spearheaded this effort. In a report written in February 1956 with geophysicist David T. Gregg, he argued that it was possible to test nuclear weapons underground to gather necessary information without a "long-term radiologic hazard."[76] After extensive research, Teller's team at Livermore successfully conducted the first fully contained underground test, code-named "Rainier," in September 1957. Teller and his supporters immediately seized on this breakthrough to claim that the Soviets could violate a test ban with impunity.[77] One scenario that they put forth was to dig a huge hole deep underground to muffle seismic waves from a nuclear explosion. For instance, even a test as large as 200 kt, if conducted in a hole of 360 feet in diameter, would generate a seismic signal of a 1-kt blast.[78] Although the cost of such an undertaking was likely to be prohibitive, those distrustful of the Soviets insisted that even the slightest risk of deception was too dangerous to take amid the deadly nuclear arms race.

As the prospect of the prohibition of all nuclear tests dimmed, the White House started considering an atmospheric ban. This fallback position was initially conceived as a way to sidestep the problem of international inspection necessary to detect underground tests. Dulles first mentioned the idea shortly before he became seriously ill, and Acting Secretary of State Christian Herter and Eisenhower briefly discussed it on February 17.[79] By the time Eisenhower returned to the subject a month later, however, the shocking news about the unusually heavy contamination of wheat and other foodstuffs had drastically changed its context. "All available evidence indicates that nuclear testing is bad," the president said in a meeting with Herter and his deputies on March 17. "The allowable dose of strontium 90 is being approached in some foods in some areas of the country." With this development, he declared, "we would no longer test atomic weapons in the atmosphere." Unlike his earlier dismissive remarks on the global-fallout problem as a matter of public relations, Eisenhower now viewed it as a sober warning about the catastrophic outcomes of the nuclear arms race. Convinced that "the scientists will say that any nuclear war would be disastrous, at least for the Northern Hemisphere," Eisenhower concluded that "we should be working toward acceptance of a test ban, which may not be so good as we want, but would test whether both sides are acting in good faith."[80]

Around the same time, the British also reached a similar conclusion. Although the reports of diet contamination from Wales apparently did not have as direct an impact on Macmillan as those from Minnesota did on Eisenhower, Macmillan clearly recognized the serious implications of the fallout controversy for the forthcoming general election. In fact, the Labour Party announced its plan to stop all nuclear tests unilaterally if it was returned to office. Party leader Hugh Gaitskell put a strong emphasis on fallout hazards, asserting that any nations which further contaminated the atmosphere by resuming tests would be guilty of a "monstrous crime against humanity."[81] On top of this domestic political pressure, Macmillan was also eager to make even small progress in nuclear disarmament to stop the nuclear dominoes in Europe following the French. For these reasons, the prime minister made his trip to Moscow in February 1959, sounding out Soviet Premier Nikita Khrushchev on a variety of ideas, including an atmospheric ban, to break the stalemate in the Geneva test-ban negotiations.[82] When he arrived in Washington a month later, he suggested an interim agreement limited to cessation of atmospheric tests to eliminate fallout while deterring countries with nuclear ambition. Eisenhower agreed, saying that his government was ready to "refrain from atmospheric tests if the Russians did."[83]

Both Eisenhower and Macmillan initially believed that an atmospheric ban would solve both the inspection and fallout problems at once. As Eisenhower's science adviser James R. Killian quickly reminded them, however, the meaning of the "atmosphere" depended on the objective of a test ban. To prevent radioactive contamination, no tests should be allowed above ground up to thousands of kilometers. If the aim was to verify compliance without international inspection, however, an agreement would have to exclude the highest layers of the atmosphere, where long-range detection was believed to be less reliable.[84] When the U.S. and British governments formally introduced a joint proposal for an atmospheric ban on April 13, the "atmosphere" was defined in the latter term: from the surface up to fifty kilometers.[85] The Soviets immediately noticed this, and Khrushchev bluntly told Eisenhower in his letter that the proposal would make little difference as to radioactive contamination.[86] By July, however, the U.S. scientific advisers had concluded that it was feasible to detect nuclear explosions well beyond the limits of the atmosphere, namely up to fifty thousand kilometers from the ground.[87] With a gap between the range of detection and that of contamination eliminated, the U.S. Congress came to overwhelmingly support an end to atmospheric tests. All JCAE members endorsed an atmospheric ban, and the Senate signaled its readiness to ratify such a treaty.[88]

The momentum for an atmospheric ban, however, soon dissipated. Part of the reason lay in a sudden change in the Soviet attitude. Vasilii S. Emel'ianov, chief of the Directorate for Peaceful Uses of Atomic Energy, explained to McCone during a meeting in December 1959 that the Soviet government had once been prepared to propose a ban on atmospheric tests, only to cancel the plan when the Western powers started showing their interest in such an idea.[89] This last-minute reversal was puzzling because an atmospheric ban would not only stop global fallout contamination that the Soviets had been decrying in public, but also alleviate their long-standing fear of espionage. One possible explanation related to a technological gap in underground testing between the nuclear rivals. While the United States had successfully conducted over two dozen fully contained underground tests before the moratorium began, the Soviet Union had not done any.[90] Undoubtedly, the techno-military advantage of an atmospheric ban for the United States must have been clear to the Soviets. Equally important, the prohibition of atmospheric tests was a far cry from a new Soviet campaign for "general and complete disarmament" that Khrushchev launched during his historic visit to the United States in September 1959. As political scientist Matthew Evangelista has pointed out, this was not mere propaganda. The Soviet premier followed up on his words by announcing a deep cut in his nation's armed forces in January 1960.[91]

The Soviets were not the only party that turned away from an atmospheric ban. The U.S. and British political leaders also resumed their bid for a comprehensive ban soon after they presented the joint proposal. George B. Kistiakowsky, a Harvard chemist who became Eisenhower's chief science adviser in August 1959, successfully convinced the president that the risk of an undetected Soviet underground test was not as serious as Teller and his associates had alleged.[92] Macmillan welcomed this development, as he had consistently looked for a meaningful measure in nuclear disarmament and also had been willing to compromise on the outstanding problems of international inspection.[93] To signal his commitment to further negotiations with the Soviets, Eisenhower extended the moratorium, which was originally set to expire at the end of October, for another two months. Both sides of the Geneva talks, however, remained as far apart as ever, with the Soviets adamantly opposed to allowing for any on-site inspections within their territory at all. Eisenhower had no choice but to let the moratorium expire, but he did not order the immediate resumption of nuclear testing, believing that it was still possible to reach a compromise with the Soviets on the matter of inspection. "Our real aim," Eisenhower told his advisers, "is to open that country up to some degree." Herter agreed with the president, saying that a comprehensive ban would serve as a test case for workable disarmament agreements in the future.[94]

The desire to achieve a genuine breakthrough in disarmament and peace, then, foiled the early momentum for an atmospheric ban. This, however, did not mean that the U.S. and British governments simply viewed the partial-ban proposal as politically expedient. On the contrary, both found it increasingly difficult to resume a type of nuclear test that would contaminate the biosphere regardless of the outcome in the test-ban negotiations. In a letter to British Foreign Minister Selwyn Lloyd dated August 7, 1959, when the Anglo-American joint offer of an atmospheric ban was all but dead, Macmillan declared that at no point in the future would he allow atmospheric tests.[95] A similar sentiment was also echoed in Washington. Despite little progress made toward a comprehensive ban, as McCone put it, "we are 'out of the business' of atmospheric testing anyhow."[96] Eisenhower strongly agreed. In a meeting with his aides in September 1959, he said that "no free country can go back to atmospheric testing," expecting that "[w]orld opinion—the adverse effect of alienating free world countries—would stop it."[97]

What decisively swung Eisenhower to a unilateral renouncement of atmospheric tests was yet another breach in the USAEC's knowledge claim. In spring of 1960, the commission reported that the bran of wheat harvested in 1958 in nine states was heavily contaminated with Sr-90, with one sample showing as much as 675 picoCi of Sr-90 per kilogram. Although the U.S. diet typically contained bran only in small amounts, the values of Sr-90 in it did indeed "*look* high."[98] When Kistiakowsky briefed Eisenhower on the "hot" bran, he found the president "obviously unhappy about the past assurances from the AEC that there was no problem from fallout."[99] Eisenhower's sense of disillusionment must have been profound because of his leadership style, in which he always chose to trust "his scientists" in the government despite his keen awareness of the deep differences of opinion among experts. The loss of confidence in the counsel of the USAEC staff scientists drove Eisenhower to the side of caution regarding the dangers of fallout. Eisenhower told McCone that "he ha[d] become more concerned about the consequences of testing on the human body." He blamed "our scientists" for being "too rosy regarding the possibilities of damage to the world." The USAEC chairman agreed, reassuring Eisenhower that he would not recommend testing nuclear weapons in the atmosphere.[100]

No longer certain about the judgment of risk by his scientific advisers, Eisenhower was determined not to resume fallout-producing nuclear tests for the rest of his presidency. By that time, his final push for a comprehensive test ban had failed. In May 1960, the Paris Summit fell apart with no agreement on international inspection in the aftermath of an incident in which a U-2 spy plane was shot down over Soviet airspace.[101] Eisenhower openly regretted this failure, but still found some comfort in his unilateral abstention from atmospheric testing. In August 1960, he declared in a press conference, "I will not allow anything to be exploded in the atmosphere that would add anything to the apprehensions of people about their health."[102] Eisenhower abided by this pledge until he left office in January 1961.

Conclusion

The regional fallout surveys conducted by concerned scientists and activists in the United States and Britain marked a local turn in the politics of risk. Radioactive hot spots within the two Western nuclear countries interacted with the pockets of political dissent in the era of the Cold War consensus. Steeped in a rich tradition of progressivism, and alarmed by the reports of unusual contamination in their communities, scientists and citizens in Minnesota and Wales as well as U.S. consumer activists started checking radioactivity in their local food and drink. Although this "information" model of antinuclear activism was in part a forced adoption to the conservative political climate of the 1950s, it proved highly effective because U.S. and British nuclear officials had always presented their safety claims as factual statements based on scientific evidence. Moreover, the evidence-based argument helped the activists in the United States attack the root cause of the problem: a conflict of interest embedded in the USAEC's dual functions as nuclear promoter and regulator. The stunning findings of unusually contaminated wheat and milk by independent fallout surveys, combined with the public revelation of the USAEC's effort to withhold radiological data, finally put an end to the commission's information monopoly by which to guide the public interpretation of fallout data toward a reassuring conclusion.

The loss of the USAEC's scientific credibility in the fallout controversy affected the course of the test-ban negotiations during the moratorium period. Test-ban opponents in the United States again displayed a combination of political savvy and technological prowess, successfully developing the technique of underground testing to turn the fallout argument against the goal of a comprehensive ban. The idea of an atmospheric ban initially emerged out of the stalemate over the detection of this latest type of nuclear test, but it soon became much more than a fallback position. As independent fallout surveys in the United States revealed extreme contamination of some foodstuffs, Eisenhower also came to view an atmospheric ban as an antipollution measure. The comparable monitoring efforts in Wales had less impact on British test-ban policy, but it added to Macmillan's domestic political calculations in support of an atmospheric ban. Although Eisenhower and Macmillan soon returned to their pursuit of a comprehensive ban, both unilaterally abandoned the option of resuming atmospheric tests even after the declared period of the moratorium expired.

Despite this powerful influence of risk politics on the test-ban negotiations, questions still lingered. Were the concessions made by the policymakers toward concerns about fallout politically motivated to the point of being meaningless? To what extent was the inner conviction against fallout felt by some political leaders like Eisenhower genuine? The strength of the emerging antipollution norms was put to its most public and crucial test when the Soviet Union abruptly ended the moratorium in August 1961, setting in motion a global upheaval which ultimately led to the conclusion of the Partial Test Ban Treaty two years later.

Chapter 7

FALLOUT

The Making of the Partial Test Ban Treaty, 1961-1963

O N AUGUST 31, 1961, the Soviet government abruptly announced the end of the moratorium on all nuclear tests. A public statement, printed on the front page of Soviet newspapers *Pravda* and *Izvestia*, explained the decision at great length. Recalling that the test-ban negotiations in Geneva remained deadlocked over the U.S. demand for the international inspection of suspect underground explosions, the Kremlin accused the Western powers of using the question of verification as a pretext for preventing the conclusion of an agreement. While forcefully defending the necessity of atmospheric tests, it did gesture toward concerns about fallout. "The harmful effects of thermonuclear weapons tests on living organisms are well known to the Soviet Union," the statement noted, "and hence every possible step is being taken to reduce those effects to a minimum." Nonetheless, it insisted that "[t]he Soviet government has been forced to take this step [to resume tests], the significance of which it fully appreciates, under the pressure of the international situation being created by the imperialist countries."[1] Recounting a number of recent events, most notably the testing of nuclear weapons by the French and an armed standoff over the status of Berlin, the Kremlin asserted the country's sovereign right to defend itself in the face of an imminent threat of war. "To discourage an aggressor from criminally playing with fire," it said in the statement, "we must let him know and see that there is in the world a force armed at all points and ready to repulse any challenge to the independence and security of peace-loving states, and that the weapon of retribution will smite the aggressor in his own lair."[2]

The Soviet statement seemed to suggest that security anxiety trumped any concerns about fallout at the time of major Cold War crises. Indeed, in the next two years, the world experienced the worst fallout pollution in history. The Soviet Union tested 136 nuclear devices in the atmosphere with a total explosive yield of approximately 220 Mt. The United States initially limited its tests to underground but eventually resumed atmospheric testing in April 1962, conducting 40 tests aboveground with about 36 Mt. Taken together, the atmospheric tests conducted during this brief period accounted for more than half of the total yield of all such tests ever conducted. Does this mean that all concessions made by the political leaders of the nuclear powers about the dangers of fallout shortly before and during the moratorium period were nothing more than a politically expedient gesture, not indicative of a fundamental shift in their private views?

A closer look at the politics of risk unfolding both within the government and without during one of the most dangerous periods of the Cold War reveals the astonishing resilience of the changed norms on global fallout contamination. Although most policymakers and advisers still believed that the need to demonstrate the country's nuclear might and resolve in the face of provocations by the enemy would clearly outweigh what appeared to be a quantitatively small risk of harm from fallout, some, most notably U.S. president John F. Kennedy and Soviet physicist Andrei D. Sakharov, repeatedly expressed strong misgivings about harming innocent bystanders. The international community also rallied in protest against fallout and in support for an atmospheric ban. Contamination was also no longer acceptable to a growing number of American and British citizens. The broken promise of timely countermeasures fed distrust of the government, especially among women who led antinuclear activism and turned children and milk into a powerful symbol of protest against fallout. Mutual distrust and suspicion between the Cold War rivals finally dashed any hope for a comprehensive ban and paved the way for an atmospheric ban. The Partial Test Ban Treaty, then, was a hybrid construct of the Cold War mind-set and global environmental consciousness.

The End of the Moratorium

U.S. nuclear physicist Harold Brown, who served in the Kennedy administration as director of Defense Research and Engineering, recalled that Kennedy entered the White House with an "instinctive feeling against nuclear testing." The new president "felt very strongly about fallout, much more strongly, I think, than was justified on purely technical grounds." This feeling, Brown said, was "a correct evaluation" of the political situation of the time.[3] Indeed, by the time Kennedy was sworn into office, radioactive fallout had become a political taboo in the United States. The 1960 presidential election, unlike that in 1956, saw no partisan debate as to the dangers of fallout to human health. Kennedy endorsed a partial test-ban proposal during his campaign, making the pledge that he would unilaterally prohibit any fallout-producing tests if another round of negotiations failed to agree on a comprehensive ban.[4] His Republican opponent, Richard Nixon, followed Eisenhower with the promise that future nuclear tests be kept underground.[5] Beyond playing politically "safe" with fallout, Kennedy also seemed to be genuinely concerned about its moral implications. During the campaign, he declared that even a very small number of individual tragedies caused by fallout "loom very large indeed in human and moral terms." He also vowed to address the issue of fallout carefully not because its dangers were well established, but rather because "there is still much that we do not know."[6]

What drove Kennedy's test-ban policy in his first months of presidency, however, was not health concerns about radioactive contamination, but rather a growing national security threat from the nuclear arms race and proliferation of nuclear weapons. The Soviet launch of the Sputnik satellite in 1957 had sparked a public uproar in the United States over the alleged "missile gap," which Kennedy as presidential candidate shrewdly exploited to attack his Republican opponent. Once in the White House, he discovered that a missile gap did exist—but overwhelmingly in his nation's favor. Nonetheless, Kennedy proceeded to order a massive nuclear arms buildup, driving the arms race with the Soviet Union into a new and more dangerous phase.[7] Moreover, the three countries that originally had entered into the test-ban negotiations no longer monopolized the world's nuclear arsenals. France conducted its first atomic bomb test in 1960, stoking the fear that Communist China, West Germany, and many more countries might follow suit.[8] Kennedy and British Prime

Minister Harold Macmillan thus gave priority to a comprehensive ban over a partial one to restrict underground testing as much as technically verifiable. For this purpose, the U.S. and British governments jointly introduced a new draft treaty in April 1961 prohibiting all tests but those underground below 4.75 on the Richter scale, combined with the offer of a three-year moratorium on underground testing of any size. To encourage the Soviets to meet halfway at the upcoming summit meeting in Vienna in June, the two governments also relaxed their demand on the annual quota of on-site inspections from a fixed number of twenty to a range of twelve to twenty depending on the number of unidentified seismic events on Soviet soil.[9]

By the time Kennedy and Khrushchev met in Vienna, however, the Soviet premier had already lost his interest in a comprehensive ban. The U-2 incident of May 1960 seemed to validate the Kremlin's suspicion that any kind of international inspection for arms control was a disguise of Western espionage. The Congo crisis, in which U.S.-backed forces toppled the Soviet-supported regime in September 1960 under the cloak of UN intervention, deepened Moscow's distrust in the appointment of a "neutral" official in charge of monitoring a test ban. The Soviets now demanded a "troika" scheme, in which the communist administrator could veto his Western and neutral counterparts in enforcing the test ban. On the eve of the summit meeting, Khrushchev told members of the Presidium that a test-ban agreement would never happen. To avoid a "harmful appearance" before world opinion, he proposed to link a test ban to an impossible demand of general and complete disarmament. "If we do not give them an espionage system" for the test ban, he said, "this will push them to agreement on disarmament."[10] In Vienna, Khrushchev acted exactly as he had promised his colleagues in Moscow. He not only made no concession on the troika demand but also insisted that a test ban was useless except as part of complete disarmament.[11]

With no immediate prospect for a breakthrough in the test-ban negotiations, both the Western and Soviet sides quietly took steps to end the moratorium. Washington and London, however, planned to limit any resumed tests to underground to avoid releasing radioactive fallout in the atmosphere. Jerome Wiesner, Kennedy's scientific adviser, proposed this partial approach. Secretary of State Dean Rusk agreed, noting that as long as there was no fallout, the resumption was "a manageable political problem."[12] The British also signaled their support for the resumption of tests, so long as the Western powers would not be "the first to create fall-out."[13] In a letter to Macmillan dated August 3, Kennedy reassured the British prime minister that any U.S. tests would remain underground unless and until the Soviets conducted atmospheric tests. Far from being defensive in the face of public opinion against fallout, the president was eager to turn it against the Soviets. He proposed to launch an international campaign for an atmospheric test ban in an attempt to "get wide public understanding of 'fall-out testing' as bad, and underground testing as reasonable."[14] Macmillan agreed, claiming that such an overture would demonstrate that "we are fully conscious of world-wide anxiety about fallout and ready to do our best to avoid it."[15]

While Kennedy and Macmillan were mindful of the political costs of fallout, Khrushchev seemed to be hardly bothered by them. His paramount concern lay in the unsettled status of divided Berlin, a hole in the Iron Curtain through which an increasing number of East Germans were escaping to the West.[16] After Kennedy refused to discuss this problem in Vienna, Khrushchev tried to force him to the negotiating table by issuing an ultimatum, threatening to sign a separate peace treaty with East Germany and hand over to the regime the control of Western access to West Berlin. Kennedy, however, stood firm. On July 25, in a speech delivered on nationwide television, he expressed his willingness to pursue a diplomatic settlement over Berlin, but he also unveiled a plan to send armed forces to West Berlin. Tensions continued to mount into August as the East Germans began to build a wall surrounding West Berlin.[17] Against the backdrop of heightened international tension, Khrushchev decided to abandon the moratorium. On July 10, a month after the Vienna summit, he suddenly called a meeting in the Kremlin with a group of nuclear weapons scientists, ordering them to prepare for the resumption of nuclear testing in all environments including the atmosphere.[18]

If Khrushchev believed that national security should take precedence over any concern about radioactive fallout, not all Soviet nuclear weapons scientists agreed. Many of them had taken active part in an international campaign against radioactive fallout along with geneticists to help the Kremlin put pressure on the Western nuclear rivals for a test ban. Their active involvement in Moscow's Cold War propaganda, however, had irreversibly changed how some of them viewed the global radiological consequences of their own invention. In the process of writing a well-publicized paper calling for a test ban, Andrei Sakharov, the chief designer of Soviet thermonuclear weapons, developed an inner conviction against imposing risks of harm upon a vast number of innocent bystanders by radioactive fallout. During the July 10 meeting with Khrushchev, Sakharov asked the premier to reconsider the decision to resume nuclear tests, arguing that it would be detrimental to both arms control and public health.[19] His defiance infuriated Khrushchev. "Leave politics to us—we're the specialists," Sakharov recalled Khrushchev's words. "You make your bombs and test them, and we won't interfere with you; we'll help you. But remember, we have to conduct our policies from a position of strength. We don't advertise it, but that's how it is!"[20]

Having failed to dissuade the premier, Sakharov tried to minimize the biological impact of the resumed tests. When his team was ordered to test a 50-Mt thermonuclear bomb popularly known as Tsar Bomba—the world's most powerful nuclear weapon ever detonated in history—Sakharov decided to develop a "clean" version to reduce the amounts of strontium-90 and other harmful fission products. Still, he must have known that it would still produce carbon-14, a radionuclide which he had once estimated would ultimately kill sixty-six hundred people per megaton worldwide over the next few thousand years.[21] Sakharov's antifallout propaganda thus came full circle: having denounced the enthusiasm of Edward Teller and other U.S. scientists for the "clean bomb" as unethical, he ironically found himself in the same position as theirs.

Although Khrushchev rejected Sakharov's plea for refraining from nuclear testing, this did not mean that he was oblivious to fallout hazards. Indeed, Sergei Khrushchev later claimed that his father was genuinely worried about the risk of overexposure for a great many people living downwind of the landlocked Semipalatinsk Test Site. For this reason, the Soviet premier ordered that as many planned tests as possible be transferred to the Arctic archipelagos of Novaya Zemlya. In the meantime, the Ministry of Medium Machine-Building was also instructed to test some weapons underground.[22] Although its minister, Efim P. Slavskii, was rather skeptical about such a need, physicist Yulii B. Khariton and his team at the Arzamas-16 nuclear weapons laboratory led the crash effort to develop the techniques to contain fallout below the ground. Like Sakharov, Khariton also opposed the resumption of nuclear testing, unsuccessfully trying to persuade Leonid Brezhnev, a Politburo member who later replaced Khrushchev in 1964 as the general secretary of the Communist Party, to cancel the test series at the eleventh hour.[23] Just as Sakharov looked to the "clean bomb" to escape the dilemma between his antifallout conviction and national security, Khariton found a similar solution in the technique of underground testing. His vigorous work resulted in the first fully contained underground test (1.2 kt) conducted in Semipalatinsk on October 11, 1961.[24] This breakthrough provided a technological basis that eventually made the partial test ban ultimately acceptable to the Soviet Union. For now, however, the Soviet nuclear weapons scientists had no choice but to rely on "technological fixes" to minimize the fallout, both literally and figuratively, from the Kremlin's decision to end the moratorium.

Fallout

On August 31, 1961, the Soviet news agency TASS broke the news on the resumption of Soviet nuclear tests. With this announcement, the Soviet Union and its Western rivals suddenly changed places in the politics of risk. Until then, the Soviets had successfully exploited the fallout argument to bring world opinion to bear on the Americans and British. The Western powers now turned the tables on the Soviets. On September 2, Rusk suggested to Kennedy that the U.S. and British governments should revive the joint offer of an atmospheric ban to prevent further pollution.[25] As Rusk explained to British ambassador David Ormsby-Gore, the proposal was more than propaganda. It was also designed to disrupt Soviet anti-ICBMs tests in the atmosphere that the U.S. government feared might alter the strategic nuclear balance.[26] Macmillan liked Rusk's idea, but he also added a vague call for peace to the draft statement in order to dilute the singular emphasis on the dangers of fallout to human health and thereby retain the option of atmospheric testing for themselves.[27] Despite this change in the wording, the Anglo-American public appeal, issued on September 3, successfully put the Kremlin on the spot. In reply, Khrushchev was forced to admit that radioactive contamination was "undesirable." Nonetheless, he insisted that the risk of harm from fallout did not match the threat to the security of the Soviet people from the catastrophic risk of nuclear warfare.[28]

If the Soviet government believed that it had the sovereign rights to pollute Earth for self-defense, other countries strongly disagreed. As it did after the discovery of unusually contaminated wheat and milk in the prairie provinces, Canada again led a voice of dissent against nuclear fallout in the United Nations. The increasing possibility of U.S. atmospheric testing in response to the Soviet action deepened Canada's dilemma as one of the closest U.S. allies and a northern high-latitude country believed to be particularly vulnerable to the influence of global fallout contamination. Moreover, the resumption of atmospheric tests came after the deployment of nuclear-capable Bomarc antiaircraft missiles in Ontario and Quebec became known in 1960, stirring a controversy over the acquisition of U.S.-supplied nuclear weapons. In this delicate political situation, Canadian foreign officials decided to focus on the problem of radioactive fallout as "a telling argument for the need to cease at least unregulated testing in the atmosphere."[29]

On October 13, Canada introduced a UN resolution calling upon the UNSCEAR to speed up its work while urging all nations and specialized agencies to share fallout data without delay. In explaining this initiative at the UN General Assembly's Special Political Committee, Canadian representative Paul Tremblay portrayed his country as a fallout victim, saying that "Canada was situated in the latitudes which seemed to have received some of the heaviest concentrations of radio-active fall-out."[30] Although the Canadian resolution was supposed to assist the U.S. publicity campaign against the Soviets, its singular emphasis on fallout annoyed some U.S. officials who were eager to keep the issue under control while preparing for the resumption of atmospheric testing.[31] Despite such misgivings, the Canadian resolution successfully rallied world opinion against the Soviet disregard for the global environmental consequences of their resumed tests. The General Assembly overwhelmingly passed the resolution on October 27 by eighty-seven to eleven, with only the Soviet bloc, Mongolia, and Cuba opposing.[32]

The Soviet government responded to the international uproar over radioactive contamination by proceeding with its atmospheric testing program with greater determination. At the time, the Berlin crisis had reached the most dangerous phase. With the Kremlin's approval, East Germany began in August to construct a wall dividing East and West Berlin. In response, Kennedy called up the National Guard to active duty, while U.S. officials in Berlin repeatedly tried to cross the border to reassert the rights of free passage.[33] In the midst of this war of nerves, the Soviets again leveraged nuclear tests as a psychological weapon to demonstrate their resolve. In his opening address delivered to the twenty-second Soviet Communist Party congress on October 17, Khrushchev announced a plan to end the current test series with the explosion of a 50-Mt bomb. Although U.S. and British officials quickly dismissed this world's largest nuclear test as a propaganda gimmick of no military value, a worldwide fallout scare nevertheless followed. Denmark, Canada, Iceland, Japan, Norway, and Sweden introduced an emergency UN resolution urging the Soviet Union to refrain from conducting the bomb test in the name of all countries located in the direct path of fallout.[34] The Western news media also universally condemned the planned Soviet test not only as a danger to peace but also as a crime against humanity. The *Washington Post* even ran an obituary notice on its editorial page for the "unnumbered hundreds of thousands" who would ultimately die because of the Soviet fallout.[35] Although the "Tsar Bomba" test, conducted on October 30 in Novaya Zemlya, turned out to be one of the "cleanest" nuclear weapons ever tested, it still generated an enormous amount of radioactive carbon, which simply deferred the potential impact of radioactive debris to the distant future.[36]

The aggressive Soviet atmospheric testing program made Kennedy's self-restraint increasingly untenable. Kennedy initially responded to the end of the moratorium by ordering the resumption of underground testing, hoping that this reserved countermeasure would give the Western powers the moral high ground while arresting a dangerous spiral of the nuclear arms race. As Soviet tests continued above ground, however, pressure mounted for Kennedy to respond in kind to show his resolve to confront the Soviets. A poll taken in winter of 1961 showed the U.S. public evenly divided over the question of whether the United States should test in the atmosphere, with 44 percent supporting and 45 percent opposing.[37] The U.S. military added to the pressure, warning that the continued unilateral moratorium on atmospheric testing would allow the Soviets to reduce and eventually overtake the U.S. nuclear technological lead. Underground testing was not only slow and costly but also incapable of testing weapons systems and also of studying weapons effects, including the electromagnetic pulse that a nuclear explosion in high altitudes could generate to knock out ballistic missiles and radar facilities.[38]

At this critical moment in the U.S. policymaking process, the norms of antipollution seemed to give way to the logic of national security. Although the British urged the Americans to refrain from testing nuclear weapons above ground, they did so out of concern about nuclear proliferation. In a letter to Kennedy after a summit meeting held in December 1961, Macmillan warned about the dangers of nuclear weapons in the hands of "dictators, reactionaries, revolutionaries, [and] madmen."[39] Kennedy's aides also debated the importance of atmospheric testing for national security, but all fell back on relative measures of risk to dismiss the health impact of the planned tests. "Except for the moral problem of inflicting damage on the environment of other nations," national security adviser McGeorge Bundy claimed in presidential briefing material, "the magnitude of the fall-out problem is smaller than that of building roads on which, statistically, many thousands of people will die."[40] Just as Khrushchev had spurned Sakharov's plea, Kennedy's advisers urged the president to set aside any concerns about fallout as a trifle that paled before the monumental national security risk.

Kennedy, however, apparently held to his antifallout conviction right up to the moment of decision. In a memo sent to the president prior to a National Security Council meeting on February 27, 1962, which gave final approval to the atmospheric testing plans, Rusk expressed his sympathy for Kennedy's loathing at "having even one individual affected by radioactive fallout." He nevertheless presented the question of atmospheric testing in a stark binary, arguing that the biological damage from U.S. tests would be "minimal as compared with the hazards which might be caused by misunderstandings about our nuclear strength."[41] Seeing no way out of the false dilemma arising from the Cold War mind-set, Kennedy chose atmospheric testing. On March 2, Kennedy issued a public statement announcing an atmospheric test series beginning in April. In explaining this decision, he echoed Rusk's binary. He declared that he would not dismiss the moral implication of risking "even one additional individual's health." "And however remote and infinitesimal those hazards may be," the president said, "I still exceedingly regret the necessity of balancing these hazards against the hazards to hundreds of millions of lives which would be created by any relative decline in our nuclear strength."[42]

Acceptable Risk?

It was one thing that Khrushchev, Kennedy, and Macmillan all deemed radioactive fallout to be an acceptable risk for national survival in the midst of the Cold War. It was another thing, however, whether citizens in their countries would actually accept such a risk. A key mechanism for the public acceptance of radiation exposure was the concept of permissible dose. Nuclear authorities had long insisted that fallout posed no undue danger to the public so long as its levels remained well below the maximum permissible dose recommended by standards organizations such as the National Committee on Radiation Protection (NCRP) and the International Commission on Radiological Protection (ICRP). What the maximum permissible dose actually denoted, however, was not an absolutely "safe" threshold but rather a limit of exposure that the experts considered as biologically and socially acceptable in exchange for the benefits of radiation use. While the maximum permissible dose for the general population was repeatedly revised downward, radioactive contamination steadily increased. As a result, the margin of safety appeared fast diminishing, an impression which drastically magnified the social impact of isolated yet unusually contaminated "hot spots" and "hot food" across the country. This created a particularly serious problem in the United States and Britain, where the governments could no longer guide the public interpretation of fallout hazards through well-controlled information management. An internal memo written by a U.S. congressional staff member for the Joint Committee on Atomic Energy in May 1959 sounded an alarm about an impending public relations disaster. Noting that confusion over the concept of permissible dose had already begun to affect nuclear policies, the memo stated, "[w]e are stuck with the problem, which is of our own making, and it is too late to go back and start over again."[43]

By the time the moratorium ended in August 1961, then, U.S. and British officials had taken steps to counter the idea of a "danger point" in radiation exposure. For this purpose, the U.S. Federal Radiation Council (FRC) and the British Medical Research Council (MRC) separately introduced a graded system of countermeasures in proportion to the magnitude of exposure. An FRC report published in September 1961 established three ranges of radioactive concentrations in food and drink regarding iodine-131 (I-131), strontium-89 (Sr-89), and strontium-90 (Sr-90) (Table 2). Range I called for no more than routine surveillance. Range II required "active surveillance and routine control," with its upper value corresponding to the annual dose limit for the thyroid in the case of I-131 and

TABLE 2. Guidance on Daily Intake (picoCi per day).

	RANGE I	RANGE II	RANGE III
Iodine-131	0 to 10	10 to 100	100 to 1,000
Strontium-89	0 to 200	200 to 2,000	2,000 to 20,000
Strontium-90	0 to 20	20 to 200	200 to 2,000

Note: The value of iodine-131 intake is for milk. Those of strontium-89 and of strontium-90 are for total diet.

Source: Adopted from Federal Radiation Council, *Report No. 2*, 4–5.

one-third of the limit for bone in the case of Sr-89 and Sr-90. If contamination levels rose to Range III, the FRC would recommend "appropriate positive control measures."[44] In Britain, the MRC adopted a similar but much simpler system, establishing an annual average value for I-131 (130 picoCi per liter in milk) and also for Sr-90 (130 picoCi per gram of calcium in the total diet) as a reference point for considering some remedial actions. The value for each element corresponded to the MRC's "warning dose" for the general population, or one one-hundredth of the maximum permissible dose.[45]

At first, it seemed that the graded system would be helpful in disabusing the public of the notion of a danger point. In reality, however, it complicated the problem in three major ways. First, it was impossible to predict the duration of radiation exposure due to radioactive fallout. The daily values shown in the FRC and MRC guides were derived from the annual limit based on the assumption that the relationship between dose and effect would be directly proportional regardless of dose rate at any given time. This meant that exposure in excess of the daily value would be permissible if it would last only for a brief period.[46] It was impossible, however, to determine whether no further contamination would occur because that question depended on the decisions made by policymakers to conduct more weapons tests. Second, the distribution of radioactive debris in the environment was highly uneven, creating numerous "hot spots" and "hot food" across the country.[47] Finally and most important, health officials were ordered to carefully compare risks and benefits before taking countermeasures, and yet it was impossible for them to balance the health impact of fallout and the national security needs in each exposure situation. Even if one considered the fallout from U.S. tests as "acceptable" for American citizens, there would be no benefit for them from Soviet tests. "Yet you're to weigh the benefits against the risks," geneticist Edward B. Lewis said in an advisory committee meeting of the U.S. Public Health Service (USPHS). "So it is a very bizarre position to be in to apply any kind of rules."[48]

Indeed, the resumption of Soviet atmospheric tests forced its Western nuclear rivals to confront the dilemma embedded in the new system of radiation protection. The British were the first who realized the looming trouble. Upon hearing the Soviet plan to detonate a 50-Mt bomb, the MRC became worried about the worst-case scenario that 830,000 infants under one year of age in Britain might consume I-131 in fresh milk up to the annual limit. In response, the Ministry of Health hastily drew up a contingency plan to switch infant diets from fresh to dried milk through its Welfare Foods Scheme.[49] During a special meeting held on the eve of the super-bomb test, however, a group of radiation protection experts advised health officials against preemptively implementing the milk substitution plan lest such an action cause a panic. Health physicist W. G. Marley of Harwell urged that "every attempt should be made to avoid impressions of criticality."[50] The Ministry of Agriculture, speaking for milk producers, also opposed the plan, arguing that any premature action might simply result in a loss of milk consumption without any gains in public health.[51] Instead of taking countermeasures to reduce radiation exposure, the British government staged a public relations campaign, with Macmillan drinking a glass of fresh milk in front of the cameras at the Royal Dairy Show held a few days prior to the Soviet test.[52] Ironically, it was the Soviets who ultimately saved British officials from a potential disaster by detonating a "clean bomb."

The U.S. government also found itself in a dilemma over Soviet fallout. The stakes were even higher for the Kennedy administration, since it had been contemplating the resumption of atmospheric tests. U.S. military and foreign officials thus strongly argued against establishing any clear-cut maximum permissible level to avoid a situation in which the fallout from Soviet tests might fill up the "quota" before the United States conducted its own tests.[53] Unlike their British counterparts, however, U.S. health officials faced a unique challenge posed by federalism, with the Surgeon General only authorized to advise state health agencies on the matter of radiation protection. On the eve of the Soviet super-bomb test, the USPHS hastily called an emergency meeting with state and territorial health officers. Insisting that hazards involved in countermeasures would almost certainly outweigh the estimated risk of harm from fallout, the service forcefully brought the participants to "understandings" that no action would be necessary to reduce radiation exposure due to the 50-Mt Soviet test.[54] This heavy-handed guidance by Washington seemed to work, as no state or municipal government took action when radioactive debris from Novaya Zemlya reached the United States.

Once the United States resumed atmospheric testing, however, the federal government quickly lost control over state authorities. When the Sedan shot, a 100-kt nuclear device, was detonated at the Nevada Test Site on July 6, 1962, its radioactive cloud passed over Utah, depositing much radioactivity on the ground. Alerted to this fact by radiobiologist Robert C. Pendleton of the University of Utah, state health officials immediately checked fresh milk. The survey showed that I-131 in some of the samples could expose fifty-three thousand children to radiation at about 10 milligray (milliGy) to the thyroid on average and up to 140 milliGy at maximum, a range well above the annual limit of the thyroid dose.[55] The FRC, however, decided to "do nothing," claiming that the risk of thyroid cancer from the estimated radioiodine intake was so small that "countermeasures would possibly have a net adverse effect rather than a net benefit."[56] Torn between Washington and the anxious Utahns, the Utah government asked milk producers in the state to "volunteer" the transfer of cattle from pasture to stored hay and hold fresh milk from sales until much of the I-131 decayed.[57] By the time countermeasures took place, however, more than 80 percent of the potential iodine intake had already taken place.[58] Upset by this failure, a Utah state health advisory board passed a unanimous resolution in May 1963 demanding that the federal government implement preventive actions in a timely manner so as to hold the levels of exposure below the annual limit.[59]

As the graded system of radiation protection failed to reassure the general public, the FRC retreated from its earlier promise of timely countermeasures, arguing that exposures many times the existing reference values for mitigating action would not result in any detectable increase in disease. During a congressional hearing held in June 1963, Paul Tompkins of the FRC reasserted the right to determine a permissible level of radiation exposure on a case-by-case basis instead of establishing "legal type maximum permissible doses."⁶⁰ Such a claim made by the government after it had repeatedly presented such guidelines as the objective definition of an acceptable risk, however, only deepened a sense of mistrust among concerned citizens. Richard Starners, a syndicated columnist, called the FRC's decision "madness that George Orwell in his wildest genius never imagined." "Like the limit on the public debt," he said, "the limit on fallout will be progressively raised until—when we take off our clothes—we all begin to glow in the dark."⁶¹

If U.S. officials still believed that the risk of harm from fallout was acceptable to the whole nation, a numerically small yet vocal segment of the population strongly disagreed. During the test moratorium, a combination of nuclear pacifism and consumerism had inspired grassroots campaigns in the United States against radioactive dust. When fresh fallout reappeared in the air, an increasing number of women began to raise their voices in dissent. On November 1, 1961, over fifty thousand women across the United States went on a "strike" for ending the testing of nuclear weapons, leaving homes and offices in protest. This remarkable show of antinuclear sentiments was a result of grassroots canvassing initiated by Dagmar Wilson, a Washington-based children's book illustrator and member of a disarmament campaign called the National Committee for a Sane Nuclear Policy (SANE).⁶² Building on this astonishing success, Wilson and her supporters organized a campaign called Women Strike for Peace (WSP).

What set WSP apart from the earlier antinuclear groups was that the activists identified themselves as housewives and mothers. Such an image belied the actual background of the WSP members, who tended to be white, middle-class, and more educated and politically active than most women of the time.⁶³ The conservative gender norm for women, however, uniquely empowered these dissidents. As historian Elaine Tyler May has shown, the national security state sought to mobilize women to fight the Cold War but only in such a manner as to reaffirm and reinforce their homebound role.⁶⁴ WSP put this traditional gender norm on its head, asserting their right to speak up for peace in the public sphere as necessary to perform their domestic duty. As was the case in Japan after the *Lucky Dragon* incident, the contamination of food and drink with radioactive fallout provided a powerful conduit to connect traditional motherhood with peace activism. Some of the most widely used slogans in WSP's first nationwide strike were "Pure Milk Not Poison" and "Let the Children Grow."⁶⁵

WSP's strong emphasis on fallout, however, was much more than a campaign tactic to mask its political message. It reflected genuine concerns among women rooted in the cultural discourse of motherhood in Cold War America. As historians Molly Ladd-Taylor and Lauri Umansky have observed, one of the popular discourses of a "bad" mother in the modern United States was one who failed to protect her children from harm.[66] It was thus no surprise that many women found it intolerable to allow even the smallest risk of injury from fallout for their children. This, however, did not mean that these concerned mothers acted irrationally. On the contrary, they embraced the model of "scientific motherhood," actively seeking expert opinions from a variety of sources and drawing their own informed conclusions.[67] Indeed, shortly after the first mass demonstration, WSP created their Committee on Radiation Problems not only to educate themselves on the problem of fallout but also to monitor contamination on their own. For this purpose, the campaign urged its members to mail the baby teeth of their children to the Baby Tooth Survey conducted by the Greater St. Louis Citizens's Committee for Nuclear Information. Each lab report was then forwarded to a U.S. senator from the child's state as a protest against contamination in the most intimate and compelling way.[68]

A combination of traditional motherhood and scientific information also led to a reemphasis on the problem of fallout in British antinuclear activism. While the largest disarmament advocacy group, the Campaign for Nuclear Disarmament (CND), concentrated its efforts on pressing the British government for the unilateral liquidation of nuclear weapons, some women scientists organized a fallout information clearinghouse called the Women's Association for Radiation Information (WARI). Although many of its members were active in the CND and other peace groups, the WARI disavowed connection with any of them, stressing its mission as "an experiment in communication" to bridge the gap between independent experts and concerned citizens. Its sponsors included not only members of the atomic scientists' movement such as crystallographer Kathleen Lonsdale and physicist Joseph Rotblat but also notable women scientists such as Hilda Lloyd, a surgeon who became the first woman elected as president of the Royal College of Obstetricians and Gynaecologists.[69] As WSP did in the United States, the WARI lobbied the government for a prompt release of fallout data to help concerned women in judging the timing of taking countermeasures for themselves.[70]

As more women on both sides of the Atlantic refused to accept the risk of harm from fallout, their dissent began to transform the broader nuclear disarmament movement. The story of SANE was a case in point. Established by U.S. journalist Norman Cousins in 1957, SANE initially drew its inspirations from Albert Schweitzer. The renowned mission doctor and humanist had publicly denounced nuclear tests as immoral for harming a countless number of innocent bystanders around the world. SANE's emphasis on the ethics of a test ban quickly turned the group into the largest antinuclear campaign in the United States, boasting about 130 branches and twenty-five thousand members in summer of 1958.[71] After the three nuclear powers temporarily suspended all tests and entered negotiation for a test-ban treaty, SANE began to broaden the scope of activities. Around the same time, it also became subject to red-baiting, plunging into a crisis over the handling of its members suspected as being communists.[72]

The reappearance of radioactive fallout in the world's atmosphere following the sudden end of the moratorium, however, helped SANE refocus on a test ban. A key to their resurgence as a leading nuclear disarmament movement in the United States was a new emphasis on the dangers of contamination to children. In January 1962, Homer Jack, the group's executive director, invited pediatrician Benjamin Spock to join SANE. Author of a bestselling childcare manual that advocated a child-centered, commonsense approach, "Dr. Spock" was a household name for countless parents struggling in childrearing. Initially, Spock declined SANE's request, convinced that the existing fallout levels were inconsequential to the health of children. He soon realized the self-reinforcing logic of the nuclear arms race, with the pursuit of superiority breeding an even more acute sense of insecurity. Convinced that nuclear testing would never stop unless people rose up and demanded it, Spock decided to join SANE as one of its sponsors.[73] In April 1962, SANE's full-page *New York Times* ad showed him looking over a little girl, with a bold headline, "Dr. Spock Is Worried." "Not so much about the effect of past tests but at the prospect of endless future ones," Spock stated in the ad. "As the tests multiply, so will the damage to children—here and around the world. Who gives us this right?" (Figure 5). Reprinted in more than eighty newspapers in the United States and abroad, the ad was a powerful statement against the notion of acceptable risk by projecting the impact of fallout into the future—both in the alarming trend of contamination and in the haunting image of innocent children.

FIGURE 5. "Dr. Spock Is Worried."
Source: SANE, Inc. Records, Swarthmore College Peace Collection, Swarthmore, PA. Reprinted with permission.

The maternal turn in the politics of risk also created another symbolism of nuclear peril: contaminated milk. U.S. and British officials had consistently refused to take action to reduce or eliminate radioactive contents in this favorite drink for children, arguing that the chance of harm from contaminated milk was extremely small compared to the risks involved in almost any countermeasure—whether it was the technical means of decontamination or the political solution of a test ban. By describing contaminated milk as a "poison," antinuclear activists firmly rejected the relative measures of risk, seeking to bring the public's attention to the presence of toxic substances, no matter how small in quantity, in the drink for children, who deserved the maximum degree of protection from harm. In Britain, the Committee of 100, a group of militant pacifists led by philosopher Bertrand Russell, launched a guerilla campaign to label hundreds of milk bottles "DANGER-RADIOACTIVE" and place them in front of the Soviet Embassy in London.[74] CND women distributed a leaflet, "Shopping in a Nuclear Age," listing various less-contaminated food items such as white bread, breast milk, and fresh produce from Australia and New Zealand.[75] In the United States, SANE took out another full-page newspaper ad featuring a bottle of milk with the mark of skull and crossbones.[76] It was WSP, however, which staged the most impressive action against contaminated milk. In April 1962, on the eve of the first U.S. atmospheric test after the moratorium, the group called a nationwide boycott of fresh milk. "The economic weapon is our greatest power," WSP stated in one leaflet, "and we should not hesitate to use it when the health and lives of our children are at stake."[77] The effect was sensational. According to historian Amy Swerdlow, WSP's slogan, "Pure Milk Not Poison" became "their most effective peace slogan" and "did much to swell the ranks of WSP in 1962 and 1963."[78]

The cult of traditional motherhood in the early Cold War era completed the development of a robust counterpoint in the United States and Britain against the notion of acceptable risk. It did so by seamlessly combining the scientific information on fallout and its disproportionate impact on children and moral absolutism against harming the innocent next generation. Although this alternative interpretation of fallout hazards did not single-handedly replace the official version, it laid the important groundwork for the overwhelming popular support of an atmospheric test ban when policymakers in the United States, Britain, and the Soviet Union decided to revisit that option in their arms control negotiations.

Toward the Partial Test Ban

The initiative to reintroduce the idea of an atmospheric ban first came from non-nuclear, neutral countries serving on the Eighteen-Nation Committee on Disarmament (ENCD). Established by the UN General Assembly in December 1961, the ENCD added eight neutral countries to the Ten-Nation Committee created a year earlier consisting of five NATO nations and five Warsaw Pact nations. The neutral countries viewed the prohibition of nuclear tests above ground, which was verifiable by national monitoring methods with high confidence, as a means to break an impasse in the tripartite test-ban negotiations over a comprehensive ban. On May 9, 1962, shortly after the United States resumed atmospheric testing, Mexico proposed a cut-off date of January 1, 1963, for all tests conducted in the open environment. Brazil also made a similar motion on July 25, calling for an immediate ban on test explosions in the atmosphere to be followed by further negotiations on the control of underground testing. Sweden merged both ideas into one package, pressing the nuclear powers to suspend all tests that would require no international inspection.[79]

The initiative from the ENCD rekindled Washington's interest in an atmospheric ban. Rusk told Kennedy that such an agreement would allow the United States to score a propaganda point and also to bypass the knotty issue of safeguards.[80] Kennedy approved the idea to prepare a draft treaty for a partial ban, but he decided to combine it with a comprehensive version. In a letter to Macmillan, he explained the reason for this dual offer. "In terms of world opinion, atmospheric fall-out may be more important than the arms race itself," he wrote. "On the other hand, in terms of the great problem of nuclear proliferation, a comprehensive treaty still seems better to me."[81] With Macmillan's endorsement, the U.S. and British representatives in Geneva introduced two draft treaties on August 27. Notably, a draft of the partial test-ban agreement presented at that time made no reference to radioactive fallout, merely declaring that the treaty would facilitate further negotiations to end nuclear testing in all environments.[82] This wording clearly reflected the priority that Kennedy and Macmillan gave to a comprehensive ban and its arms control effects. Neither of the two drafts, however, seemed acceptable to Khrushchev. The Soviet premier flatly rejected a comprehensive ban by reiterating his objection to any on-site inspections by international inspectors within Soviet borders. He gestured toward agreeing on a partial option, but only on condition that all countries voluntarily refrain from underground testing. Given the Soviet breach of the moratorium in the past, the Western powers were in no mood to suspend tests voluntarily once again.[83]

Khrushchev's apparent indifference to a test ban, however, belied his continued interest in reaching agreement with the West on this important subject. In fact, part of the reason for his negative reactions might have been a matter of timing. When the Western powers tabled their draft treaties in August, the Soviet Union had just started a new test series early that month. Seeing that the United States had been vigorously conducting atmospheric tests since April to widen its technological lead, Khrushchev had no intention of allowing any Western peace initiative to interrupt his country's pursuit of nuclear strength. He told U.S. ambassador Llewellyn E. Thompson that both the United States and the USSR would become ready to conclude a test ban following the completion of their respective nuclear test series then in progress.[84] He did not completely ignore fallout hazards, either. His son Sergei later claimed that the premier spurned the request made by Slavskii for exploding a 100-Mt warhead, fearing that it would provoke another international uproar in Scandinavia and elsewhere as the super-bomb test had done a year earlier.[85] Despite this intervention, the 1962 Soviet test series, which lasted until Christmas Day, turned out to be the dirtiest of all. Its estimated fission yield of over 50 Mt far exceeded the highest amount recorded by any nuclear country until then.[86]

While the Soviet political leaders continued to prioritize the logic of national security over health concerns, Sakharov maintained his moral conviction against fallout, struggling to keep its amounts to the absolute minimum. Upon learning the plan to test two almost identical devices designed by different laboratories, Sakharov directly contacted Slavskii and Khrushchev among others, urging them to cancel or at least postpone one of the planned tests. He later recalled having told the atomic minister that up to a six-figure number of people "are going to die for no reason."[87] His intervention nevertheless failed. When the duplicate tests took place, Sakharov put his face down on his desk and wept. This experience, he recalled, made him all the more determined to work toward ending biologically harmful tests.[88]

Sakharov was not alone in such a conviction. In summer of 1962, with the Soviet test series still in progress, Viktor B. Adamskii, a young theoretical physicist at Arzamas-16 working under Sakharov, handed his boss a draft letter to Khrushchev. The message included a proposal for a partial ban. Sakharov later remembered that, like him, Adamskii harbored a strong concern regarding the harmful effects of fallout. In the letter, he explained the merits of the partial ban in eliminating fallout and slowing nuclear proliferation without abandoning the option of continued tests underground. Sakharov informed Slavskii of Adamskii's idea, and the atomic minister forwarded it to the foreign ministry.[89] It was not the first time Sakharov confronted the Kremlin: he had failed to dissuade Khrushchev from ordering the resumption of atmospheric testing. This time, however, the Soviet political leadership proved unusually welcoming. A few months after Sakharov handed the letter to Slavskii, the minister called back, telling Sakharov that the Kremlin was interested and was about to take action in line with his proposal.[90]

If Sakharov's recollections are correct, the timing of Slavskii's phone call must have occurred in fall of 1962 or a little later. Two international crises happening that fall had drastically changed the context of a test ban for U.S., British, and Soviet political leaders. The Cuban Missile Crisis in October pushed the whole world to the brink of a nuclear holocaust. Around the same time, the Chinese went to war against India over their disputed borders in Himalaya, dramatizing the danger of Beijing's nuclear ambition.[91] Combined, these geopolitical events rekindled the interests of Washington, London, and Moscow in a test ban as a first step to promote nuclear disarmament, reduce international tensions, and check nuclear proliferation.

Shortly after the Cuban Missile Crisis, Khrushchev told British ambassador Frank Roberts that he was ready to sign a partial test-ban agreement. He added, however, that he still preferred stopping underground tests as a better brake upon the arms race.[92] To give the comprehensive ban another chance, Khrushchev persuaded the Presidium to permit two to three annual on-site inspections within Soviet territory by members of an international control organization. In a letter to Kennedy dated December 19, he formally made this offer, asking the U.S. president to accept it as a basis for a comprehensive ban.[93] Khrushchev's bold initiative, however, was based on a serious misunderstanding. While U.S. officials informally signaled their willingness to meet halfway on a smaller number than its original demand for twelve inspection visits to the USSR, their Soviet contacts seemed to believe that the United States would need no more than a token number as a purely political gesture to persuade the reluctant Senate.[94] This miscommunication publicly embarrassed Khrushchev at the time when he faced a strong pushback against his initiative for détente from some Presidium members as well as from China.[95] Once it was evident that his overture had failed to move the Americans, the Soviet premier vowed not to make any more concessions.

If the Cuban Missile Crisis spurred the U.S., British, and Soviet political leaders to find an elusive compromise essential for a comprehensive ban, it had the opposite effect on the U.S. Senate. For many fiercely anticommunist senators, Khrushchev's secret attempt at installing nuclear missiles in Cuba seemed to vindicate their fear that the Soviets would cheat on a test ban unless there was a robust international verification regime. As soon as Kennedy's decision to lower the demand for an annual on-site inspection quota from twelve to seven leaked to the press, Democratic senators Henry M. Jackson, Stuart Symington, and Richard Russell Jr., all ranking members on the Armed Services Committee, issued a strong warning to the White House against any compromise. Senator Thomas J. Dodd, another anticommunist Democrat and member of the Senate Foreign Relations Committee, also joined the chorus, sharply criticizing Kennedy's test-ban policy both from the Senate floor and in open letters.[96] Although the Democrats managed to keep control of the Senate by a comfortable margin of sixty-six to thirty-four after the 1962 midterm elections, Kennedy could ill afford to lose these and other powerful senators capable of blocking the two-thirds majority necessary for the ratification of a test-ban treaty.[97]

When Kennedy and Macmillan sent a joint letter to Khrushchev on April 15 with the request to receive their special envoys on the matter of on-site inspection, the Kremlin knew the U.S. president's deep trouble with the Senate very well. Three days earlier, Norman Cousins had met Khrushchev in his capacity as the Vatican's emissary, informing him that his offer of three inspections would not help Kennedy in Congress.[98] Shortly after this conversation, the Soviet premier made a curious move. In his note to the Presidium, Khrushchev recounted his meeting with Cousins, arguing that he no longer considered a test ban an important issue in arms control. He then went on to downplay the need for continued atmospheric testing, citing defense officials as claiming that it was unnecessary for the Soviet military. With this rhetoric, Khrushchev successfully lowered the political hurdle for a partial test ban.[99] He did so despite a persistent fear that the partial ban might simply take the arms race underground in a much more costly manner.[100] A session of the Presidium, held on April

25, duly endorsed Khrushchev's proposal of exploring a partial option in the test-ban talks.[101]

With the domestic political groundwork for a policy shift completed, Khrushchev finally returned to the theme of radioactive fallout, skillfully drawing on the emerging norms of antipollution. In a letter to Kennedy and Macmillan dated May 8, Khrushchev agreed to receive their personal emissaries. He called the cessation of fallout contamination a problem of "an incontestable significance from the moral-humane point of view." "As we see it," he wrote, "this alone is a sufficient incentive motive to agree on a test ban," aside from the test ban's positive effect of relaxing Cold War tensions.[102] Kennedy immediately understood its import. In a speech delivered at American University in Washington on June 10, he declared that the United States would not resume atmospheric testing as long as the Soviet Union refrained alike.[103] As Rusk observed, this unilateral declaration of a moratorium also aimed to rally Dodd and other skeptical senators behind the White House at this crucial moment for the test ban.[104]

For London and Washington, however, the prize was still a comprehensive ban. Kennedy's envoy W. Averell Harriman was instructed to pursue a compromise in Moscow as close to a total ban as possible.[105] Khrushchev, however, was determined to seek a partial ban only. In a July 2 speech in East Berlin, the Soviet premier for the first time publicly declared his readiness to sign a partial ban treaty. Lord Hailsham, Macmillan's science minister, who accompanied Harriman to Moscow, immediately realized the import of Khrushchev's speech. It essentially "slammed the door on a total ban," he said, "and opened the door to a partial ban."[106] Indeed, when Harriman and Hailsham finally met Khrushchev in Moscow on July 15, the Soviet premier reiterated his decision to withdraw his earlier offer of accepting a token number of on-site inspections on Soviet soil.[107]

Having eliminated any prospect for a comprehensive ban, Khrushchev then took an unusual initiative. Unlike the draft of a partial test ban agreement proposed by the United States and Britain in August 1962, the Soviet text made an explicit reference to radioactive fallout, declaring that one of the treaty's objectives was "to put an end to the contamination of man's environment by radioactive substances."[108] It is conceivable that this additional rationale for an atmospheric ban closely reflected one of Khrushchev's political aims, namely to boost Moscow's image in the world. A closer look at the agreed text, however, showed that the essence of the treaty was still linked to the principle of arms control verification. The agreement covered not only the atmosphere and under water but also outer space, where contamination was unlikely to affect Earth. It also prohibited nuclear tests in any other environment "if such explosion causes radioactive debris to be present outside the territorial limits of the State under whose jurisdiction or control such explosion is conducted."[109] This was a nod to the possibility of accidental venting from an underground nuclear explosion for both military and peaceful purposes that might result in radioactive contamination around the proving ground. All of these qualifications indicated that the three nuclear powers, which signed the Partial Test Ban Treaty on August 5, still considered the agreement primarily from the point of view of national security rather than environmental health.

Nonetheless, the last-minute addition of antipollution language to its preamble helped the U.S. political leaders sell the PTBT to wavering lawmakers. In his request to the Senate for ratification, Kennedy stressed the treaty's contribution to "freeing the world from the fears and dangers of radioactive fallout." He further linked this benefit of pollution control with nuclear disarmament, arguing that the continuation of an unrestricted nuclear arms race, soon to be joined by many more countries, would exacerbate radioactive contamination far more than the existing levels. Even then, he argued, the effects of fallout on humans would be likely to remain small compared to those from natural background radiation. "But," he said, "this is not a natural health hazard—and it is not a statistical issue." Breaking from the notion of acceptable risk that had long shaped the outlook of U.S. officials, the president recast fallout as morally unacceptable to the world's present and future generations:

> The loss of even one human life, or the malformation of even one baby—who may be born long after we are gone—should be of concern to us all. Our children and grandchildren are not merely statistics toward which we can be indifferent. Nor does this affect the nuclear powers alone. These tests befoul the air of all men and all nations, the committed and the uncommitted alike, without their knowledge and without their consent. That is why the continuation of atmospheric testing causes so many countries to regard all nuclear powers as equally evil; and we can hope that its prevention will enable those countries to see the world more clearly, while enabling all the world to breathe more easily.[110]

One might consider Kennedy's eloquent account of the global-fallout problem as politically expedient. As noted earlier, however, he had voiced strong misgivings about the ethics of harming innocent bystanders on the eve of his decision to resume atmospheric testing, even when his policy advisers were all united in dismissal of such concerns as trifling compared to the allegedly clear and present danger of lagging behind the Soviets in the nuclear arms race. Although he ultimately failed to overcome the trap of the zero-sum logic that characterized the Cold War, it was likely that Kennedy spoke to the Senate as much from his own inner conviction as from political necessity.

Once Congress began its deliberations on the PTBT, however, the White House toned down its antipollution rhetoric. When Glenn Seaborg briefed the Senate Foreign Relations Committee on the treaty's merits, Senator Russell did not hide his "surprise" that the chairman of the U.S. Atomic Energy Commission "did not refer to any advantages to be gained from lessened fallout" in view of the importance of the "mother vote."[111] The Kennedy administration's conspicuous silence on the problem of fallout, however, seemed to be a calculated political move to avoid opening the Pandora's Box of uncertainty in knowledge about the biological effects of low-dose radiation. Indeed, test-ban opponents all cited Edward Teller and others who insisted that a small dose of radiation was inconsequential or even beneficial to health.[112]

Instead of engaging in the endless debate over the nature of scientific proof on the Senate floor, the White House chose to let its grassroots partners operating outside Congress underline the moral imperative to stop pollution. The Citizens' Committee for a Nuclear Test Ban, organized by Cousins and other antinuclear activists, solicited favorable messages from doctors, biologists, and other life scientists in support of the treaty.[113] Spock lent his public fame to the test-ban cause, appearing on TV and issuing a public statement to "get at the so-called mothers' vote."[114] Public opinion was generally favorable for ratification but far from unanimous. Some regarded anticommunism as a moral issue. A man from Pennsylvania angrily wrote to Harriman, "Fallout, Yes, Sellout, No!"[115] Nor did all women agree that mothers should naturally support the treaty. In a letter sent to the White House, a woman from California wrote as one of "some stout-hearted wives with 'patient wisdom' who have the courage to declare that this treaty solves very little."[116] These voices of dissent, however, failed to swing the Senate, as the paramount concern among the legislators was not fallout but rather safeguards. Although Teller and his allies still claimed that the Soviets were certain to cheat the ban by all means possible, the Senate accepted the claim by the White House that it would be nearly impossible to test nuclear weapons above ground anywhere in the world without being detected by the United States. With this reassurance, the Senate overwhelmingly approved the PTBT on September 24, 1963, by a vote of eighty to nineteen.

Conclusion

Contrary to its declared objective as a step toward disarmament, it became immediately clear that the PTBT did little to check the nuclear arms race. Under strong pressure from the U.S. Congress, Lyndon B. Johnson, who succeeded Kennedy as president after the latter's assassination in November 1963, was forced to make an announcement in April 1964 that the United States would conduct "comprehensive, aggressive, and continuing underground nuclear test programs."[117] The rate of U.S. testing greatly accelerated thereafter. The United States carried out as many as 338 nuclear tests underground from 1964 to 1971, a number nearly equal to all the tests that it had conducted through 1963. The Soviet Union reciprocated with 133 subsurface test explosions during the same period.[118] Neither did the PTBT reverse the tide of nuclear proliferation. France refused to sign the treaty, and China conducted its first atomic bomb test in October 1964 as the newest member of the nuclear club. As a diplomatic follow-up for a comprehensive ban quickly lost its momentum, the nuclear powers pursued a nonproliferation treaty instead to perpetuate what historian Shane J. Maddock has aptly called "nuclear apartheid."[119]

As the nuclear arms race continued unabated, Johnson found it difficult to name any positive consequence of the PTBT when he delivered a message to commemorate the first anniversary of the treaty. The only concrete outcome was the significant reduction of fallout emissions in the world. "We can live in strength without adding to the hazards of life on this planet," Johnson stated in the address. "We need not relax our guard in order to avoid unnecessary risks. This is the legacy of the nuclear test ban treaty and it is a legacy of hope."[120] Johnson's backdated rebranding of the PTBT as a global environmental agreement pointed to an ultimate irony: the three Cold War rivals, having failed to overcome mutual distrust and suspicion, simply managed to make the nuclear arms race environmentally "sustainable." As Edward R. Murrow, a famous U.S. broadcast journalist and Kennedy's information adviser, put it, the PTBT represented "a partial recognition of the realities of the world in which we live."[121]

Conclusion

"WE CAN LIVE IN STRENGTH WITHOUT ADDING TO THE HAZARDS OF LIFE ON THIS PLANET"

THE PARTIAL TEST BAN TREATY (PTBT) concluded in 1963 is typically viewed as a first step toward nuclear arms control and the easing of hostility between two superpowers. The most important aspects of the PTBT, however, came as a result of humans' accelerating impact on Earth. Atmospheric nuclear testing directly brought the Cold War and the Anthropocene, two parallel historical processes in the latter half of the twentieth century, into synchronization. The relationship between the two, however, was fundamentally asymmetrical. While the global dispersion of the bombs' debris raised radiation levels worldwide by a small degree, the spatial and temporal scale of radioactive contamination systematically obscured its public health and environmental effects from view. For the slightly more radioactive planet to affect the course of the Cold War leading to the PTBT, then, it had to undergo a worldwide struggle to determine its biological effects, social acceptability, and policy implications. This politics of risk transformed the meaning of radioactive fallout from a harmless side effect to an unacceptable hazard that demanded international regulation.

At the beginning of the nuclear age, the U.S., British, and Soviet governments monitored the radioactive fallout of their nuclear tests and judged the potential risks for themselves. The technopolitical notion of "emergency" guided this self-regulation of fallout hazards. Keenly aware of the possibility of a severe accident caused by a nuclear explosion, but equally determined to minimize disruptions in their atomic projects, the Cold War rivals similarly checked radioactivity only in the vicinity of the test sites and took little or no protective action until exposure was projected to reach severe and injurious levels. The exigencies of the Cold War also led them to secretly track the global dispersion of radioactive debris as a useful tracer for nuclear intelligence and war planning. The compartmentalization of knowledge production further concealed from both government officials and the general public the widespread and long-term health effects of global fallout contamination.

The power of the nuclear security states to define the radiological risks of weapons testing, however, was far from absolute. The coverage of fallout monitoring had many gaps and holes, and the exposure limits in use excluded consideration of genetic damage that might have no absolutely safe threshold and could be quite large if a great many individuals were exposed to radiation. These epistemic vulnerabilities embedded in the safety claims of the nuclear powers led to two concurrent openings for reassessing risk following the Castle Bravo hydrogen-bomb test in 1954. Reassured by U.S. officials, the Japanese government checked radioactivity in tuna from the Pacific against the most stringent standard, only to discover more "atomic-bomb tuna." The subsequent consumer panic sparked the rise of antinuclear activism led by concerned housewives. Meanwhile, Washington's categorical denial of any harm whatsoever from fallout provoked an uproar among geneticists in the United States. The ensuing controversy crossed the Iron Curtain and reached the Soviet Union, where Nikolai P. Dubinin, a leading geneticist purged by his rival Trofim Lysenko, attempted to convince the Soviet scientific leadership that genetic knowledge could assist the Kremlin's test-ban propaganda.

In each case, dissenters faced the asymmetric relations of definitional power in favor of government officials and their scientific allies. Washington and Tokyo managed to introduce the concept of permissible dose and called off tuna inspections despite strong opposition from many Japanese scientists. U.S. geneticists had no institutional mechanism through which to forge a united front against the government's disregard for genetic damage and its social implications. Soviet geneticists also found their arguments effectively sidelined by the Lysenkoites who controlled key scientific bodies. Nonetheless, the ensuing stalemate in the fallout controversy caused the first crack in the relations of definition. The atomic scientists' movement in the United States played a key role in this development. Instead of pushing its political and moral arguments against nuclear weapons, the Federation of American Scientists astutely highlighted the uncertainty of knowledge about fallout and called for an independent safety review under the auspice of the United Nations as a step toward the international regulation of nuclear testing. To keep the issue of arms control apart from that of fallout, U.S. and British officials made successful maneuvers, commissioning studies from the politically reliable National Academy of Sciences (NAS) and the Medical Research Council (MRC) as well as the politically constrained United Nations Scientific Committee on the Effects of Atomic Radiation (UNSCEAR).

Contrary to what Washington and London had expected, however, the review of fallout hazards conducted by these expert bodies from 1956 to 1958 further undermined the scientific and ethical basis of their safety claim. The appointed experts were not faithful agents of state power but rather actors in their own right, seeking to advance their professional agendas within the constraints of their political mandate. Some experts reached a conclusion similar to that of their sponsors. With much hesitation, the geneticists in the NAS and MRC panels accepted the notion of permissible dose in an attempt to offer a practical guide for radiation protection. Others, however, took a more cautious view. The NAS and MRC medical researchers alike also applied the same principle of permissible dose to the cancer risks of strontium-90, but the British members unilaterally lowered the acceptable level of contamination to keep its potential impact insignificant. Some UNSCEAR delegations went even further, seeking to connect the issue of fallout to that of a nuclear-test ban. The Japanese scientists adroitly used the concept of permissible dose and radiological data to downplay the immediate dangers of fallout while presenting Asian people as exceptionally vulnerable to the long-term effects of global radioactive contamination caused by the Western nuclear powers. Meanwhile, a coalition of Soviet nuclear physicists and radiation geneticists came to the assistance of the Kremlin's test-ban propaganda to combat the nuclear threat abroad and Lysenkoism at home. All these conflicts within the scientific institutions were linked to a broader dispute over the notion of acceptable risk, as the transnational movement involving intellectuals from both sides of the Iron Curtain recast global radioactive contamination as a humanitarian crisis that put countless innocent bystanders around the world in harm's way. Ultimately, a combination of these scientific and moral arguments pushed the UNSCEAR toward a position that radioactive fallout was an unacceptable hazard due to its deeply uncertain, uncontrollable, and indiscriminate effects on the world population.

While U.S. and British officials found their definition of risk soundly rejected at the international level by the time the nuclear-test moratorium began in fall of 1958, what decisively undermined the credibility of their safety claim was the troubling outcomes of grassroots fallout surveys in North America and the British Isles reported during the moratorium. This local turn in the politics of risk was crucial because the impacts of global radioactive contamination varied widely across different regions due to the complex interplay of natural processes and human activities. Some of the major radioactive hot spots in the United States and Britain happened to overlap with pockets of political dissent in the era of the Cold War consensus, most notably Minnesota and Wales. Unlike those who brought political and ethical arguments against nuclear testing to the forefront, concerned scientists and citizens conducted independent radiological checks to local food and drink in an attempt to break government control of fallout information. The results of these community-based fallout surveys had a particularly strong impact in the United States, where the findings of heavily contaminated foodstuffs in some locales forced nuclear officials to admit notable gaps in their fallout monitoring program and to disclose all radiological data without interpretation or judgment. U.S. and British officials tried to regain the public's confidence by promising protective action in a timely manner to minimize radiation exposure. When the resumption of atmospheric testing in 1961 pushed fallout levels dangerously close to the new benchmarks set for countermeasures, however, Washington and London refused to take action for fear of a panic, thus deepening the public's distrust in their safety assurances.

The transformed understanding of radioactive fallout as an unacceptable hazard, however, did not prescribe any specific solution to the identified problem. The definitional struggles over global radioactive contamination changed the course of the test-ban debate in a number of subtle yet important ways. Some policymakers actively exploited the potential dangers of fallout to advance their Cold War agenda. In pursuit of a test ban as part of a peace offensive, Soviet leader Nikita Khrushchev encouraged scientists in his country to publicly denounce fallout in an attempt to bring world opinion to bear on their Western rivals. His opportunism was on full display when Khrushchev suddenly reversed his position and ordered the resumption of atmospheric testing in 1961. Meanwhile, those who did not necessarily agree with the idea of fallout as an unacceptable hazard nevertheless viewed it as a political fact to be reckoned with. British Prime Minister Anthony Eden declared his support for a test limitation after the MRC released its report in 1956 to blunt pressure for a total ban. Likewise, U.S. President Dwight Eisenhower and his secretary of state John Foster Dulles fully recognized the significant impact of the 1958 UNSCEAR report on world opinion, so much so that the U.S.-British joint announcement on the nuclear-test moratorium specifically omitted any reference to the problem of fallout lest it might make it impossible to resume tests in case of a failure in the subsequent test-ban talks.

While many policymakers continued to prioritize national security and treated any gesture toward growing public concern about fallout as politically expedient, the struggles over the definition of risk in the nuclear Anthropocene also profoundly shaped the beliefs of some politicians, scientists, and activists. Alsatian philosopher and mission doctor Albert Schweitzer recast fallout as a humanitarian crisis, and his call for action against the evil of harming innocent bystanders inspired many around the world, including U.S. chemist Linus Pauling and Soviet physicist Andrei Sakharov. Contamination of food and drink also drove many women around the world into antinuclear activism. By identifying themselves as concerned housewives and mothers, these women successfully made subversive use of the conservative gender norms of the early Cold War period. Some political leaders also came to see the health risks of fallout as morally objectionable. President Eisenhower's view drastically changed when he realized that his scientific advisers had consistently downplayed the problem of fallout despite the uncertainty of their claims. His successor, John F. Kennedy, came to the White House with deep concern about the effects of fallout on children. Even Khrushchev became anxious about at least some of the fallout hazards and ordered the transfer of nuclear tests from the Kazakh steppe to the remote corners of the Soviet Arctic. Although the logic of national security still prevailed in the thinking of these political leaders, their changed opinions about fallout ultimately left their mark in the environmental objective inserted into the preamble of the PTBT.

Some of the most surprising and consequential policy impacts of the shifting definition of risk, however, were found among the staunchest opponents to a test ban. Some senior officials and scientists of the U.S. Atomic Energy Commission, most notably chairman Lewis L. Strauss, commissioner Willard F. Libby, and physicist Edward Teller, never changed their dismissive opinions about fallout and continued to cast doubt on the scientific basis of arguments against contamination. Nonetheless, these cold warriors displayed a remarkable degree of flexibility in appropriating the rhetoric of antipollution to arrest the growing momentum for the international regulation of nuclear testing. Behind the scenes, Libby and Teller repeatedly raised the possibility of a political agreement to reduce fallout emissions without stopping all tests in order to divide and conquer test-ban advocates. Most significantly, both scientists tried to offer a "technological fix" to the problem of fallout as an alternative to an international agreement. Libby and Teller actively promoted the idea the "clean bomb," and Strauss successfully used this technical argument to confuse the Soviets and to force U.S. allies to rescind their proposals for a test limitation. Another innovation, underground testing, brought the test-ban negotiations to a standstill over the on-site inspection of suspicious seismic events within the Soviet Union. The Soviet scientists, torn between the Kremlin's demand for tests and fallout hazards, also developed their own version of the "clean bomb" and underground testing techniques. This parallel technological development in the United States and the Soviet Union ultimately made the PTBT feasible.

Legacies of the Politics of Risk

The redefinition of radioactive fallout from a harmless phenomenon to an unacceptable hazard, a shift codified in the 1958 UNSCEAR report and also in the PTBT, had many different and often conflicting implications beyond its contributions to the making of the landmark arms control treaty. To understand these mixed legacies of risk politics, it is important to recall that understanding of fallout changed through the politics of risk. The result was thus neither a triumph of science over politics nor a political distortion of science, but specific to its own historical context and dynamic. For this reason, the ways in which the redefinition of fallout unfolded fundamentally determined its broader implications for the nuclear age, the Cold War, and the Anthropocene.

The most immediate outcome of the redefinition of fallout was the continuing global efforts to end all atmospheric nuclear testing. France and China refused to sign the PTBT and went on to test their weapons above ground until 1974 and 1980, respectively. Their defiance, however, did not go unchallenged. Japan and New Zealand, located downwind from the Chinese and French-Pacific test sites, routinely monitored fallout and kept pressure on the non-PTBT signatories.[1] An international outcry against atmospheric testing became even stronger as modern environmentalism went global. When the United Nations hosted its first environmental summit in 1972, the draft declaration introduced for discussion included a Japanese proposal for the immediate suspension of all test explosions that caused contamination across borders. Although the conference eventually adopted a modified version that simply condemned all weapons of mass destruction, fallout continued to strain French and Chinese foreign relations and taint their international images.[2]

Notably, all countries that have acknowledged developing nuclear weapons after France and China have conducted underground testing only. Although the needs for security and secrecy related to clandestine nuclear activities undoubtedly explain this choice, the newly declared nuclear countries invariably made a point of their conscious efforts to contain fallout. When India, a signatory to the PTBT, tested what it claimed to be a "peaceful nuclear explosive" in 1974, Indian officials insisted that the test did not violate its treaty obligations.[3] Pakistan, another PTBT party which detonated its first nuclear devices in 1998, declared that "[t]here was no release of radioactivity."[4] Even North Korea, which allegedly conducted an H-bomb test in September 2017, announced that "there were neither emission through ground surface nor leakage of radioactive materials nor did it have any adverse impact on the surrounding ecological environment."[5] However empty it may sound, such a statement attests to the remarkable strength of the international norms against fallout even to this day.

By driving nuclear weapons tests underground, however, the redefinition of fallout rendered one of the most direct and intimate connections between the Cold War and the Anthropocene imperceptible once again. Throughout their life cycle from uranium mining to radioactive waste disposal, nuclear weapons involved the handling of radioactive substances in massive amounts. Yet the thick shroud of secrecy and security surrounding nuclear facilities, combined with the complex system of reward and discipline for workers and residents, systematically obscured the health and environmental effects of nuclear weapons production from public view.[6] In the case of uranium mines, typically located in remote regions far removed from the seats of power in the global nuclear order, there was even a question as to whether these sites should be classified as "nuclear" workplaces worthy of a special political, economic, safety, and health status.[7] A notable exception to these "regimes of imperceptibility," to borrow from historian Michelle Murphy, was atmospheric nuclear testing.[8] This technopolitical spectacle, often staged for public consumption through photography and film, not only visualized the destructive power of nuclear weapons but also scattered the radioactive debris that literally brought the nuclear threat to home for people around the world. The end of atmospheric testing disconnected the rising awareness of the peacetime hazards of nuclear weapons from the continuous nuclear arms buildup. Although the leakage and venting of radioactive gases from underground tests occasionally made the headlines and provoked a public uproar, few people paused to consider the human and environmental costs of nuclear weapons that many believed would prevent war between the superpowers.

The redefinition of fallout contributed to another kind of disconnection: a perceived distinction between military and civilian uses of atomic energy. Materially speaking, there is no difference whether radiation comes from nuclear weapons or nuclear power plants if its type, energy level, quantity, and mode of exposure are the same. In declaring nuclear test debris as an unacceptable hazard, however, many scientists drew a sharp contrast between that source of radiation and those in civilian applications. This reflected a modernist belief, reinforced by the discourse of redemption from the original sin of Hiroshima and Nagasaki, that atomic energy, if redirected toward peaceful purposes, would bring a boon to humanity.[9] While explaining that effects of radioactive fallout were worldwide, uncertain, and uncontrollable, the 1958 UNSCEAR report stated that the industrial, research, and medical applications affected only part of the population and "are for the benefit of mankind and can be controlled." This contrast led the committee to make two starkly different recommendations to minimize irradiation of human populations: one was "the cessation of contamination of the environment by explosions of nuclear weapons," and the other was "the avoidance of *unnecessary* exposure resulting from medical, industrial and other procedures for peaceful uses [emphasis added]."[10] The arguments made by scientists to reject the acceptability of radioactive fallout even in the smallest amounts thus gave a license to higher exposures to radiation related to civilian uses of atomic energy as necessary to reap the rewards of the nuclear age. It was no coincidence that the issue of permissible dose flared up again in the 1970s when the controllability and benefits of nuclear power also came into question.[11]

The most important consequence of the redefinition of fallout was perhaps the most paradoxical: while it contributed to the end of atmospheric nuclear testing, it drew little public attention to the suffering of those who were present at the test sites or nearby during the atmospheric tests. This may seem puzzling, given the fact that those so-called atomic veterans and downwinders were likely to be exposed to radiation in much larger amounts than those located farther away. The reasons for the disconnect between the global and local risk perceptions lay in the ways in which fallout came to be seen as harmful during the atmospheric-testing period. A central piece of knowledge that helped to redefine the meaning of fallout was the linear non-threshold hypothesis. Predicting a potentially significant impact that a tiny increase in radiation-related health risks might have if applied to billions of people around the world, this proposition initially led scientists to recognize fallout hazards only on much larger demographic scales than those of the test-site regions. At the same time, the focus of fallout monitoring also shifted from external to internal exposure. The fallout surveillance, however, was primarily concerned with strontium-90 and other longer-lived elements scattered worldwide. This lopsided attention to the global and long-term effects of fallout systematically obscured the intakes of short-lived and prolific radionuclides, most notably iodine-131, by the test participants and residents near the test sites.[12] Once the issue of fallout changed from the prevention of global contamination to recognition of and restitution to those directly affected by atmospheric tests, the intrinsic and manufactured uncertainty of their exposure continued to hamper their quest for justice.[13]

The redefinition of radioactive fallout through the politics of risk, then, subtly yet powerfully influenced the course of the nuclear age, the Cold War, and environmental consciousness after the PTBT by unraveling emerging awareness of their intimate and reciprocal connections. Recently, however, the enduring legacies of atmospheric nuclear testing, materially inscribed into our bodies and Earth's crust, have moved into the limelight again. Since 2016, the Anthropocene Working Group of the Subcommission on Quaternary Stratigraphy has been developing a proposal to formalize "Anthropocene" as the latest geologic epoch within the Quaternary Period (the last 2.58 million years). One of the ideas that has guided this effort is that the artificial radionuclides scattered by atmospheric tests may constitute a primary marker of Anthropocene strata due to the durability, sharpness, and global synchronicity of their signals.[14] Meanwhile, in July 2017, a United Nations conference made a major breakthrough toward a world free of nuclear weapons by adopting the Treaty on the Prohibition of Nuclear Weapons. In its preamble, this landmark international convention made a specific reference to "the unacceptable suffering of and harm caused to the victims of the use of nuclear weapons (hibakusha), as well as of those affected by the testing of nuclear weapons."[15] By providing compelling evidence of the nuclear peril and the accelerating impact of humans on the global environment, radioactive fallout has outlived the Cold War that unleashed it and has continuously shaped our understanding of the world that we live in.

NOTES

Introduction

1. Treaty Banning Nuclear Weapon Tests in the Atmosphere, in Outer Space and Under Water, August 5, 1963, accessed October 1, 2019, http://disarmament.un.org/treaties/t/test_ban/text.

2. Henry Tanner, "Test Ban Treaty Signed in Moscow," *New York Times*, August 6, 1963.

3. For U.S. test-ban policy, see Robert A. Divine, *Blowing on the Wind: The Nuclear Test Ban Debate, 1954–1960* (New York: Oxford University Press, 1978); Benjamin P. Greene, *Eisenhower, Science Advice, and the Nuclear Test-Ban Debate, 1945–1963* (Stanford, CA: Stanford University Press, 2007); Gregg Herken, *Cardinal Choices: Presidential Science Advising from the Atomic Bomb to SDI* (New York: Oxford University Press, 1992), 101–126; Shane J. Maddock, *Nuclear Apartheid: The Quest for American Atomic Supremacy from World War II to the Present* (Chapel Hill: University of North Carolina Press, 2010), 92–216; Paul Rubinson, *Redefining Science: Scientists, the National Security State, and Nuclear Weapons in Cold War America* (Amherst; Boston: University of Massachusetts Press, 2016), 93–116; and Martha Smith-Norris, "The Eisenhower Administration and the Nuclear Test Ban Talks, 1958–1960: Another Challenge to 'Revisionism'," *Diplomatic History* 27, no. 4 (2003): 503–541. For British policy, see Kendrick Oliver, *Kennedy, Macmillan and the Nuclear Test-Ban Debate, 1961–63* (London: Macmillan, 1998); and John R. Walker, *British Nuclear Weapons and the Test Ban, 1954–73: Britain, the United States, Weapons Policies and Nuclear Testing: Tensions and Contradictions* (Farnham, UK;

Burlington, VT: Ashgate, 2010). For Soviet policy, see Matthew Evangelista, *Unarmed Forces: The Transnational Movement to End the Cold War* (Ithaca, NY: Cornell University Press, 1999), 45–89; and Vojtech Mastny, "The 1963 Nuclear Test Ban Treaty: A Missed Opportunity for Détente?" *Journal of Cold War Studies* 10, no. 1 (2008): 3–25. For the role of the nuclear disarmament movement in the nuclear-test ban debate, see Lawrence S. Wittner, *The Struggle Against the Bomb, vol. 2: Resisting the Bomb: A History of the World Nuclear Disarmament Movement, 1954–1970* (Stanford, CA: Stanford University Press, 1997).

4. *UNSCEAR 2000 Report, Volume 1: Sources* (New York: United Nations, 2000), 160–162, 213. The unofficial nuclear weapons states, namely Israel, India, Pakistan, South Africa, and North Korea, presumably have never conducted a weapon test above ground, with a possible exception of South Africa. While the debris of underground testing at times vented into the air, UNSCEAR believes that the contributions of underground tests to global radioactive contamination are essentially nil.

5. Treaty Banning Nuclear Weapon Tests.

6. For Rachel Carson and her role in the birth of modern environmentalism, see Mark H. Lytle, *The Gentle Subversive: Rachel Carson, Silent Spring, and the Rise of the Environmental Movement* (New York: Oxford University Press, 2007). For the recent scholarly effort to recover the movement's earlier origins, see Chad Montrie, *The Myth of Silent Spring: Rethinking the Origins of American Environmentalism* (Oakland: University of California Press, 2018).

7. For the global development of environmentalism in the 1970s, see John R. McNeill, "The Environment, Environmentalism, and International Society in the Long 1970s," in *The Shock of the Global: The 1970s in Perspective*, ed. Niall Ferguson, Charles S. Maier, Erez Manela, and Daniel J. Sargent, 263–278 (Cambridge, MA: The Belknap Press of Harvard University Press, 2010).

8. For the centrality of nuclear weapons in Cold War history, see David Holloway, "Nuclear Weapons and the Escalation of the Cold War, 1945–1962," in *The Cambridge History of the Cold War, vol. 1: Origins*, ed. Melvyn P. Leffler and Odd Arne Westad, 376–397 (Cambridge: Cambridge University Press, 2010).

9. For the U.S. case, see Howard Ball, *Justice Downwind: America's Atomic Testing Program in the 1950s* (New York: Oxford University Press, 1986); Holly M. Barker, *Bravo for the Marshallese: Regaining Control in a Post-Nuclear, Post-Colonial World*, 2nd ed. (Belmont, CA: Wadsworth, 2013); Barton C. Hacker, *Elements of Controversy: The Atomic Energy Commission and Radiation Safety in Nuclear Weapons Testing, 1947–1974* (Berkeley: University of California Press, 1994); and Laura J. Harkewicz, "'The Ghost of the Bomb': The Bravo Medical Program, Scientific Uncertainty, and the Legacy of

U.S. Cold War Science, 1954–2005" (PhD diss., University of California-San Diego, 2010). For the British case, see Lorna Arnold and Mark Smith, *Britain, Australia and the Bomb: The Nuclear Tests and Their Aftermath* (Basingstoke, UK; New York: Palgrave Macmillan, 2006); Roger Cross, "British Nuclear Tests and the Indigenous People of Australia," in *The British Nuclear Weapons Programme, 1952–2002*, ed. Douglas Holdstock and Frank Barnaby, 75–88 (London: Frank Cass, 2003); and Sue Rabbitt Roff, "Long-Term Health Effects in UK Test Veterans," in *The British Nuclear Weapons Programme, 1952–2002*, ed. Holdstock and Barnaby, 99–112. For the Soviet case, see Susanne Bauer, "Radiation Science After the Cold War: The Politics of Measurement, Risk, and Compensation in Kazakhstan," in *Health, Technologies, and Politics in Post-Soviet Settings: Navigating Uncertainties*, ed. Olga Zvonareva, Evgeniya Popova, and Klasien Horstman, 225–249 (Basingstoke, UK: Palgrave Macmillan, 2017); Magdalena E. Stawkowski, "Radioactive Knowledge: State Control of Scientific Information in Post-Soviet Kazakhstan" (PhD diss., University of Colorado-Boulder, 2014); and Cynthia Werner and Kathleen Purvis-Roberts, "After the Cold War: International Politics, Domestic Policy and the Nuclear Legacy in Kazakhstan," *Central Asian Survey* 25, no. 4 (2006): 461–480.

10. Paul J. Crutzen, "Geology of Mankind," *Nature* 415, no. 6867 (2002): 23; and Crutzen and Eugene F. Stoermer, "The 'Anthropocene'," *IGBP Global Change Newsletter* 41 (2000): 17–18.

11. John R. McNeill and Peter Engelke, *The Great Acceleration: An Environmental History of the Anthropocene Since 1945* (Cambridge, MA: The Belknap Press of Harvard University Press, 2014).

12. John R. McNeill and Corinna R. Unger, "Introduction: The Big Picture," in *Environmental Histories of the Cold War*, ed. McNeill and Unger, 1–18 (Cambridge: Cambridge University Press, 2010); and Simo Laakkonen, Viktor Pál, and Richard Tucker, "The Cold War and Environmental History: Complementary Fields," *Cold War History* 16, no. 4 (2016): 377–394.

13. For example, see Nick Cullather, *The Hungry World: America's Cold War Battle Against Poverty in Asia* (Cambridge, MA: Harvard University Press, 2010); Bob H. Reinhardt, *The End of a Global Pox: America and the Eradication of Smallpox in the Cold War Era* (Chapel Hill: University of North Carolina Press, 2015); and Richard Tucker, "Containing Communism by Impounding Rivers: American Strategic Interests and the Global Spread of High Dams in the Early Cold War," in *Environmental Histories of the Cold War*, ed. McNeill and Unger, 139–163.

14. Lisa M. Brady, "Life in the DMZ: Turning a Diplomatic Failure into an Environmental Success," *Diplomatic History* 32, no. 4 (2008): 585–611;

Kai Hünemörder, "Environmental Crisis and Soft Politics: Détente and the Global Environment, 1968–1975," in *Environmental Histories of the Cold War*, ed. McNeill and Unger, 257–276; and Tuomas Räsänen and Simo Laakkonen, "Cold War and the Environment: The Role of Finland in International Environmental Politics in the Baltic Sea Region," *Ambio* 36, no. 2–3 (2007): 229–236.

15. Jacob Darwin Hamblin, *Arming Mother Nature: The Birth of Catastrophic Environmentalism* (New York: Oxford University Press, 2013).

16. David Zierler, *The Invention of Ecocide: Agent Orange, Vietnam, and the Scientists Who Changed the Way We Think about the Environment* (Athens: University of Georgia Press, 2011). Also see Evelyn Frances Krache-Morris, "Into the Wind: The Kennedy Administration and the Use of Herbicides in South Vietnam" (PhD diss., Georgetown University, 2012).

17. *UNSCEAR 2000 Report, Volume 1*, 207.

18. Jan Zalasiewicz et al., "When Did the Anthropocene Begin? A Mid-Twentieth Century Boundary Level Is Stratigraphically Optimal," *Quaternary International* 385 (2015): 196–203.

19. Jan Zalasiewicz, Will Steffen, Reinhold Leinfelder, Mark Williams, and Colin Waters, "Petrifying Earth Process: The Stratigraphic Imprint of Key Earth System Parameters in the Anthropocene," *Theory, Culture & Society* 34, no. 2–3 (2017): 86.

20. For the critical view of the term *Anthropos*, see Andreas Malm and Alf Hornborg, "The Geology of Mankind? A Critique of the Anthropocene Narrative," *The Anthropocene Review* 1, no. 1 (2014): 62–69.

21. Divine, *Blowing on the Wind*; and Spencer R. Weart, *Nuclear Fear: A History of Images* (Cambridge, MA: Harvard University Press, 1988), 183–214.

22. Silvio O. Funtowicz and Jerome R. Ravetz, "Science for the Post-Normal Age," *Futures* 25, no. 7 (1993): 744.

23. Ulrich Beck, *Risk Society: Towards a New Modernity*, trans. Mark Ritter (London: SAGE, 1992).

24. Ulrich Beck, *World at Risk*, trans. Ciaran Cronin (Cambridge; Malden, MA: Polity Press, 2009), 30.

25. Ibid., 32.

26. Karin Zachmann, "Risk in Historical Perspective: Concepts, Contexts, and Conjunctions," in *Risk: A Multidisciplinary Introduction*, ed. Claudia Klüppelberg, Daniel Straub, and Isabell M. Welpe, 3–35 (Cham, Switzerland: Springer International Publishing, 2014). For a similar development in the United States, see Arwen P. Mohun, *Risk: Negotiating Safety in American Society* (Baltimore: Johns Hopkins University Press, 2013).

27. Daniel Paul Serwer, "The Rise of Radiation Protection: Science, Medicine and Technology in Society, 1896–1935" (PhD diss., Princeton University, 1976), 173–182, 212–236.

28. Ibid., 226–236; and Gilbert F. Whittemore Jr., "The National Committee on Radiation Protection, 1928–1960: From Professional Guidelines to Government Regulation" (PhD diss., Harvard University, 1986), 57–90.

29. Arthur Mutscheller, "Physical Standards of Protection Against Roentgen-Ray Dangers," *American Journal of Roentgenology and Radium Therapy* 13 (1925): 65.

30. National Bureau of Standards, *Permissible Dose from External Sources of Ionizing Radiation: Recommendations of the National Committee on Radiation Protection* (National Bureau of Standards Handbook 59) (Washington, DC: Government Printing Office, 1954), 26–27. For the adoption of "permissible dose" by the radiation protection committees after World War II, see Whittemore Jr., "National Committee on Radiation Protection," 322–326.

31. Soraya Boudia, "Global Regulation: Controlling and Accepting Radioactivity Risks," *History and Technology* 23, no. 4 (2007): 389–406; Jacob Darwin Hamblin, "'A Dispassionate and Objective Effort': Negotiating the First Study on the Biological Effects of Atomic Radiation," *Journal of the History of Biology* 40, no. 1 (2007): 147–177; Hamblin, *Poison in the Well: Radioactive Waste in the Oceans at the Dawn of the Nuclear Age* (New Brunswick, NJ: Rutgers University Press, 2008); Gabrielle Hecht, *Being Nuclear: Africans and the Global Uranium Trade* (Cambridge, MA: MIT Press, 2012); Hiroshi Ichikawa, *Soviet Science and Engineering in the Shadow of the Cold War* (London: Routledge, 2018), 143–178; J. Christopher Jolly, "Thresholds of Uncertainty: Radiation and Responsibility in the Fallout Controversy" (PhD diss. Oregon State University, 2004); Olga Kuchinskaya, *The Politics of Invisibility: Public Knowledge About Radiation Health Effects After Chernobyl* (Cambridge, MA: MIT Press, 2014); Laura Sembritzki, "Maiak 1957 and Its Aftermath: Radiation Knowledge and Ignorance in the Soviet Union," *Jahrbücher für Geschichte Osteuropas* 66, no. 1 (2018): 45–64; and J. Samuel Walker, *Permissible Dose: A History of Radiation Protection in the Twentieth Century* (Berkeley: University of California Press, 2000).

32. For the republic of science and its Cold War construction, see Mary Jo Nye, *Michael Polanyi and His Generation: Origins of the Social Construction of Science* (Chicago: University of Chicago Press, 2011); and Audra J. Wolfe, *Freedom's Laboratory: The Cold War Struggle for the Soul of Science* (Baltimore: Johns Hopkins University Press, 2018). For the classic interpretation of the Lysenko Affair, see David Joravsky, *The Lysenko Affair* (Chicago: University of Chicago Press, 1986 [1970]; and Zhores A. Medvedev, *The Rise and Fall of*

T. D. Lysenko, trans. I. Michael Lerner (New York: Columbia University Press, 1969). The recent scholarship has examined this historiography as a historical product of the Cold War. See William deJong-Lambert and Nikolai Krementsov, "On Labels and Issues: The Lysenko Controversy and the Cold War," *Journal of the History of Biology* 45, no. 3 (2012): 373–388; and Michael D. Gordin, "Lysenko Unemployed: Soviet Genetics After the Aftermath," *Isis* 109, no. 1 (2018): 56–78.

33. Naomi Oreskes, "Introduction," in *Science and Technology in the Global Cold War*, ed. Oreskes and John Krige, 1–9 (Cambridge, MA; London: MIT Press, 2014). For the United States, see, e.g., Paul Forman, "Behind Quantum Electronics: National Security as Basis for Physical Research in the United States, 1940–1960," *Historical Studies in the Physical and Biological Sciences* 18, no. 1 (1987): 149–229; Rubinson, *Redefining Science*; Mark Solovey, ed., "Science in the Cold War," special issue, *Social Studies of Science* 31, no. 2 (2001): 165–297; and Jessica Wang, *American Science in an Age of Anxiety: Scientists, Anticommunism, and the Cold War* (Chapel Hill: University of North Carolina Press, 1999). For Britain, see, e.g., Jon Agar and Brian Balmer, "British Scientists and the Cold War: The Defence Research Policy Committee and Information Networks, 1947–1963," *Historical Studies in the Physical and Biological Sciences* 28, no. 2 (1998): 209–252; David Edgerton, *Warfare State: Britain, 1920–1970* (Cambridge: Cambridge University Press, 2006); and Alexandros-Panagiotis Oikonomou, "The Hidden Persuaders: Government Scientists and Defence in Post-War Britain" (PhD diss., Imperial College London, 2011).

34. For example, Loren R. Graham, *What Have We Learned About Science and Technology from the Russian Experience?* (Stanford, CA: Stanford University Press, 1998); David Holloway, "Physics, the State, and Civil Society in the Soviet Union," *Historical Studies in the Physical and Biological Sciences* 30, no. 1 (1999): 173–192; and Ethan Pollock, *Stalin and the Soviet Science Wars* (Princeton, NJ: Princeton University Press, 2006).

35. For the case of the United States, see, e.g., John Krige, *American Hegemony and the Postwar Reconstruction of Science in Europe* (Cambridge, MA: MIT Press, 2006). For the case of Britain, see, e.g., Néstor Herran, "Spreading Nucleonics: The Isotope School at the Atomic Energy Research Establishment, 1951–67," *The British Journal for the History of Science* 39, no. 4 (2006): 569–586; and Shinsuke Tomotsugu, "Decolonization and the United Kingdom's Atoms for Peace Program in the Middle East," presented at the Annual Meeting of the Society for Historians of American Foreign Relations, Lexington, Kentucky, June 2014. For the case of the Soviet Union, see, e.g., William deJong-Lambert and Nikolai Krementsov, eds., *The Lysenko Controversy as a Global Phenomenon*, 2 vols. (New York: Palgrave Macmillan, 2017); Izabella

Goikhman, "Soviet-Chinese Academic Interactions in the 1950s: Questioning the 'Impact-Response' Approach," in *China Learns from the Soviet Union, 1949–Present*, ed. Thomas P. Bernstein and Hua-Yu Li, 275–302 (Lanham, MD: Lexington, 2010); and Sonja D. Schmid, "Nuclear Colonization? Soviet Technopolitics in the Second World," in *Entangled Geographies: Empire and Technopolitics in the Global Cold War*, ed. Gabrielle Hecht, 125–154 (Cambridge, MA; London: MIT Press, 2011).

36. Sven Ove Hansson, "Risk and Ethics: Three Approaches," in *Risk: Philosophical Perspectives*, ed. Tim Lewens, 21–35 (London; New York: Routledge, 2007).

Chapter 1

1. Ferenc Morton Szasz, *The Day the Sun Rose Twice: The Story of the Trinity Site Nuclear Explosion* (Albuquerque: University of New Mexico Press, 1984), 134–135.

2. *UNSCEAR 2000 Report, Volume 1: Sources* (New York: United Nations, 2000), 196–201.

3. "Invisible Atom Blast Debris Blankets City," *Chicago Tribune*, June 6, 1952, 25.

4. Hansard (UK), House of Commons Debate, October 23, 1952, vol. 505, col. 1268.

5. "Government Announcement of Test of a Hydrogen Bomb in the Soviet Union," *Current Digest of the Soviet Press*, September 5, 1953, 3; "TASS Report on Tests of New Types of Atomic Bombs in the Soviet Union," *Current Digest of the Soviet Press*, October 31, 1953, 15.

6. Szasz, *The Day the Sun Rose Twice*, 63.

7. Joseph O. Hirschfelder, "The Scientific-Technological Miracle at Los Alamos," in *Reminiscences of Los Alamos, 1943–1945*, ed. Lawrence Badash, Joseph O. Hirschfelder, and Herbert P. Broida, 67–88 (Dordrecht, the Netherlands: D. Reidel, 1980), 74.

8. Joseph O. Hirschfelder and John Magee to Kenneth Bainbridge, "Danger from Active Material Falling from Cloud," June 16, 1945, The Department of Energy OpenNet System (DOE OpenNet), NV0059784.

9. Quoted in Martin J. Sherwin, *A World Destroyed: Hiroshima and the Origins of the Arms Race* (New York: Vintage Books, 1987), 220.

10. Thomas E. Widner and Susan M. Flack, "Characterization of the World's First Nuclear Explosion, the Trinity Test, as a Source of Public Radiation Exposure," *Health Physics* 98, no. 3 (2010): 481; and Valerie L. Kuletz, *The Tainted Desert: Environmental Ruin in the American West* (New York: Routledge, 1998), 12.

11. Stafford L. Warren, interview by Adelaide Tusler for the Center for Oral History Research, University of California, Los Angeles, September 15, 1966, vol. 2. Accessed October 1, 2019. http://oralhistory.library.ucla.edu/viewItem.do?ark =21198/zz0008zcbp&title=Interview%20of%20Stafford%20L.%20 Warren%20(1983).

12. Barton C. Hacker, *The Dragon's Tail: Radiation Safety in the Manhattan Project, 1942–1946* (Berkeley: University of California Press, 1987), 93.

13. Ibid., 91–92.

14. Ibid., 103–105.

15. Quoted in Widner and Flack, "Characterization of the World's First Nuclear Explosion," 495.

16. For more detail about the role of the U.S. nuclear intelligence program in framing the issue of radioactive fallout, see Michael R. Lehman, "Nuisance to Nemesis: Nuclear Fallout and Intelligence as Secrets, Problems, and Limitations on the Arms Race, 1940–1964" (PhD diss., University of Illinois at Urbana-Champaign, 2016).

17. Quoted in Charles A. Ziegler and David Jacobson, *Spying Without Spies: Origins of America's Secret Nuclear Surveillance System* (Westport, CT: Praeger, 1995), 38. For the 1883 eruption of Krakatoa, see Simon Winchester, *Krakatoa: The Day the World Exploded, August 27, 1883* (New York: Harper-Collins, 2003).

18. J. M. Blair, D. H. Frisch, and S. Katcoff, "Detection of Nuclear-Explosion Dust in the Atmosphere," October 2, 1945, Los Alamos National Laboratory.

19. Ziegler and Jacobson, *Spying Without Spies*, 127–129.

20. Ibid., 204–205.

21. Barton C. Hacker, *Elements of Controversy: The Atomic Energy Commission and Radiation Safety in Nuclear Weapons Testing, 1947–1974* (Berkeley: University of California Press, 1994), 47–48.

22. Operation Plan, Rad Safe Group for Operation Ranger, January 20, 1951, Military Research & Appl. 7-1, Nevada Proving Grounds (Report), Box 31, Central Subject Files (CSF), Division of Biomedical and Environment Research (DBER), Records of the Atomic Energy Commission (RG 326), The National Archives at College Park (NACP), Maryland.

23. Merril Eisenbud, *An Environmental Odyssey: People, Pollution, and Politics in the Life of a Practical Scientist* (Seattle: University of Washington Press, 1990), 64–66.

24. Ibid., 27.

25. Ibid., 36, 48.

26. Thomas L. Shipman to John C. Clark, "Special Rad Safe Problems—Operation Bungle [*sic*]," July 11, 1951, Military Research & Appl. 7-1, Nevada Proving Grounds, August 1951–December 1951, Box 30, CSF, DBER, RG 326, NACP.

27. Merril Eisenbud and John H. Harley, "Radioactive Dust from Nuclear Detonations," *Science* 117, no. 3033 (1953): 141–147.

28. Robert K. Plumb, "Increased Radiation Found in East," *New York Times*, February 3, 1951, 1.

29. Hacker, *Elements of Controversy*, 78.

30. Ibid., 104.

31. Herbert M. Clark, "The Occurrence of an Unusually High-Level Radioactive Rainout in the Area of Troy, N.Y.," *Science* 119, no. 3097 (1954): 619–622.

32. U.S. Atomic Energy Commission, *Thirteenth Semiannual Report of the Atomic Energy Commission* (Washington, DC: Government Printing Office, 1953), 114.

33. Ibid., 123.

34. Lyle B. Borst to Gordon Dean, May 16, 1953, enclosed with AEC 604/1, July 13, 1953, Military Research & Appl. 7-4, Test Personnel Exposures, 1953, Box 31, CSF, DBER, RG 326, NACP.

35. Clark, "Occurrence," 620.

36. John C. Bugher to Lyle B. Borst, July 1, 1953, enclosed with AEC 604/1, July 13, 1953.

37. Norris Bradbury to Elmo Morgan, April 3, 1953, Medicine, Health & Safety 3-1, Effects, 1/53–6/53, Box 36, CSF, DBER, RG 326, NACP.

38. Jacob Darwin Hamblin, *Arming Mother Nature: The Birth of Catastrophic Environmentalism* (New York: Cambridge: Oxford University Press, 2013).

39. Nicholas M. Smith Jr., "Gabriel Project, Reopened," November 8, 1951, Organization & Management 11, Project Gabriel 1953, Box 34, CSF, DBER, RG 326, NACP.

40. Status of Gabriel Studies as of February 1953, n.d., Organization & Management 11, Project Gabriel 1953, Box 34, CSF, DBER, RG 326, NACP.

41. Willard F. Libby, interview by Greg Marlowe, Session II, April 16, 1979, Niels Bohr Library & Archives, American Institute of Physics, College Park, Maryland. Accessed October 1, 2019. www.aip.org/history-programs/niels-bohr-library/oral-histories/4743-1.

42. Ibid.

43. Ibid.

44. Ibid.

45. Complete Transcription of Project SUNSHINE Conference Held on July 21–23, 1953, 1, Box 39, CSF, DBER, RG 326, NACP.

46. RAND Corporation, "Worldwide Effects of Atomic Weapons: Project Sunshine," R-251–AEC, August 6 (Oak Ridge, TN: Technical Information Service Extension, 1953), 7.

47. Complete Transcription of Project SUNSHINE Conference, 188.

48. Ibid., 188.

49. Ibid., 76

50. Jane C. Loeffler, *The Architecture of Diplomacy: Building America's Embassies* (New York: Princeton Architectural Press, 1998).

51. Melvyn P. Leffler, *A Preponderance of Power: National Security, the Truman Administration, and the Cold War* (Stanford, CA: Stanford University Press, 1992).

52. Lyle E. Alexander, "Samples Collected on Trip to Europe, Asia, and Africa, February 1955," n.d., USDA (Beltsville)—Soil Program (Sunshine), Box 8, Records Relating to Fallout Studies (RRFS), RG 326, NACP.

53. U.S. Congress, Joint Committee on Atomic Energy, Special Subcommittee on Radiation, *Fallout from Nuclear Weapons Tests* (Washington, DC: Government Printing Office, 1959), 493, 512.

54. U.S. Atomic Energy Commission, Health and Safety Laboratory, *Environmental Contamination from Weapon Tests* (HASL-42) (Oak Ridge, TN: Technical Information Service Extension, 1958), 49.

55. See, e.g., John P. DiMoia, *Reconstructing Bodies: Biomedicine, Health, and Nation-Building in South Korea Since 1945* (Stanford, CA: Stanford University Press, 2013); Jonathan Zimmerman, *Innocents Abroad: American Teachers in the American Century* (Cambridge, MA: Harvard University Press, 2008).

56. J. C. Bugher to Andrew Warren, December 30, 1953, Gabriel 1951 thru 1953, Box 1, RRFS, RG 326, NACP.

57. M. Susan Lindee, "The Repatriation of Atomic Bomb Victim Body Parts to Japan: Natural Objects and Diplomacy," *Osiris* 13 (1998): 376–409.

58. J. Laurence Kulp to W. F. Bulle, March 14, 1955, 13, Radiation and Fallout, 13.8 Projects—Gabriel, 1954–56, Box 235, Records Relating to Atomic Energy Matters, 1948–1962 (RRAEM), Special Assistant to the Secretary of State for Atomic Energy and Outer Space (AEOS), Records of the Department of State (RG 59), NACP.

59. J. Laurence Kulp et al., "Project Sunshine, Annual Report, Period March 31, 1955–April 1, 1956," April 15, 1956, Box 10, Office Files of Commissioner Lewis L. Strauss, RG 326, NACP.

60. Robert A. Dudley, "Report on Chicago Trip, September 14–15, 1953," September 18, 1953, Gabriel 1951 thru 1953, Box 1, RRFS, RG 326, NACP.

61. Walter D. Claus, "Reconsideration of Objectives of Project Sunshine," May 1, 1956, Administrative File—Fallout Program Policy, Box 10, RRFS, RG 326, NACP.

62. "Sunshine Meeting at New York Operations Office," May 23, 1955, Box 5, RRFS, RG 326, NACP.

63. V. A. Logachev, *Semipalatinskii poligon: Obespechenie obschchei i*

radiatsionnoi bezopasnosti iadernykh ispytanii [Semipalatinsk Proving Ground: Securing General and Radiation Safety of Nuclear Tests] (Moscow: Medbioekstrem RF, 1997), 17; Ministerstvo Oborony Rossiiskoi Federatsii, *Rozhdennye atomnoi eroi: Istoriia sozdaniia i razvitiia 12-e Glavnogo Upravleniia Ministerstva Oborony Rossiiskoi Federatsii* [Birth of the Atomic Era: History of the Creation and Development of the 12th Chief Directorate of the Ministry of Defense of Russian Federation], vol. 1 (Moscow: Nauka, 2007), 21–22.

64. Quoted in David Holloway, *Stalin and the Bomb: The Soviet Union and Atomic Energy, 1939–1956* (New Haven, CT: Yale University Press, 1994), 213.

65. Steven Sabol, *Russian Colonization and the Genesis of Kazak National Consciousness* (Basingstoke, UK; New York: Palgrave Macmillan, 2003), 34–51; A. S. Elagin, ed., *Semipalatinsk* (Alma-Ata, Kazakh SSR: Nauka, 1984), 151–172.

66. E. L. Iakubovskaiia, V. I. Nagibin, and V. P. Sislin, *Semipalatinskii ispytatel'nyi poligon: Nezavisimyi analiz problemy* [Semipalatinsk Proving Ground: An Independent Problem Analysis] (Novosibirsk: Novosibirskii Poligrafkombinat, 2003), 3; and Logachev, *Semipalatinskii poligon*, 193.

67. A. Lavrent'eva, "Stroiteli novogo mira" [Builders of the New World], *V mire knig* [In the World of Books] 9 (1970): 4. Also see Michael D. Gordin, *Red Cloud at Dawn: Truman, Stalin, and the End of the Atomic Monopoly* (New York: Farrar, Straus and Giroux, 2009), 7–10; and Holloway, *Stalin and the Bomb*, 132.

68. Logachev, *Semipalatinskii poligon*, 61–62.

69. Ibid., 121–122.

70. Ibid., 37–38; and L. D. Riabev, ed., *Atomnyi proekt SSSR: Dokumenty i materialy* [Atomic Project of the USSR: Documents and Materials], t. 2, kn. 1 (Moscow: Nauka, 1999), 648.

71. A. I. Burnazian, "O radiatsionnoi bezopasnosti" [On Radiation Safety], in *Vospominaniia ob Igore Vasil'eviche Kurchatove* [Reminiscences on Igor Vasil'vich Kurchatov], ed. A. P. Aleksandrov, 305–311 (Moscow: Nauka, 1988), 311.

72. Riabev, *Atomnyi proekt SSSR*, t. 2, kn. 1, 649.

73. "Truman Statement on Atom," *New York Times*, September 24, 1949, 1.

74. A. P. Vasil'ev, ed., *Rozhdennaia atomnym vekom: Sbornik istoricheskikh ocherkov, dokumentov i vospominanii veteranov k 40–letiiu sozdaniia v SSSR sluzhby spetsial'nogo kontrolia Ministerstva oborony* [Birth of the Atomic Age: Collection of Historical Assessments, Documents, and Reminiscences of Veterans toward 40 Years of the Establishment of the USSR Service of Special Control of the Ministry of Defense] (Moscow: Sluzhba spetsial'nogo kontrolia, 1998), 5–6, 194–197; Vasil'ev, "Sistema dal'nego obnaruzheniia iadernykh vzryvov i

Sovetskii atomnyi proekt" [System of the Distant Observation of Nuclear Explosions and the Soviet Atomic Project], in *Istoriia Sovetskogo atomnogo proekta: Dokumenty, vospominaniia, issledovaniia* [History of the Soviet Atomic Project: Documents, Reminiscences, and Investigations], ed. V. P. Vizgin, vol. 2, 237–278 (St. Petersburg: Izdatel'stvo Russkogo Khristianskogo Gumanitarnogo Instituta, 2002), 240; and V. N. Natal'ina, "Aerozol'nyi (radionuklidnyi) metod kontrolia ispytanii"[Aerosol (Radionuclide) Method of Control of Tests], in Vasil'ev, *Rozhdennaia atomnym vekom*, 183–184.

75. Vasil'ev, *Rozhdennaia atomnym vekom*, 6; and Vasil'ev, "Sistema dal'nego obnaruzheniia," 244.

76. Vasil'ev, "Sistema dal'nego obnaruzheniia," 243, 245.

77. Riabev, *Atomnyi proekt SSSR*, t. 3, kn. 2, 127.

78. L. M. Mezelev, *Oni byli pervymi: Iz istorii iadernykh ispytanii* [They Were the First: From the History of Nuclear Tests], vol. 1 (Moscow: n.p., 2001), 78–82; and Riabev, *Atomnyi proekt SSSR*, t. 3, kn. 2, 142–143, 320.

79. Logachev, *Semipalatinskii poligon*, 39, 52.

80. Andrei D. Sakharov, *Memoirs*, trans. Richard Lourie (London: Hutchinson, 1990), 172. For the interpretation of exposure levels, see Samuel Glasstone, ed., *The Effects of Atomic Weapons* (Washington, DC: Government Printing Office, 1950), 342.

81. Sakharov, *Memoirs*, 172.

82. V. N. Mikhailov, ed., *Iadernye ispytaniia SSSR* [Nuclear Tests of the USSR], vol. 1 (Sarov, Russia: RFIaTs-VNIIEF, 1997), 228; and Riabev, *Atomnyi proekt SSSR*, t. 3, kn. 1, 34–35, 52–53, 633–634.

83. Riabev, *Atomnyi proekt SSSR*, t. 3, kn. 1, 53.

84. Mikhailov, *Iadernye ispytaniia SSSR*, 230. The rate of background radiation is a worldwide average.

85. Sakharov, *Memoirs*, 172.

86. Margaret Gowing, *Independence and Deterrence: Britain and Atomic Energy, 1945–1952*, vol. 1 (New York: St. Martin's Press, 1974), 113; and Norman Spoor, "Dr W. G. Marley," *Annals of Occupational Hygiene* 24, no. 2 (1981): 258–259.

87. Gowing, *Independence and Deterrence*, vol. 1, 183.

88. Ibid., 306–307; and Septimus H. Paul, *Nuclear Rivals: Anglo-American Atomic Relations, 1941–1952* (Columbus: Ohio State University Press, 2000), 184–185.

89. Lorna Arnold, *A Very Special Relationship: British Atomic Weapon Trials in Australia* (London: H.M.S.O., 1987), 89–90, 97; and Robert Milliken, *No Conceivable Injury* (Ringwood, Australia: Penguin, 1986), 161–162.

90. Department of Fisheries and Wildlife, Western Australia, *The Pearling Industry of Western Australia, 1850–1979* (Perth: Extension and Publicity Service, Department of Fisheries & Wildlife, Western Australia, 1979).

91. Peggy Brock, "South Australia," in *Contested Ground: Australian Aborigines Under the British Crown*, ed. Ann McGrath, 208–239 (St. Leonards, Australia: Allen & Unwin, 1995), 216–217, 223.

92. Robert Menzies to E. J. Williams, January 24, 1952, "Epicure," FCO 1/3, The National Archives of the United Kingdom (TNA), Kew.

93. Gowing, *Independence and Deterrence*, vol. 1, 335–337.

94. E. L. Sykes to Ben Cockram, n.d., AB 16/571, TNA.

95. "British Atomic Weapons," *Times* (London), February 18, 1952, 4.

96. Arnold, *A Very Special Relationship*, 39.

97. Ibid., 59–61.

98. "Melbourne Rain Radio-Active," *The News* (Adelaide), October 11, 1952, 12.

99. "Public Safety and the Woomera Tests," *Sydney Morning Herald*, October 9, 1953, 2.

100. Gowing, *Independence and Deterrence*, vol. 2, 114–115; and N. G. Stewart and E. M. R. Fisher, "The Deposition of Sr90, Cs137, and Pu239 on U.K. from Atomic Bombs Exploded Before 31st December 1952," July 1953, AB 15/3751, TNA.

101. Richard J. Aldrich, *The Hidden Hand: Britain, America and Cold War Secret Intelligence* (London: John Murray, 2001), 229; and Jeffrey T. Richelson, *Spying on the Bomb: American Nuclear Intelligence from Nazi Germany to Iran and North Korea* (New York: W. W. Norton, 2006), 86–87.

102. Hansard (Australia), House of Representatives Debate, October 8, 1953.

103. Milliken, *No Conceivable Injury*, 95.

104. *The Report of the Royal Commission into British Nuclear Tests in Australia*, vol. 1 (Canberra: Australian Government Publishing Service, 1985), 194.

Chapter 2

1. For the full description of the Lucky Dragon incident, see Ralph E. Lapp, *The Voyage of the Lucky Dragon* (New York: Harper and Brothers, 1958).

2. Barton C. Hacker, *The Dragon's Tail: Radiation Safety in the Manhattan Project, 1942–1946* (Berkeley: University of California Press, 1987), 116–118; and Hacker, *Elements of Controversy: The Atomic Energy Commission and Radiation Safety in Nuclear Weapons Testing, 1947–1974* (Berkeley: University of California Press, 1994), 14–19.

3. Chikashi Kataoka, *Nanyō no Nihonjin gyogyō* [Japanese Fisheries in South Sea] (Tokyo: Dōbunkan Shuppan, 1991).

4. U.S. Congress, House Committee on Ways and Means, Subcommittee on Tuna Imports, *Tuna Imports* (Washington, DC: Government Printing Office, 1951), 45–46; and Andrew F. Smith, *American Tuna: The Rise and Fall of an Improbable Food* (Berkeley: University of California Press, 2012), 107–111.

5. Memo of Conversation, June 21, 1955, 894.245/6-2155, Central Decimal Files (CDF), RG 59, NACP.

6. Sayuri Shimizu, *Creating People of Plenty: The United States and Japan's Economic Alternatives, 1950–1960* (Kent, OH: Kent State University Press, 2001), 102–109.

7. David Bradley et al. to Chief of Radiological Safety Section, "Report of Radioactivity of Migratory Fish Caught at Wotho, Ailingiane, Rongelap, and Rongerik Atolls," October 4, 1946, 6/4 Bikini 1946–47, University of Washington Laboratory of Radiation Biology Records (UWLRB), University of Washington Special Collections, Seattle.

8. Neal O. Hines, *Proving Ground: An Account of the Radiobiological Studies in the Pacific, 1946–1961* (Seattle: University of Washington Press, 1963), 65–66.

9. John C. Bugher, "Effects of Pacific Tests on Commercial Tuna Fishing," March 31, 1954, DOE OpenNet, NV0404580.

10. E. Jerry Jessee, "Radiation Ecologies: Bombs, Bodies, and Environment During the Atmospheric Nuclear Weapons Testing Period, 1942–1965" (PhD diss., Montana State University, 2013), 226–230.

11. Bugher, "Effects of Pacific Tests on Commercial Tuna Fishing."

12. "Notes on Conference of Applied Fisheries Laboratory Staff Members with Dr. John C. Bugher, Chief, Division of Biology & Medicine US AEC, Washington, DC, held on November 1, 1953, in Seattle," n.d., 39, Box 1, UWLRB.

13. "Anzen ni kueruyō taisaku kōzu" [Countermeasures Taken to Ensure Safe Diet], *Asahi Shimbun*, March 18, 1954, evening edition, 3.

14. Memo, March 21, 1954, enclosed with Tokyo to Washington, Desp. no. 1482, April 30, 1954, 711.5611/4-3054, CDF, RG 59, NACP; and Tokyo to Washington, no. 2260, March 21, 1954, 711.5611/3-2154, CDF, RG 59, NACP.

15. Bugher, "Effects of Pacific Tests on Commercial Tuna Fishing."

16. Richard Hirsch to the Executive Officer, Operations Coordinating Board, "AEC Action on Japanese Fish Problem," March 17, 1954, Japan (1) September 58–June 57 (2), Box 46, OCB Central Files Series, National Security Council Staff Papers (NSCS), White House Office (WHO), Dwight D. Eisenhower Presidential Library (DDEL), Abilene, Kansas.

17. Washington to Tokyo, March 23, 1954, 711.5611/3-2254, CDF, RG 59, NACP.

18. Eisenbud to Bugher, "Contamination of the Fukuryu Maru and Associated Problems in Japan: Preliminary Report," n.d., enclosed with Tokyo to Washington, Desp. no. 1481, April 30, 1954, 711.5611/4-3054, Box 3174, CDF, RG 59, NACP.

19. Yasuo Miyake, Yoshio Hiyama, and Nobuo Kusano, eds., *Bikini suibaku hisai shiryōshū* [Collected Data from the Bikini Hydrogen Bomb Disaster], rev. ed. (Tokyo: Tokyo Daigaku Shuppankai, 2014), 422-423.

20. Ibid., 433.

21. Kōseishō, "Hōshanō osen gyorui ni kansuru shiryō" [Materials Relating to Fish Contaminated with Radioactivity], November 14, 1954, C'4.2.1.5, vol. 2, Gaikō Shiryōkan (GS), Tokyo.

22. "Mada mada tsuzuku 'maguro junan'" ["Tuna's Suffering" Goes on], *Asahi Shimbun*, May 16, 1954, evening edition, 3.

23. Memo, "Report of Radioactive Fish on Formosan Market," July 14, 1954, Japan d. Fukuryu Maru 1954, part 1, Box 422, RRAEM, AEOS, RG 59, NACP; and Taipei to Washington, no. TOUSFO 42, July 12, 1954, Japan d. Fukuryu Maru 1954, part 1, Box 422, RRAEM, AEOS, RG 59, NACP.

24. Kōchiken Bikini Suibaku Jikken Hisai Chōsadan, *Mōhitotsu no Bikini jiken: 1000 seki o koeru hisaisen o ou* [Another Bikini Incident: Searching for over 1,000 Ships Affected by the Disaster] (Tokyo: Heiwabunka, 2004), 12-13.

25. "Radioactive Tuna," July 6, 1954, 711.5611/7-654, CDF, RG 59, NACP; and Washington to Tokyo, no. 2636, May 24, 1954, DOE OpenNet, NV0410249.

26. Yasuo Miyake, *Shi no hai to tatakau kagakusha* [Scientists Who Fight the Ashes of Death] (Tokyo: Iwanami Shoten, 1972), 128.

27. USAEC New York Operations Office, *Operation Troll*, p. 10, Troll, 1955-56, Box 425, RRAEM, AEOS, RG 59, NACP.

28. K. D. Nichols to Robert B. Carney, January 31, 1955, Troll, 1955-56, Box 425, RRAEM, AEOS, RG 59, NACP.

29. Miyake, Hiyama, and Kusano, *Bikini suibaku hisai shiryōshū*, 178, 180-187.

30. "Tenya wanya no Ōshima" [Oshima in Utter Confusion]," *Asahi Shimbun*, May 26, 1954, evening edition, 3.

31. "Tsuyu hitoashi hayameni" [The Rainy Season Came a Little Early], *Asahi Shimbun*, May 23, 1954, 7.

32. Miyake, Hiyama, and Kusano, *Bikini suibaku hisai shiryōshū*, 180-187.

33. Miyake, *Shi no hai to tatakau kagakusha*, 87-89, 95-98.

34. "Mada tsuzuku hōshanōu" [Radioactive Rain Still Continues], *Yomiuri Shimbun*, June 16, 1954, 7. For the respective values, see U.S. Atomic Energy Commission, *Thirteenth Semiannual Report of the Atomic Energy Commission* (Washington, DC: Government Printing Office, 1953), 116; and National Bureau of Standards, *Maximum Permissible Amounts of Radioisotopes in the Human Body and Maximum Permissible Concentrations in Air and Water* (National Bureau of Standards Handbook 52) (Washington, DC: Government Printing Office, 1953), 11.

35. Miyake, Hiyama, and Kusano, *Bikini suibaku hisai shiryōshū*, 189.

36. Yasuo Miyake, *Nihon no ame* [Rain in Japan] (Tokyo: Hōsei Daigaku Shuppankai, 1956), 20–24.

37. Tōru Kobayashi, ed., *Gensuibaku kinshi undō shiryōshū* [The Collection of Materials on the Ban-the-Bomb Movement] (Tokyo: Ryokuin Shobō, 1995), vol. 2, 81.

38. Miyake, *Nihon no ame*, 56.

39. Yasuo Miyake, *Kaere Bikini e: Gensuibaku kinshi undō no genten o kangaeru* [Return to Bikini: A Reflection on the Origins of the A- and H-Bomb Ban Movement] (Tokyo: Suiyōsha, 1984), 102.

40. Kobayashi, *Gensuibaku kinshi undō shiryōshū*, vol. 1, 47, 69–70; and Miura-shi, ed., *Bikini jiken Miura no kiroku* [The Bikini Incident: The Records of Miura] (Miura, Japan: Miura-shi, 1996), 32, 54.

41. Washington to Tokyo, no. 566, May 29, 1954, C'4.2.1.5-1, GS.

42. *Gyoson* [Fishing Village] (May 1954): 30.

43. Nihon Katsuo-Maguro Gyogyō Kyōdō Kumiai Rengōkai, ed., *Nikkatsuren shi* [The History of Nikkatsuren], vol. 1 (Tokyo: Nihon Katsuo-Maguro Gyogyō Kyōdō Kumiai Rengōkai, 1966), 438–439.

44. Miura-shi, *Bikini jiken Miura no kiroku*, 241–242.

45. Yasuo Kondō, ed., *Suibaku jikken to Nihon gyogyō* [H-Bomb Tests and Japan's Fisheries] (Tokyo: Tokyo Daigaku Shuppankai, 1958), 63.

46. Nisseikyō Sōritsu 50 Shūnen Kinen Rekishi Hensan Iinkai, ed., *Gendai Nihon seikyō undō shi: Shiryōshū* [The History of the Contemporary Japanese Co-operative Movement: The Collection of Materials] (Tokyo: Nihon Seikatsu Kyōdō Kumiai Rengōkai, 2001), 326.

47. Nisseikyō Sōritsu 50 Shūnen Kinen Rekishi Hensan Iinkai, ed., *Gendai Nihon seikyō undō shi* [The History of the Contemporary Japanese Co-operative Movement], vol. 1 (Tokyo: Nihon Seikatsu Kyōdō Kumiai Rengōkai, 2002), 203.

48. Kathleen S. Uno, "The Death of 'Good Wife, Wise Mother'?" in *Postwar Japan as History*, ed. Andrew Gordon, 293–322 (Berkeley: University of California Press, 1993), 308–309.

49. Yōko Matsuoka, "Machi no sumi kara sekai e" [From the Street Corner to the World], *Asahi Shimbun*, May 20, 1954, evening edition, 2.

50. Kobayashi, *Gensuibaku kinshi undō shiryōshū*, vol. 3, 73.
51. See Mari Yamamoto, *Grassroots Pacifism in Post-War Japan: The Rebirth of a Nation* (London: RoutledgeCurzon, 2004), 131–137.
52. Kobayashi, *Gensuibaku kinshi undō shiryōshū*, vol. 3, 73.
53. Ibid., vol. 1, 409.
54. Eriko Maruhama, *Gensuikin shomei undō no tanjō: Tokyo Suginami no jūmin pawā to suimyaku* [Birth of the Petition Campaign for a Ban on A- and H-Bombs: Power of Citizens in Suginami, Tokyo, and Its Context] (Tokyo: Gaifūsha, 2011), 280–282, 287–304.
55. Kobayashi, *Gensuibaku kinshi undō shiryōshū*, vol. 1, 132.
56. Ibid., vol. 1, 131.
57. Ibid., vol. 2, 5–11, quote from 11.
58. Maruhama, *Gensuikin shomei undō no tanjō*, 311–337.
59. Tokyo to Washington, no. 2466, April 10, 1954, 711.5611/4-1054, CDF, RG 59, NACP.
60. Ajia Kyoku Dai Go Ka, "Beikoku Genshiryoku Iinkai Bosu, Donarudoson ryōshi to Nihon gawa senmonka tono kondan ni kansuru ken" [On the meeting of Japanese experts with Messer. Boss and Donaldson from the U.S. Atomic Energy Commission], July 1, 1954, FOIA No. 2004–00197, courtesy of Daigo Fukuryū Maru Kinenkan, Tokyo.
61. Washington to Tokyo, no. 2471, May 10, 1954, 711.5611/4-1054, CDF, RG 59, NACP; Tokyo to Washington, Desp. no. 1130, page 89, 711.5611/3-2855, CDF, RG 59, NACP; and "Fundamental Considerations in Determining Significant Levels of Radioactive Contamination of Food and Water," Japan d. Fukuryu Maru 1954, part 1, Box 422, RRAEM, AEOS, RG 59, NACP.
62. Washington to Tokyo, no. 2412, May 10, 1954, DOE OpenNet, NV0410249.
63. Yoshikuma Nagasawa, "Shokuhin eisei bukai hōkoku" [Report of the Food Sanitation Panel], October 15, 1954, C'4.2.1.5, vol. 2, GS.
64. Toshihiro Higuchi, "The Strange Career of Dr. Fish: Yoshio Hiyama, Radioactive Fallout, and Nuclear Fear Management in Japan, 1954–1958," *Historia Scientiarum* 25, no. 1 (2015): 57–77.
65. Yoshio Hiyama to Arthur D. Welander, March 23, 1954, 13 Hiyama, Box 12, UWLRB.
66. Hiyama to Lauren Donaldson, April 12, 1954, 11 Hiyama, Box 4, UWLRB.
67. Tokyo to Washington, no. 2912, May 25, 1954, 711/5611/5-2554, CDF, RG 59, NACP.
68. Memo of Hōshasen eikyō chōsa tokubetsu iinkai yakuinkai, n.d., FOIA No. 2004–00197.

69. Press Release, November 15, enclosed with Tokyo to Washington, Desp. no. 696, December 10, 1954, 994.7138/12–1054, CDF, RG 59, NACP; and "Hōshasen busshitsu no eikyō to riyō ni kansuru Nichi-Bei kaigi hōkoku," November 15–19, 1954, enclosed with Tokyo to Washington, Desp. no. 790, January 4, 1955, 994.8138/1–455, CDF, RG 59, NACP.

70. "Chikaku ōhaba ni kanwa, genshi maguro no haiki kijun" [The Significant Relaxation of the Disposal Standards of Atomic Tuna Will Come Soon], *Yomiuri Shimbun*, November 19, 1954, 7.

71. "Nichi-Bei hōshanō kaigi no shūkaku" [Outcomes of the Japan-U.S. Radioactivity Conference], *Yomiuri Shimbun*, November 30, 1954, 6.

72. Ibid.

73. For the "Atoms for Peace" campaign in Japan as a propaganda counterweight to the rise of antinuclear weapons sentiments, see Yuka Tsuchiya, "Kōhō bunka gaikōto shiteno genshiryoku riyō kyanpe'en to 1950 nendai no Nichi-Bei kankei" [The Campaign for Atomic Energy Use as Public-Relations Cultural Diplomacy and U.S.-Japan Relations in the 1950s], in *Nichi-Bei dōmei ron: Rekishi, kinō, shūhen shokoku no shiten* [On the U.S.-Japan Alliance: Perspective of History, Function, and Neighboring Countries], ed. Toshitaka Takeuchi, 180–209 (Kyoto: Mineruba Shobō, 2011); Masakatsu Yamazaki and Kenzō Okuda, "Bikini jiken go no genshiro dōnyūron no taitō" [Rise of the Call for the Introduction of Atomic Power Reactors After the Bikini Incident], *Kagakusi kenkyū*, no. 230 (2004): 83–93; and Ran Zwigenberg, "'The Coming of a Second Sun': The 1956 Atoms for Peace Exhibit in Hiroshima and Japan's Embrace of Nuclear Power," *Asia-Pacific Journal: Japan Focus* 10, no. 6 (February 4, 2012), https://apjjf.org/2012/10/6/Ran-Zwigenberg/3685/article.html.

74. *Shōwa 29 nen ni okeru Bikini kaiiki no hōshanō eikyō chōsa hōkoku* [Report of the Study of the Effects of Radioactivity in Waters in the Vicinity of Bikini in Showa 29], vol. 2 (Tokyo: Suisanchō Chōsa Kenkyū Bu, 1955), 81–99.

75. Miyake, Hiyama, and Kusano, *Bikini suibaku hisai shiryōshū*, 143–148; "Maguro wa mō daijōbu" [Tuna Is Fine Now], *Asahi Shimbun*, December 23, 1954, 7.

76. For details of the compensation negotiations, see Roger Dingman, "Alliance in Crisis: The Lucky Dragon Incident and Japanese-American Relations," in *The Great Powers in East Asia, 1953–1960*, ed. Warren I. Cohen and Akira Iriye, 187–214 (New York: Columbia University Press, 1990); Kazuya Sakamoto, "Kakuheiki to Nichi-Bei kankei: Bikini jiken no gaikō shori" [Nuclear Weapons and Japan-U.S. Relations: Diplomatic Settlement of the Bikini Incident], *Nenpō kindai Nihon kenkyū* 16 (1994): 243–271; and John Swenson-Wright, *Unequal Allies? United States Security and Alliance Policy Toward Japan, 1945–1960* (Stanford, CA: Stanford University Press, 2005), 150–186.

Chapter 3

1. "Text of Statement and Comments by Strauss on Hydrogen Bomb Tests in the Pacific," *New York Times*, April 1, 1954, 20.

2. Edward B. Lewis, "Remembering Sturtevant," in *Genes, Development and Cancer: The Life and Work of Edward B. Lewis*, ed. Howard D. Lipshitz, 511–516 (Dordrecht, the Netherlands: Springer, 2007), 514.

3. Alfred H. Sturtevant, "Social Implications of the Genetics of Man," *Science* 120, no. 3115 (1954): 407.

4. Phillip R. Sloan and Brandon Fogel, eds., *Creating a Physical Biology: The Three-Man Paper and Early Molecular Biology* (Chicago: University of Chicago Press, 2011).

5. Douglas E. Lea, *Actions of Radiations on Living Cells*, 2nd ed. (Cambridge: Cambridge University Press, 1955), 69.

6. Elof Axel Carlson, *Genes, Radiation, and Society: The Life and Work of H. J. Muller* (Ithaca, NY: Cornell University Press, 1981), 160.

7. Lauriston S. Taylor, *Organization for Radiation Protection: The Operations of the ICRP and NCRP, 1928–1974* (Springfield, VA: National Technical Information Service, 1979), 5.013.

8. Ibid., 5.016.

9. Ibid., 5.017.

10. Ibid., 5.019.

11. Gilbert F. Whittemore Jr., "The National Committee on Radiation Protection, 1928–1960: From Professional Guidelines to Government Regulation" (PhD diss., Harvard University, 1986), 282–287.

12. Tolerance Doses Panel of the Protection Sub-committee, Minutes, November 29, 1946, FD 1/7106, TNA.

13. M. Susan Lindee, *Suffering Made Real: American Science and the Survivors at Hiroshima* (Chicago: University of Chicago Press, 1994); and Karen A. Rader, *Making Mice: Standardizing Animals for American Biomedical Research, 1900–1955* (Princeton, NJ: Princeton University Press, 2004), 221–250.

14. Donald Falconer, "Quantitative Genetics in Edinburgh: 1947–1980," *Genetics* 133, no. 2 (1993): 137–142; and Rader, *Making Mice*, 246.

15. Agreed Statement of Meeting of Geneticists and Others at the London Hospital on 25th January 1947, FD 1/465, TNA; and D. G. Catcheside, "Genetic Effects of Irradiation with Reference to Man," February 6, 1947, FD 1/479, TNA.

16. J. S. Mitchell, "Report on Visit to USA, 25th March to 14th April, 1948," May 4, 1948, F192, Joseph S. Mitchell Papers (JSMP), Cambridge University Library Special Collections, Cambridge.

17. W. Binks to Members of the Protection Subcommittee, March 25, 1947, F293, JSMP.

18. Minutes of the 13th Meeting of the Tolerance Doses Panel of the Protection Subcommittee, August 5, 1948, FD 1/7106, TNA.

19. Lauriston S. Taylor, *The Tripartite Conferences on Radiation Protection: Canada, United Kingdom, United States (1949–1953)* (Washington, DC: Office of Scientific and Technical Information, U.S. Department of Energy, 1984), 18.10.

20. Taylor, *Organization for Radiation Protection*, 7.124. Also see J. Christopher Jolly, "Thresholds of Uncertainty: Radiation and Responsibility in the Fallout Controversy" (PhD diss., Oregon State University, 2003), 105–112.

21. Taylor, *Organization for Radiation Protection*, 7.124–7.125. For more, see Whittemore Jr., "The National Committee on Radiation Protection," 431–478.

22. Taylor, *The Tripartite Conferences*, 18.10.

23. Hermann J. Muller, "The Manner of Dependence of the 'Permissible Dose' of Radiation on the Amount of Genetic Damage," *Acta Radiologica* 41, no. 1 (1954): 11–12.

24. G. Failla to C. Stern, January 17, 1953, SC-1 Correspondence 1953–54, Box 20, Lauriston S. Taylor Papers, the Countway Library of Medicine, Boston; and Jolly, "Thresholds of Uncertainty," 195.

25. Taylor, *The Tripartite Conferences*, 23.11–23.12.

26. H. J. Muller to A. H. Sturtevant, September 22, 1954, Box 11, Alfred H. Sturtevant Papers (AHSP), the Caltech Archives, Pasadena, California.

27. Michael David-Fox, *Showcasing the Great Experiment: Cultural Diplomacy and Western Visitors to the Soviet Union, 1921–1941* (New York; Oxford: Oxford University Press, 2012).

28. Mark B. Adams, "The Founding of Population Genetics: Contributions of the Chetverikov School 1924–1934," *Journal of the History of Biology* 1, no. 1 (1968): 23–39.

29. Carlson, *Genes, Radiation, and Society*, 244–273.

30. Nils Roll-Hansen, *The Lysenko Effect: The Politics of Science* (Amherst, NY: Humanity Books, 2005).

31. Nikolai P. Dubinin, "Work of Soviet Biologists: Theoretical Genetics," *Science* 105, no. 2718 (1947): 109; and Nikolai P. Dubinin, *Vechnoe dvizhenie* [Perpetual Motion] (Moscow: Politizdat, 1989), 254.

32. Dubinin, *Vechnoe dvizhenie*, 250–254.

33. Ethan Pollock, *Stalin and the Soviet Science Wars* (Princeton, NJ: Princeton University Press, 2006), 41–71. Also see Nikolai Krementsov, *Stalinist Science* (Princeton, NJ: Princeton University Press, 1997), 158–183; and Valery N. Soyfer, *Lysenko and the Tragedy of Soviet Science* (New Brunswick, NJ: Rutgers University Press, 1994), 172–204.

34. *The Situation in Biological Science: Proceedings of the Lenin Academy of Agricultural Sciences of the U.S.S.R. Session: July 31–August 7, 1948, Verbatim Report* (Moscow: Foreign Languages Publishing House, 1949), 31.

35. Ethan Pollock, "From *Partiinost'* to *Nauchnost'* and Not Quite Back Again: Revisiting the Lessons of the Lysenko Affair," *Slavic Review* 68, no. 1 (2009): 95–115.

36. Michael D. Gordin, "Lysenko Unemployed: Soviet Genetics After the Aftermath," *Isis* 109, no. 1 (2018): 56–78.

37. "History of the Genetics Conflict," *Bulletin of the Atomic Scientists* 5, no. 5 (1949): 138–140. For Orbeli and the physiology controversy, see Pollock, *Stalin and the Soviet Science Wars*, 136–167; and George Windholz, "The 1950 Joint Scientific Session: Pavlovians as the Accusers and the Accused," *Journal of the History of the Behavioral Sciences* 33, no. 1 (1997): 61–81.

38. Pollock, "From *Partiinost'* to *Nauchnost'*," 110–111.

39. William deJong-Lambert, "Hermann J. Muller, Theodosius Dobzhansky, Leslie Clarence Dunn, and the Reaction to Lysenkoism in the United States," *Journal of Cold War Studies* 15, no. 1 (2013): 78–118; deJong-Lambert and Nikolai Krementsov, eds., *The Lysenko Controversy as a Global Phenomenon*, 2 vols. (New York: Palgrave Macmillan, 2017); Kaori Iida, "A Controversial Idea as a Cultural Resource: The Lysenko Controversy and Discussions of Genetics as a 'Democratic' Science in Postwar Japan," *Social Studies of Science* 45, no. 4 (2015): 546–569; Greta Jones, "British Scientists, Lysenko and the Cold War," *Economy and Society* 8, no. 1 (1979): 26–58; Diane B. Paul, "A War on Two Fronts: J. B. S. Haldane and the Response to Lysenkoism in Britain," *Journal of the History of Biology* 16 (1983): 1–37; and Audra J. Wolfe, "What Does It Mean to Go Public? The American Response to Lysenkoism, Reconsidered," *Historical Studies in the Natural Sciences* 40, no. 1 (2010): 48–78.

40. Trofim D. Lysenko, *Agrobiology: Essays on Problems of Genetics, Plant Breeding and Seed Growing* (Moscow: Foreign Language Publishing House, 1954), 492.

41. Ibid., 326.

42. Ibid., 548.

43. V. V. Babkov and E. S. Sakanian, *Nikolai Vladimirovich Timofeev-Resovskii, 1900–1981* (Moscow: Pamiatniki Istoricheskoi Mysli, 2002), 168–169; and B. M. Emel'ianov and V. S. Gavril'chenko, *Laboratoriia "B": Sungul'skii fenomen* (Snezhinsk, Russia: RFIaTs—VNIITF, 2000), 33–40.

44. Mira M. Kosenko, *Where Radiobiology Began in Russia: A Physician's Perspective* (Fort Belvoir, VA: Defense Threat Reduction Agency, 2011), 6; H. Bentley Glass, "Timofeeff-Ressovsky, Nikolai Vladimirovich," *Complete Dictionary of Scientific Biography*, accessed October 1, 2019, https://www.encyclopedia.com/science/dictionaries-thesauruses-pictures-and-press-releases

/timofeeff-ressovsky-nikolai-vladimirovich; and Nikolaus Riehl and Frederick Seitz, *Stalin's Captive: Nikolaus Riehl and the Soviet Race for the Bomb* (Washington, DC: American Chemical Society, 1996), 202.

45. Kosenko, *Where Radiobiology Began in Russia*, 9.

46. Robert A. Divine, *Blowing on the Wind: The Nuclear Test Ban Debate, 1954–1960* (New York: Oxford University Press, 1978), 27–29; and Martha J. Smith, "The Nuclear Testing Policies of the Eisenhower Administration, 1953–60" (PhD diss., University of Toronto, 1997), 68–77, 215–221.

47. Jolly, "Thresholds of Uncertainty," 199.

48. Bugher to Sturtevant, June 16, 1954, Box 11, AHSP.

49. Joseph and Stewart Alsop, "Matter of Fact . . . The Radiological Hazard I," *Washington Post*, November 22, 1954, 11.

50. Willard F. Libby, "Dosages from Natural Radioactivity and Cosmic Rays," *Science* 122, no. 3158 (1955): 57–58.

51. Libby to Strauss, October 27, 1955, Muller, Herman, Box 5, Office Files of Commissioner Willard F. Libby (OFWL), RG 326, NACP.

52. Peter Thorsheim, "Interpreting the London Fog Disaster of 1952," in *Smoke and Mirrors: The Politics and Culture of Air Pollution*, ed. E. Melanie DuPuis, 154–169 (New York: New York University Press, 2004), 161–163.

53. Alfred H. Sturtevant, "The Genetic Effects of High Energy Irradiation of Human Populations," *Engineering and Science* 18 (1955): 12.

54. John Beatty, "Weighing the Risks: Stalemate in the Classical/Balance Controversy," *Journal of the History of Biology* 20, no. 3 (1987): 289–319.

55. T. C. Carter, "The Genetic Problem of Irradiated Human Populations," *Bulletin of the Atomic Scientists* 11, no. 10 (1955): 363.

56. Sturtevant, "Genetic Effects of High Energy Irradiation," 12.

57. Muller to E. L. Green, April 13, 1955, Correspondence, March-July 1955, Box 11, AHSP.

58. Minutes, Advisory Committee for Biology and Medicine, May 5, 6, 7, 1955, 17–18, DOE OpenNet, NV0708697.

59. Wolfe, "What Does It Mean to Go Public?"

60. Muller to R. R. Lanier, March 28, 1955, 1955, March 21–31, Box 6, Correspondence, Papers of Hermann Joseph Muller (Muller mss.), Lilly Library, Indiana University, Bloomington.

61. Curt Stern, "One Scientist Speaks Up," *Science* 120, no. 3131 (1954): 5A.

62. For the cult of productivity in the Stalin era, see Lewis H. Siegelbaum, *Stakhanovism and the Politics of Productivity in the USSR, 1935–1941* (Cambridge; New York: Cambridge University Press, 1988).

63. Paul R. Josephson, *Red Atom: Russia's Nuclear Power Program from Stalin to Today* (New York: W.H. Freeman, 2000), 20–21.

64. Dubinin, *Vechnoe dvizhenie*, 336.

65. "Stenogramma zasedaniia Biuro otdeleniia biologicheskikh nauk AN SSSR ot 5 aprelia 1955 g" [Transcript of the Session of the Bureau of the Division of Biological Sciences, the Academy of Sciences of the USSR from April 5, 1955], f. 534, op. 1 (1950–1956), d. 226, ll. 3–4, Arkhiv Rossiiskoi akademii nauk (ARAN), Moscow.

66. Dubinin, *Vechnoe dvizhenie*, 336–338.

67. K. V. Kosikov, "Zakliuchenie na tematisheskii plan rabot laboratorii eksperimental'noi genetiki na 1955 god, predstavlennyi v OBN AN SSSR chlenom-korrespondentom AN SSSR N. P. Dubininym" [Conclusion on the Thematic Work Plan of the Laboratory of Experimental Genetics in 1955, Presented at the Division of Biological Sciences, the USSR Academy of Sciences by Corresponding Member of the Academy N. P. Dubinin], f. 534, op. 1 (1950–1956), d. 154, ll. 17–20, ARAN.

68. "Stenogramma zasedaniia Biuro otdeleniia biologicheskikh nauk AN SSSR ot 5 aprelia 1955 g," l. 44.

69. N. I. Nuzhdin, N. I. Shapiro, O. N. Petrova, and O. N. Kitaeva, "Effect of Ionizing Radiations on the Fertility of Mice and the Viability of Their Progeny," in *Conference of the Academy of Sciences of the USSR on the Peaceful Uses of Atomic Energy, July 1–5, 1955, Session of the Division of Biological Science*, English trans. (Washington, DC: U.S. Atomic Energy Commission, 1956), 21–30.

70. Mark B. Adams, "The Soviet Nature-Nurture Debate," in *Science and the Soviet Social Order*, ed. Loren R. Graham, 94–138 (Cambridge, MA: Harvard University Press, 1990), 109–110; and Loren R. Graham, *Science in Russia and the Soviet Union: A Short History* (New York: Cambridge University Press, 1993), 121–134.

71. "Stenogramma, zasedaniia uchenogo komiteta meditsinskoi radiologii pri ministerstve zdravookhraneniia Soiuza SSR" [Transcript of the Session of the Scientific Committee of Medical Radiology Under the USSR Ministry of Health], October 7, 1955, f. R-8009, op. 40, d. 27, l. 70, Gosudarstvennyi Arkhiv Rossiiskoi Federatsii, Moscow.

72. I. K. Zakharov and V. K. Shumnyi, "K 50-letiiu 'pis'ma trekhsot'" [Toward the 50th Year of the "Letter of 300"], *Vestnik VOGuS* 9, no. 1 (2005): 19. Also see V. Ia. Aleksandrov, *Trudnye gody sovetskoi biologii: Zapiski sovremennika* [The Tragic Years of Soviet Biology: Memoirs of a Contemporary] (St. Petersburg: Nauka, 1993), 165–170.

73. "Proposal for a United Nations Commission to Study the Problem of H-Bomb Tests," February 16, 1955, 13.5b Studies on Effects of, 1955, Box 232, RRAEM, AEOS, RG 59, NACP.

74. Lawrence S. Wittner, *The Struggle Against the Bomb, vol. 1: One World*

or None: A History of the World Nuclear Disarmament Movement Through 1953 (Stanford, CA: Stanford University Press, 1993), 59–66, 263–268, 326; and Wittner, *The Struggle Against the Bomb, vol. 2: Resisting the Bomb: A History of the World Nuclear Disarmament Movement, 1954–1970* (Stanford, CA: Stanford University Press, 1997), 5–7.

75. "Proposal for a United Nations Commission to Study the Problem of H-Bomb Tests," February 16, 1955.

76. Philip M. Williams, *Hugh Gaitskell: A Political Biography* (London: Jonathan Cape, 1979), 295–308, 337–339.

77. Hugh Dalton, *The Political Diary of Hugh Dalton, 1918–40, 1945–60* (London: Cape, 1986), 648.

78. Hansard (UK), House of Commons Debate, March 22, 1955, vol. 538, cc. 1881–1948. For Summerskill, see John Stewart, "Summerskill, Edith Clara, Baroness Summerskill (1901–1980)," in *Oxford Dictionary of National Biography*, ed. H. C. G. Matthew and Brian Harrison, vol. 53, 321–323 (Oxford: Oxford University Press, 2004); and Penny Summerfield, "'Our Amazonian Colleague': Edith Summerskill's Problematic Reputation," in *Making Reputations: Power, Persuasion and the Individual in Modern British Politics*, ed. Richard Toye and Julie Gottlieb, 135–150 (London: I.B. Tauris, 2005).

79. Benjamin P. Greene, *Eisenhower, Science Advice, and the Nuclear Test-Ban Debate, 1945–1963* (Stanford, CA: Stanford University Press, 2007), 68–74.

80. Lorna Arnold, *Britain and the H-Bomb* (Basingstoke, UK; New York: Palgrave, 2001), 54.

81. "Moratorium on Testing of Large Thermonuclear Weapons," May 23, 1955, Moratorium on Tests (2), Box 6, Executive Secretary's Subject File Series, NSCS, WHO, DDEL.

82. "Radiation Study," May 5, 1955, 13.5b Studies on Effects of, 1955, Box 232, RRAEM, AEOS, RG 59, NACP.

83. Draft Reply to Foreign Secretary, n.d., enclosed with E. Plowden to Lord President, March 29, 1955, AB 16/1541, TNA.

84. Note by Sir Richard Powell, March 17, 1955, AB 16/1655, TNA.

85. R. N. Quirk to Lord President, March 19, 1955, AB 16/1655, TNA; and Note of Meeting, March 18, 1955, AB 16/1655, TNA.

86. Hansard (UK), House of Commons Debate, March 29, 1955, vol. 539, col. 197.

87. Jacob Darwind Hamblin, "'A Dispassionate and Objective Effort': Negotiating the First Study on the Biological Effects of Atomic Radiation," *Journal of the History of Biology* 40, no. 1 (2007): 147–148; and Jolly, "Thresholds of Uncertainty," 314–316.

88. Thomas Jonter, *The Key to Nuclear Restraint: The Swedish Plans to*

Acquire Nuclear Weapons During the Cold War (London: Palgrave Macmillan, 2016), 130–138; and Richard Hirsch to Elmer B. Staats, "Continued Swedish Demand for International Radiation Survey," May 25, 1955, OCB 000.9 (File #3) (3), Box 9, OCB Central File Series, NSCS, WHO, DDEL.

89. Robin Hankey to P. H. Dean, May 31, 1955, AB 16/1656, TNA. For Undén's opposition to the acquirement of nuclear weapons, see Jonter, *The Key to Nuclear Restraint*, 138–140.

90. *Foreign Relations of the United States, 1955–1957*, vol. 20 (Washington, DC: Government Printing Office, 1990), 73–74.

91. Ibid., 91–93.

92. "Nuclear Radiation," July 9, 1955, CAB 124/1653, TNA; and Roger Makins to Harold Caccia, August 18, 1955, CAB 124/1653, TNA.

93. United Nations General Assembly Resolution 913(X), adopted on December 3, 1955, accessed October 1, 2019, http://www.unscear.org/unscear/en/general_assembly.html#Resolution1.

94. *Documents on Canadian External Relations*, vol. 21 (Ottawa: Department of Foreign Affairs and International Trade, 1999), 73–75.

95. United Nations General Assembly, Tenth Session, Verbatim Record of the Seven Hundred and Seventy-Fifth Meeting, November 1, 1955, 36, UNSCEAR Library (UNSCEARL), Vienna.

96. New York to London, no. 1028, October 22, 1955, CAB 124/1654, TNA; New York to London, no. 1068, October 29, 1955, CAB 124/1654, TNA; and New York to London, no. 1096, November 2, 1955, CAB 124/1654, TNA.

97. New York to London, no. 1124, November 4, 1955, CAB 124/1654, TNA.

98. Quoted in Soraya Boudia, "Global Regulation: Controlling and Accepting Radioactivity Risks," *History and Technology* 23, no. 4 (2007): 396.

99. Brief for United Kingdom Delegation to the UNSCEAR, March 9, 1956, CAB 124/2496, TNA.

100. John W. Garver, *Protracted Contest: Sino-Indian Rivalry in the Twentieth Century* (Seattle: University of Washington Press, 2001), 117–120.

101. Lindesay Parrott, "U.N. Nonmembers Barred from Atom Effects Study," *New York Times*, December 4, 1955, 1.

102. P. E. Ramsbotham to I. F. Porter, December 7, 1955, CAB 124/2496, TNA.

Chapter 4

1. Press Release, April 8, 1955, Beginning of Program 1954–1955, Committees on Biological Effects of Atomic Radiation (CBEAR), the National Academies Archives (NAA), Washington, DC.

2. Hansard (UK), House of Commons Debate, March 22, 1955, vol. 538, col. 1916.

3. For the origins of the U.S. National Academy of Sciences, see Rexmond C. Cochrane, *The National Academy of Sciences: The First Hundred Years, 1863–1963* (Washington, DC: The National Academy of Sciences, 1978); and A. Hunter Dupree, "The Founding of the National Academy of Sciences: A Reinterpretation," *Proceedings of the American Philosophical Society* 101, no. 5 (1957): 434–440.

4. For the British Medical Research Council, see Arthur Landsborough Thomson, *Half a Century of Medical Research, vol. 1: Origins and Policy of the Medical Research Council (U.K.)* (London: H.M.S.O., 1973).

5. Jacob Darwin Hamblin, "'A Dispassionate and Objective Effort': Negotiating the First Study on the Biological Effects of Atomic Radiation," *Journal of the History of Biology* 40, no. 1 (2007): 147–177.

6. Genetics Panel, Second Meeting, February 5–6, 1956, 8, National Academy of Sciences, Subcommittee on the Genetic Effects, 1956, March, Box 9, Organizations, Muller mss.

7. Ibid., 2.
8. Ibid., 2.
9. Ibid., 7.
10. Ibid., 7.
11. Ibid., 2–3.
12. Ibid., 14.

13. Lauriston S. Taylor, *Organization for Radiation Protection: The Operations of the ICRP and NCRP, 1928–1974* (Springfield, VA: National Technical Information Service, 1979), 8.302.

14. Ibid., 8.324.
15. Ibid., 8.325.
16. Ibid., 8.306.

17. Panel on Genetic Effects, Minutes of the Third Meeting, 9th August, 1955, n.d., FD 1/8127, TNA.

18. Himsworth to J. S. Mitchell, April 28, 1956, FD 1/8131, TNA.

19. Medical Research Council, *The Hazards to Man of Nuclear and Allied Radiations* (London: H.M.S.O., 1956), 64.

20. H. Bentley Glass, "The Hazards of Atomic Radiations to Man: British and American Reports," *Bulletin of the Atomic Scientists* 12, no. 8 (1956): 317.

21. John Laughlin to Bronk, July 13, 1956, Genetic Laughlin-Pullman Report General 1956, CBEAR, NAA; and Full Committee, Minutes of the Second Meeting, February 8, 1956, FD 1/8116, TNA.

22. Proceedings, Conference on Genetics, February 5–6, 1956, 293, Genetic Conference Transcript Feb 1956, CBEAR, NAA.

23. Genetics Panel, Second Meeting, 8.

24. Sturtevant to Weaver, April 9, 1956, National Academy of Sciences, Subcommittee on the Genetic Effects, 1956, April, Box 9, Organizations, Muller mss.

25. Sturtevant to Weaver, May 18, 1956, National Academy of Sciences, Subcommittee on the Genetic Effects, 1956, May, Box 9, Organizations, Muller mss; and National Academy of Sciences, *The Biological Effects of Atomic Radiation: Summary Reports* (Washington, DC: National Academy of Sciences-National Research Council, 1956), 30.

26. Willard F. Libby, "Radioactive Fallout and Radioactive Strontium," *Science* 123, no. 3199 (1956): 657–660.

27. For an example of dosimetry used at the time, see Masanobu Tomonaga, "Leukemia in Nagasaki Atomic Bomb Survivors from 1945 through 1959," *Bulletin of the World Health Organization* 26 (1962): 619–631.

28. J. Christopher Jolly, "Thresholds of Uncertainty: Radiation and Responsibility in the Fallout Controversy" (PhD diss., Oregon State University, 2003), 388.

29. National Academy of Sciences, *Biological Effects of Atomic Radiation*, 34.

30. Pathologic Meetings Transcript vol. II, January 1956, 162, CBEAR, NAA.

31. Ibid., 163.

32. The rep is the same as the rem (Roentgen equivalent man). See Glossary.

33. National Academy of Sciences, *Biological Effects of Atomic Radiation*, 39.

34. Claudia Clark, *Radium Girls: Women and Industrial Health Reform, 1910–1935* (Chapel Hill: University of North Carolina Press, 1997).

35. U.S. Congress, Joint Committee on Atomic Energy, Special Subcommittee on Radiation, *Radiation Protection Criteria and Standards: Their Basis and Use* (Washington: Government Printing Office, 1960), 239.

36. Robley D. Evans, "Acceptance of the Coolidge Award," *Medical Physics* 11, no. 5 (1984): 580.

37. L. H. Gray, "A Consideration of the Hazard Associated with the Ingestion of Radium Together with a Reconsideration of the Derived Estimates of the Permissible Concentration of Plutonium and Strontium in Drinking Water," n.d., FD 1/465, TNA.

38. Minutes of the 22nd Meeting of the Tolerance Doses Panel of the Protection Subcommittee, August 11, 1949, FD 1/7106, TNA.

39. Lauriston S. Taylor, *The Tripartite Conferences on Radiation Protection: Canada, United Kingdom, United States (1949–1953)* (Washington, DC: Office of Scientific and Technical Information, U.S. Department of Energy, 1984), 18.10–18.11.

40. "An Attempt to Estimate the Hazard from Fission Products," n.d., FD 1/8132, TNA.

41. Medical Research Council, *The Hazards to Man*, 16; and Panel on Individual Effects, Minutes of the Second Meeting, June 27, 1955, FD 1/8098, TNA.

42. A. Haddow to H. Himsworth, February 16, 1956, B. 128, Alexander Haddow Papers, Wellcome Library, London.

43. Panel of Genetic Effects, Minutes of the Eleventh Meeting, April 24, 1956, FD 1/8127, TNA.

44. W. G. Marley to W. V. Mayneord, J. S. Mitchell, and J. Loutit, May 3, 1956, FD 1/8131, TNA.

45. G. G. Manov to W. F. Libby, "British Sr90 Measurements," June 18, 1956, Sunshine General, NAS Correspondence, Box 8, OFWL, RG 326, NACP.

46. W. G. Marley, "Comments on Libby's Remarks at Northwestern University, January 19, 1956," February 2, 1956, AB 6/1202, TNA; and Cockcroft to Libby, February 21, 1956, Sunshine General, British Correspondence, Box 8, OFWL, RG 326, NACP.

47. "Amendments Relating to the Hazard of Strontium 90," n.d., FD 1/8132, TNA.

48. Walter Johnson, ed., *The Papers of Adlai E. Stevenson*, vol. 6 (Boston: Little, Brown, 1976), 118.

49. "Nuclear Tests and Radiation," May 28, 1956, CAB 130/115, TNA.

50. "Nuclear Tests: Radio-Strontium Hazard," June 2, 1956, CAB 124/1651, TNA.

51. "Nuclear Tests and Radiation."

52. Hansard (UK), House of Commons Debate, July 23, 1956, vol. 557 cc. 46–47.

53. The Society of Friends to Members of Quarterly Meeting Peace Committee, January 17, 1957, CND/2008, 1/1, Campaign for Nuclear Disarmament Papers (CNDP), LSE Archives and Special Collections, London. For the role of the Atomic Scientists' Association in the fallout controversy in Britain, see Christoph Laucht, "Scientists, the Public, the State, and the Debate Over the Environmental and Human Health Effects of Nuclear Testing in Britain, 1950–1958," *The Historical Journal* 59, no. 1 (2016): 221–251.

54. Lawrence S. Wittner, *The Struggle Against the Bomb, vol. 2: Resisting the Bomb: A History of the World Nuclear Disarmament Movement, 1954–1970* (Stanford, CA: Stanford University Press, 1997), 44–45.

55. "Stop These Tests NOW!," n.d., [May 1957], CND/2008, 1/3, CNDP; and "Stop the Tests . . . ," n.d., CND/2008, 1/3, CNDP.

56. John H. Harley to Charles L. Dunham, May 28, 1956, Sunshine

Data—New York, Box 10, OFWL, RG 326, NACP; and Harley to Dunham, June 4, 1956, Sunshine Data—New York, Box 10, OFWL, RG 326, NACP.

57. Oslo to Washington, no. 748, December 22, 1956, Nuclear Weapons Tests, September-December 1956 (6), Box 6, Records of the Office of the Special Assistant for Disarmament, WHO, DDEL.

58. Ottawa to Washington, no. V-265, April 17, 1957, 50271-B-40 part 2.2, Records of the Department of External Affairs (RG 25), National Library and Archives of Canada (NLAC), Ottawa.

59. Joseph Levitt, *Pearson and Canada's Role in Nuclear Disarmament and Arms Control Negotiations, 1945–1957* (Montreal: McGill-Queen's University Press, 1993), 225–251.

60. *Documents on Canadian External Relations*, vol. 22 (Ottawa: Department of Foreign Affairs and International Trade, 2001), 697.

61. Ibid., 703–705.

62. Alvin Schuster, "Strauss Asserts Fall-Out Danger Can Be Localized," *New York Times*, July 20, 1956, 1.

63. Ralph E. Lapp, "The 'Humanitarian' H-Bomb," *Bulletin of the Atomic Scientists* 12, no. 7 (1956): 264.

64. Peter Goodchild, *Edward Teller: The Real Dr. Strangelove* (Cambridge, MA: Harvard University Press, 2004).

65. For the perceived strategic importance of the clean bomb, see Toshihiro Higuchi, "'Clean' Bombs: Nuclear Technology and Nuclear Strategy in the 1950s," *Journal of Strategic Studies* 29, no. 1 (2006): 83–116.

66. Paul Rubinson, *Redefining Science: Scientists, the National Security State, and Nuclear Weapons in Cold War America* (Amherst; Boston: University of Massachusetts Press, 2016), 63–92.

67. Teller to Alfred Starbird, June 5, 1956, enclosed with W. B. McCool to the Commissioners, June 7, 1956, DOE OpenNet, NV0074094.

68. "Study Proposal by Commissioner Libby Concerning Limitation on Weapons Testing," enclosed with AEC 859/3, August 6, 1956, Medicine, Health and Safety 3 Radioactive Fallout, Box 48, Office of the Secretary General Correspondence (OSGC), 1951–1958, RG 326, NACP.

69. Ibid.

70. Johnson, *The Papers of Adlai E. Stevenson*, vol. 6, 248.

71. Benjamin P. Greene, *Eisenhower, Science Advice, and the Nuclear Test Ban Debate, 1945–1963* (Stanford, CA: Stanford University Press, 2007), 100–102.

72. Telephone Calls, October 23, 1956, October 56 Phone Calls, Box 18, DDE Diary Series (DDS), Ann Whitman File (AWF), DDEL.

73. "Texts of Statement by Eisenhower on Nuclear Weapons and 2 Official Memoranda," *New York Times*, October 24, 1956, 18.

74. Walter Silov [sic], October 24, 1956, Folder 2, Box 372, Adlai E. Stevenson Papers (AESP), Seeley G. Mudd Manuscript Library, Princeton University, Princeton, New Jersey.

75. "Stevenson-Kefauver Campaign Committee Regarding Stopping of H-Bomb Tests," October 18, 1956, Folder 2, Box 372, AESP.

76. H-Bomb Memorandum by Adlai E. Stevenson, October 29, 1956, Folder 3, Box 372, AESP.

77. Dan Jacob to Senator Humphrey, October 11, 1956, Folder 2, Box 372, AESP.

78. "Record of Technical Discussions on the Limitation of Nuclear Weapons Tests Held at the United States Atomic Energy Commission on January 30, 1957," February 1, 1957, DEFE 7/2289, TNA; and Thomas Murray to Libby, March 18, 1957, Medicine, Health and Safety—3, Radiation vol. 3, Box 47, OSGC, 1951–1958, RG 326, NACP.

79. Gerald C. Smith to Strauss, March 21, 1957, Weapons—Test Limitation 1957–58, Box 347, RRAEM, AEOS, RG 59, NACP; and Handwritten Memo, March 23, 1957, Weapons—Test Limitation 1957–58, Box 347, RRAEM, AEOS, RG 59, NACP.

80. *Foreign Relations of the United States, 1955–1957*, vol. 27 (Washington, DC: Government Printing Office, 1992), 741.

81. For the Suez crisis and its effects on U.S.-U.K. relations, see David Carlton, *Britain and the Suez Crisis* (Oxford: Basil Blackwell, 1988).

82. Joint Statement with Prime Minister Macmillan Following the Bermuda Conference, March 24, 1957, accessed October 1, 2019, http://www.presidency.ucsb.edu/ws/index.php?pid=11002.

83. *Foreign Relations of the United States, 1955–1957*, vol. 27, 767.

84. "Technical Problems of Nuclear Test Limitation and Registration," April 30, 1957, 600.0012/4-3057, CDF, RG 59, NACP.

85. "Nuclear Test Limitation Problems," April 9, 1957, Weapons—Test Limitation 1957–58, Box 347, RRAEM, AEOS, RG 59, NACP.

Chapter 5

1. "Statement by the Secretary-General at the Opening Meeting of the Scientific Committee on the Effects of Atomic Radiation, March 14, 1956," Series 262, Box 16, File 1, ACC. A/706 75/7, United Nations Archives (UNA), New York.

2. *Report of the United Nations Scientific Committee on the Effects of Atomic Radiation* (New York, 1958), 100.

3. Ibid., 14.

4. Ibid., 42.

5. Ibid., 41.

6. Néstor Herran, "'Unscare' and Conceal: The United Nations Scientific Committee on the Effects of Atomic Radiation and the Origin of International Radiation Monitoring," in *The Surveillance Imperative: Geosciences during the Cold War and Beyond*, ed. Simone Turchetti and Peder Roberts, 69–84 (New York: Palgrave Macmillan, 2014).

7. Position Paper, March 8, 1956, 13.10, UN Scientific Committee on Radiation 1956, part 1 of 2, Box 237, RRAEM, AEOS, RG 59, NACP.

8. "Suggested Procedures for Collection of Fallout Samples," March 20, 1956, April 20, 1956, 4–5, Series 262, Box 17, File 10, ACC. A/706 75/7, UNA.

9. Letter received from the representative of the United States to the United Nations offering technical assistance in the measurement of radioactive fallout, June 13, 1956, Series 262, Box 17, File 10, ACC. A/706, 75/7, UNA.

10. "Politicheskaia storona podkhoda SShA k mezhdunarodnomu sotrudnichestvu po izucheniiu atomnoi radiatsii" [A Political Side of the Approach of the USA to International Cooperation on the Study of Atomic Radiation], enclosed with G. Saksin to A. A. Gromyko, no. 743, September 1, 1956, f. Otdela mezhdunarodnykh orzanizatsii, op. 2, p. 46, d. (116) 2 Mirnoe ispol'zovanie atomnoi energii i atomnaia radiatsiia, 11.II.56 to 23.X.56, l. 9, Arkhiv Vneshnei Politiki Rossiiskoi Federatsii, Moscow.

11. For the comparison, see Annex IV, Report of Technical Discussions and Recommendations of the Committee Concerning Methods of Measurement of Radioactive Fallout and Other Sources of Radiation, enclosed with PO 131/241, December 11, 1956, Series 262, Box 17, File 11, ACC. A/706, 75/7, UNA.

12. New York to London, no. 56, March 25, 1956, CAB 124/2496, TNA.

13. New York to London, no. 978, October 30, 1956, CAB 124/2496, TNA; and New York to London, no. 251, November 5, 1956, CAB 124/2496, TNA.

14. Shields Warren, Confidential Report, n.d., enclosed with Warren to G. C. Spiegel, November 23, 1956, UN Scientific Committee on Radiation 1956, 2 of 2, Box 237, RRAEM, AEOS, RG 59, NACP.

15. E. E. Pochin, "Discussion of Radiostrontium Hazards," December 20, 1956, enclosed with B. Lush to G. Brown, January 2, 1957, CAB 124/2496, TNA.

16. "Draft Remarks Made by W. F. Libby to Sunshine Group on Monday, Feb 4, 1957," February 11, 1957, Sunshine General, Sunshine Meetings, Box 8, OFWL, RG 326, NACP.

17. Harrison Brown, "What Is a 'Small' Risk?" *The Saturday Review* (August 3, 1957): 9–10.

18. U.S. Congress, House Committee on Appropriations, Subcommittee on Public Works Appropriations, *Atomic Energy Appropriations for 1958* (Washington, DC: Government Printing Office, 1957), 110; U.S. Congress, Joint Committee

on Atomic Energy, Special Subcommittee on Radiation, *Fallout from Nuclear Weapons Tests* (Washington, DC: Government Printing Office, 1959), 17; Statement by John A. McCone, March 24, 1959, Atomic Energy—AEC, JCAE Statements (Fallout Hearings), May 1959, Box 822, Clinton Presba Anderson Papers (CPAP), Manuscript Division, Library of Congress (LOC), Washington, DC.

19. Statement by John A. McCone, March 24, 1959; J. Laurence Kulp, "World-wide Distribution of Strontium-90 and Its Uptake in Man," *Bulletin der Schweizerischen Akademie der Medizinischen Wissenschaften* 14, no. 5–6 (1958): 421–422; and "Fallout Collection Activities," December 4, 1957, Sunshine General 1957, Box 1, RRFS, RG 326, NACP.

20. U.S. Congress, Joint Committee on Atomic Energy, Special Subcommittee on Radiation, *The Nature of Radioactive Fallout and Its Effects on Man* (Washington, DC: Government Printing Office, 1957), 690.

21. "Moscow Says Amazon Is U.S. Bomb Target," *New York Times*, June 12, 1957, 12.

22. Libby to Kulp, February 11, 1958, Reading File—W. F. Libby, January–June 1958, Box 1, OFWL, RG 326, NACP.

23. Kulp, "World-wide Distribution of Strontium-90," 420–421.

24. Hal Hollister, "Project Sunshine: Its Aims and Operation," April 7, 1958, Administrative Files Fallout Program Policy, Box 10, RRFS, RG 326, NACP.

25. M. Susan Lindee, *Suffering Made Real: American Science and the Survivors at Hiroshima* (Chicago: University of Chicago Press, 1994).

26. James J. Orr, *The Victim as Hero: Ideologies of Peace and National Identity in Postwar Japan* (Honolulu: University of Hawai'i Press, 2001).

27. "Effects of Establishment of Danger Area and Nuclear Tests on Japanese Fishing Interests," February 29, 1956, Weapons-Testing Redwing (Pacific 2 Japan Test), January–June 1956, Box 345, RRAEM, AEOS, RG 59, NACP.

28. "Hōshanō wa dono teido made anzen ka, Kōseishō de kijun o happyō, Bei suibaku jikken no eikyō" [How Much Radioactivity Is Safe? The Ministry of Health Announces Standards on the Effects of U.S. H-Bomb Tests], *Asahi Shimbun*, May 5, 1956, 7. For the explanation of the standard, see Hiyama to Willis Boss, May 16, 1956, #270 Yoshio Hiyama 1956, Box 17, John C. Bugher Papers, Rockefeller Archive Center, Sleepy Hollow, New York.

29. Masao Tsuzuki, "Hōshanō jintai eno kyoyōdo" [Permissible Dose of Radioactivity for Humans], *Asahi Shimbun*, May 12, 1956, 5.

30. Kokkyō 4, "Hōshasen eikyō chosa tokubetsu iinkai hōkoku" [Report of the Special Committee for the Study of Radiation Effects], March 3, 1956, B'.2.1.0.3-2, vol. 2, GS; and Kokkyō 3, "Genshiryoku kyoku shusai Kokuren Kagaku Iinkai junbi uchiawasekai (hōkoku) [A Preparatory Meeting for the

United Nations Scientific Committee Called by the Bureau of Atomic Energy (Report)], May 12, 1956, B'.2.1.0.3-2, vol. 3, GS.

31. Tokyo to London, no. 161, March 6, 1957, C'4.2.1.1-2, vol. 2, GS. For more detail about Kishi's political motive, see Toshihiro Higuchi, "An Environmental Origin of Antinuclear Activism in Japan, 1954–1963: The Government, the Grassroots Movement, and the Politics of Risk," *Peace & Change* 33, no. 3 (2008): 333–367.

32. "Dai nijukkai kanjikai (April 3) kiroku" [Record of the Twentieth Executive Meeting], n.d., C'4.2.1.2, GS.

33. New York to Tokyo, no. 287, March 26, 1956, B'.2.1.0.3-2, vol. 2, GS.

34. Nakayama, "Kokuren Kagaku Iinkai Nihon teishutsu shiryō naiyō hōkokukai" (A Briefing Session on Materials Submitted by Japan to the United Nations Scientific Committee), n.d. [March 23, 1957], B'.2.1.0.3-2, vol. 4, GS.

35. "Konomama jikken tsuzukeba 12–17 nen de kyoyō gendo, Hiyama Tōdai kyōju ga happyō" [If Testing Continues as Is Now, Permissible Limits Will Be Reached in 12–17 Years; Announced by Professor Hiyama of Tokyo University], *Asahi Shimbun*, May 27, 1957, 7.

36. Matthew Jones, *After Hiroshima: The United States, Race and Nuclear Weapons in Asia, 1945–1965* (Cambridge; New York: Cambridge University Press, 2010).

37. Masao Tsuzuki, "Kokusai Rengō Kagaku Iinkai dai 4 kai kaigi hōkokusho (betsuhen)" [A Report on the Fourth Session of the United Nations Scientific Committee (Supplement)], April 19, 1958, B'.2.1.0.3-2, vol. 5, GS.

38. Charles L. Dunham to Bourke B. Hickenlooper, October 22, 1958, 13.10: UN Scientific Committee on Radiation 1958, part 1 of 4, Box 237, RRAEM, AEOS, RG 59, NACP.

39. *Report of the United Nations Scientific Committee on the Effects of Atomic Radiation*, 14.

40. Arthur L. Richards to Philip Farley, June 11, 1958, 13.10: UN Scientific Committee on Radiation 1958, part 2 of 4, Box 237, RRAEM, AEOS, RG 59, NACP.

41. Shane J. Maddock, *Nuclear Apartheid: The Quest for American Atomic Supremacy from World War II to the Present* (Chapel Hill: University of North Carolina Press, 2010), 104–106; and Kenneth Osgood, *Total Cold War: Eisenhower's Secret Propaganda Battle at Home and Abroad* (Lawrence: University Press of Kansas, 2006), 189–199.

42. A. A. Fursenko, ed., *Prezidium TsK KPSS, 1954–1964, Tom 2, Postanovleniia, 1954–1958* [Presidium of the Central Committee of the CPSU, Book 2, Resolutions] (Moscow: ROSSPEN, 2006), 337.

43. V. Iu. Afiani and V. D. Esakov, eds., *Akademiia nauk v resheniiakh TsK KPSS: Biuro prezidiuma, prezidium, sekretariat TsK KPSS, 1952–1958* [Academy of Sciences in Resolutions of the Central Committee of the Communist Party of the Soviet Union: Bureau of the Presidium, Presidium, and Secretariat, 1952–1958] (Moscow: ROSSPEN, 2010), 608.

44. Welles Hangen, "Lysenko, Stalin's Protégé, Out as Soviet's Scientific Chieftain," *New York Times*, April 10, 1956, 1. For the views of Muller and other U.S. scientists on Lysenkoism, see Michael D. Gordin, "How Lysenkoism Became Pseudoscience: Dobzhansky to Velikovsky," *Journal of the History of Biology* 45, no. 3 (2012): 443–468; William deJong-Lambert, "Hermann J. Muller, Theodosius Dobzhansky, Leslie Clarence Dunn, and the Reaction to Lysenkoism in the United States," *Journal of Cold War Studies* 15, no. 1 (2013): 78–118; and Audra J. Wolfe, "What Does It Mean to Go Public? The American Response to Lysenkoism, Reconsidered," *Historical Studies in the Natural Sciences* 40, no. 1 (2010): 48–78.

45. Second Memorandum on Reports Received Concerning the Genetic Aspects of Irradiation, March 22, 1957, Annex 1, 1, UNSCEARL.

46. Daniel P. Todes, *Pavlov's Physiology Factory: Experiment, Interpretation, Laboratory Enterprise* (Baltimore: Johns Hopkins University Press, 2002).

47. Ethan Pollock, *Stalin and the Soviet Science Wars* (Princeton, NJ: Princeton University Press, 2006), 136–167; and George Windholz, "The 1950 Joint Scientific Session: Pavlovians as the Accusers and the Accused," *Journal of the History of the Behavioral Sciences* 33, no. 1 (1997): 61–81.

48. L. A. Orbeli, "Action of Ionizing Radiations on the Animal Organism," in *Conference of the Academy of Sciences of the USSR on the Peaceful Uses of Atomic Energy, July 1–5, 1955, Session of the Division of Biological Science*, English trans. 15–19 (Washington, DC: U.S. Atomic Energy Commission, 1956), 16.

49. V. V. Sedov, "Pervaia vsesoiuznaia konferentsiia po meditsinskoi radiologii" [The First All-Union Conference on Medical Radiology], *Atomnaia energiia* [Atomic Energy] 1, no. 2 (1956) : 98–100.

50. A. M. Kuzin, "Some Current Problems in Radiobiology," trans. A. Pirscenok, *Bulletin of the Atomic Scientists* 14, no. 1 (1958): 49.

51. Roger Clarke and Jack Valentin, "A History of the International Commission on Radiological Protection," *Health Physics* 88, no. 6 (2005): 720.

52. "Radiation Effects on the Nervous System," *IAEA Bulletin* 3, no. 4 (1961): 6–7.

53. Warren, Confidential Report, n.d. Lebedinskii himself admitted this problem. See Stenogramma raschirennogo zasedaniia radiobiologicheskoi komissii pri prezidiume AN SSSR, July 11, 1958, f. 1522, op. 1, d. 159, l. 72, ARAN.

54. Valery N. Soyfer, *Lysenko and the Tragedy of Soviet Science* (New Brunswick, NJ: Rutgers University Press, 1994), 255–258.

55. A. F. Andreev, ed., *Kapitsa, Tamm, Semenov: V ocherkakh i pis'makh* [Kapitsa, Tamm, Semenov: In Surveys and Letters] (Moscow: Vagris-Priroda, 1998), 29; V. A. Tsukerman and Z. M. Azarkh, *Liudi i vzryvy* [People and Explosions] (Arzamas-16, Russia: VNIIEF, 1994), 132; Nikolai V. Timofeev-Resovskii, "Iz istorii dialoga biologov i fizikov" [From the History of the Dialogue of Biologists and Physicists], in *Vospominaniia o I. E. Tamme* [Reminiscences of I. E. Tamm], ed. E. L. Feinberg, 282–285 (Moscow: IZDAT, 1995), 284–285; and Soyfer, *Lysenko and the Tragedy of Soviet Science*, 233, 241–242.

56. Michael D. Gordin and Karl Hall, "Introduction: Intelligentsia Science Inside and Outside Russia," *Osiris* 23 (2008): 1–19.

57. Kendall E. Bailes, *Technology and Society Under Lenin and Stalin: Origins of the Soviet Technical Intelligentsia, 1917–1941* (Princeton, NJ: Princeton University Press, 1978).

58. Benjamin Tromly, *Making the Soviet Intelligentsia: Universities and Intellectual Life Under Stalin and Khrushchev* (Cambridge; New York: Cambridge University Press, 2013); and Vladislav M. Zubok, *Zhivago's Children: The Last Russian Intelligentsia* (Cambridge, MA: Harvard University Press, 2009).

59. David Holloway, "Physics, the State, and Civil Society in the Soviet Union," *Historical Studies in the Physical and Biological Sciences* 30, no. 1 (1999): 173–192.

60. Stenogramma zasedaniia presidiuma akademii nauk SSSR, March 29, 1957, f. 2, op. 3a, d. 207, ll. 106–107, ARAN.

61. Norman Cousins, *Dr. Schweitzer of Lambaréné* (Westport, CT: Greenwood Press, 1973), 167–168.

62. Albert Schweitzer, "A Declaration of Conscience," *The Saturday Review* (May 18, 1957): 20.

63. Paul Rubinson, *Redefining Science: Scientists, the National Security State, and Nuclear Weapons in Cold War America* (Amherst; Boston: University of Massachusetts Press, 2016), 35–62.

64. Linus Pauling, *No More War!* (New York: Dodd, Mead, 1958), 170.

65. "An Appeal by American Scientists to the Governments and People of the World," n.d., 1.2, Box 5.001, Ava Helen and Linus Pauling Papers (AHLPP), Oregon State University Libraries Special Collections, Corvallis.

66. Pauling to Hammarskjold, January 13, 1958, 2.1, Box 5.002, AHLPP.

67. For the origins of the Pugwash Conference, see Eugene Rabinowitch, "Pugwash—History and Outlook," *Bulletin of the Atomic Scientists* 13, no. 7 (1957): 243–248.

68. Afiani and Esakov, *Akademiia nauk v resheniiakh TsK KPSS*, 734–735.

69. "The Pugwash Conference: Hazards Arising from the Use of Atomic Energy in Peace and War, Committee I," *Bulletin of the Atomic Scientists* 13, no. 7 (1957): 251.

70. Ibid.

71. Afiani and Esakov, *Akademiia nauk v resheniiakh TsK KPSS*, 818–819.

72. "Ob'edinit' usiliia uchenykh v bor'be za nemedlennoe zapreschenie iadernogo oruzhiia" [Unite the Efforts of Scientists in the Struggle for the Immediate Prohibition of Nuclear Weapons], *Pravda*, August 13, 1957, 4.

73. Pauling to Hammarskjold, January 13, 1958.

74. Afiani and Esakov, *Akademiia nauk v resheniiakh TsK KPSS*, 872–873.

75. Ovsei I. Leipunskii, "Radioaktivnaia opasnost' vzryvov chistovodorodnoi bomby i obychnoi atomnoi bomby" [The Radiological Danger of the Explosions of a Clean Hydrogen Bomb and an Ordinary Atomic Bomb], *Atomnaia energiia* [Atomic Energy] 3, no. 12 (1957): 530–539.

76. Andrei D. Sakharov, *Memoirs*, trans. Richard Lourie (London: Hutchinson, 1990), 201.

77. Andrei D. Sakharov, "Radioactive Carbon in Nuclear Explosions and Nonthreshold Biological Effects," in *Soviet Scientists on the Danger of Nuclear Tests*, ed. A. V. Lebedinsky, 39–49 (Moscow: Foreign Languages Publishing House, 1960), 48–49.

78. Afiani and Esakov, *Akademiia nauk v resheniiakh TsK KPSS*, 872–873; and Sakharov, *Memoirs*, 204.

79. Stenogramma raschirennogo zasedaniia radiobiologicheskoi komissii pri prezidiume AN SSSR, July 11, 1958, ll. 5-8.

80. Ibid., l. 15.

81. New York to Washington, no. 740, January 28, 1958, Radiation General January-March 1958, Box 236, RRAEM, AEOS, RG 59, NACP.

82. New York to Washington, no. 905, February 21, 1958, Radiation General January–March, 1958, Box 246, RRAEM, AEOS, RG 59, NACP.

83. *Report of the United Nations Scientific Committee on the Effects of Atomic Radiation*, 41.

84. Matthew Evangelista, *Unarmed Forces: The Transnational Movement to End the Cold War* (Ithaca, NY: Cornell University Press, 1999), 53–58; and James G. Richter, *Khrushchev's Double Bind: International Pressures and Domestic Coalition Politics* (Baltimore: Johns Hopkins University Press 1994), 101–111.

85. Statement by Prof. A. V. Lebedinsky at the Press Conference on June 17, 1958, 3-4, Series 262, Box 16, File 5, ACC. A/706 75/7, UNA.

86. "Suspension of Nuclear Testing," May 6, 1958, CF 1012—Chronology CPH, Box 149, Executive Secretariat—Conference Files (ES/CF), RG 59, NACP; London to Washington, no. 4583, July 16, 1958, FO 371/135550, TNA; and Susanna Schrafstetter and Stephen Twigge, *Avoiding Armageddon: Europe, the United States, and the Struggle for Nuclear Nonproliferation, 1945–1970* (Westport, CT: Praeger, 2004), 93.

87. Macmillan to Lloyd, July 8, 1958, PREM 11/2566, TNA.

88. "Exchange of Views on the Limitation of Nuclear Testing," June 9, 1958, Macmillan Talks, Washington, Chronology, June 7–11, 1958, Box 150, ES/CF, RG 59, NACP.

89. Benjamin P. Greene, *Eisenhower, Science Advice, and the Nuclear Test Ban Debate, 1945–1963* (Stanford, CA: Stanford University Press, 2007), 139–140, 143–144.

90. "Exchange of Views on the Limitation of Nuclear Testing," June 9, 1958.

91. Greene, *Eisenhower, Science Advice*, 131–135, 144–153.

92. Ibid., 144–164.

93. Memo of Conversation with John A. McCone, July 30, 1958, July 1958 (1), Box 16, JFD Chronological Series (CS), John Foster Dulles Papers (JFDP), DDEL.

94. A. A. Tomei, "GAC Paper on Atomic Tests," May 9, 1958, AEC, Box 5, Records of the Office of the Special Assistant for Science and Technology, WHO, DDEL.

95. W. F. Libby and Edward Teller to McCone, n.d., enclosed with C. A. Borg to Breithut, July 31, 1958, Weapons—Test Moratorium—Other World Opinion on Suspension of Testing, 1958, Box 432, RRAEM, AEOS, RG 59, NACP.

96. London to Washington, no. 5946, August 20, 1958, PREM 11/2566, TNA; Dulles to Macmillan, August 20, 1958, August 1958 (1), Box 16, CS, JFDP, DDEL.

Chapter 6

1. Donovan Bess, "Radioactive Test Delay Admitted," *San Francisco Chronicle*, April 4, 1958, 1, 5; Hale Champion, "Rains Boost Radioactivity in Vegetables," *San Francisco Chronicle*, April 6, 1958, 1, 7; "Radioactive Vegetables," *New York Times*, April 7, 1958, 12; and Robert S. Roe to Commissioner, April 16, 1958, 400.22, Box 2471, Subject Files, Records of the Food and Drug Administration (RG 88), NACP.

2. Kelly Moore, *Disrupting Science: Social Movements, American Scientists, and the Politics of the Military, 1945–1975* (Princeton, NJ: Princeton University Press, 2008), 96–129.

3. Michael Egan, *Barry Commoner and the Science of Survival: The Remaking of American Environmentalism* (Cambridge, MA: MIT Press, 2007). For further discussion of the emerging consumerism-environmentalism nexus in Cold War America forged through radioactive fallout, see Thomas Jundt, *Greening the Red, White, and Blue: The Bomb, Big Business, and Consumer Resistance in Postwar America* (New York: Oxford University Press, 2014), 93–100; and Kendra Smith-Howard, *Pure and Modern Milk: An Environmental History Since 1900* (Oxford; New York: Oxford University Press, 2014), 128–137.

4. Norman Lloyd, "A-Tests Pollute Midwest," *Chicago Tribune*, October 13, 1956, 1; and Willard F. Libby, "Current Research Findings on Radioactive Fallout," *Proceedings of the National Academy of Sciences* 42, no. 12 (1956): 945–962.

5. U.S. Congress, Joint Committee on Atomic Energy, Special Subcommittee on Radiation, *The Nature of Radioactive Fallout and Its Effects on Man* (Washington, DC: Government Printing Office, 1957), 557.

6. Lawrence Goodwyn, *The Populist Moment: A Short History of the Agrarian Revolt in America* (New York: Oxford University Press, 1978). For the progressive tradition of politics in Minnesota, see Daniel J. Elazar, Virginia Gray, and Wyman Spano, *Minnesota Politics & Government* (Lincoln: University of Nebraska Press, 1999).

7. For isolationism in the Midwest during the 1930s, see Peter Trubowitz, *Defining the National Interest: Conflict and Change in American Foreign Policy* (Chicago: University of Chicago Press, 1998), 96–168. For the history of peace activism in Minnesota, see Rhoda R. Gilman, *Stand Up!: The Story of Minnesota's Protest Tradition* (St. Paul: Minnesota Historical Society Press, 2012).

8. David E. Nye, *Electrifying America: Social Meanings of a New Technology, 1880–1940* (Cambridge, MA: MIT Press, 1990), 299.

9. Ibid., 307–326. For the relationship between conservation and democracy in rural America, see Sarah T. Phillips, *This Land, This Nation: Conservation, Rural America, and the New Deal* (Cambridge; New York: Cambridge University Press, 2007).

10. ANS Nuclear Café, "Elk River—Rural America's First Atomic Power Plant," July 28, 2016, accessed October 1, 2019, http://ansnuclearcafe.org/2016/07/28/elk-river-rural-americas-first-atomic-power-plant/#sthash.cpxGm4UI.dpbs.

11. Interim Report of Minnesota Atomic Development Problems Committee, Atomic Energy Development Problems in MN, February–May 1958, Box 5, Maurice B. Visscher Papers (MBVP), University Archives, University of Minnesota, Minneapolis.

12. Visscher to Lee Loevinger, March 10, 1958, Atomic Energy Development Problems in MN, February–May 1958, Box 5, MBVP.

13. "Stevenson-Kefauver Campaign Committee Regarding Stopping of H-Bomb Tests," October 18, 1956, Folder 2, Box 372, AESP.

14. Maurice B. Visscher, "Radioactive Fall-Out," *New York Times*, March 19, 1958, 30.

15. "Report by the Director, Division of Biology and Medicine, U.S. Atomic Energy Commission on Allegation by Maurice B. Visscher in Letter to Editor, New York Times, That U.S. Atomic Energy Commission Is Withholding Data on

Strontium 90 in U.S. Milk," April 1, 1958, enclosed with Dunham to Visscher, May 5, 1958, Atomic Energy Development Problems in MN, February–May 1958, Box 5, MBVP.

16. Visscher, "Radioactive Fall-Out."

17. Paul Rubinson, *Redefining Science: Scientists, the National Security State, and Nuclear Weapons in Cold War America* (Amherst; Boston: University of Massachusetts Press, 2016), 35–62.

18. Visscher to Pauling, March 3, 1958, Folder 1.9, Box 6.001, AHLPP.

19. "Program on Strontium-90 in Milk—Project C-98," August 21, 1958, Folder 11.10, Box 7.011, AHLPP.

20. "The Milk All of Us Drink—And Fallout," *Consumer Reports* 24, no. 3 (1959): 102–103.

21. Lawrence B. Glickman, *Buying Power: A History of Consumer Activism in America* (Chicago: University of Chicago Press, 2009), 212–213, 258–259; and U.S. Congress, Joint Committee on Atomic Energy, Special Subcommittee on Radiation, *Radiation Protection Criteria and Standards: Their Basis and Use* (Washington, DC: Government Printing Office, 1960), 571.

22. Glickman, *Buying Power*, 259–260, 263–264, 271.

23. "Radiation Hazards from Fallout and X Rays," *Consumer Reports* 23, no. 9 (1958): 484.

24. For the rights-based discourse of post–World War II consumerism, see Lizabeth Cohen, *A Consumers' Republic: The Politics of Mass Consumption in Postwar America* (New York: Alfred A. Knopf, 2003), 133. Also see U.S. Congress, Joint Committee on Atomic Energy, Special Subcommittee on Radiation, *Radiation Protection Criteria and Standards*, 575.

25. "Program on Strontium-90 in Milk—Project C-98," August 21, 1958, Folder 11.10, Box 7.011, AHLPP; and Letter, John Hanc to CU Shopper, July 23, 1958, Folder 11.10, Box 7.011, AHLPP.

26. Medical Research Council, *The Hazards to Man of Nuclear and Allied Radiations* (London: H.M.S.O., 1956), 124–125; and F. J. Bryant, A. C. Chamberlain, A. Morgan, and G. S. Spicer, *Radiostrontium Fallout in Biological Materials in Britain* (Harwell, UK: Atomic Energy Research Establishment, 1956).

27. Hansard (UK), House of Commons Debate, May 13, 1957, vol. 570, cc 5–6; May 16, 1957, vol. 570, cc 570–1; and May 30, 1957, vol. 571, cc 585–6.

28. Kenneth O. Morgan, *Modern Wales: Politics, Places and People* (Cardiff: University of Wales Press, 1995), 114. For the CND in Wales and its connections with the rise of "left-of-center" nationalism in Wales, see Christopher R. Hill, "Nations of Peace: Nuclear Disarmament and the Making of National Identity in Scotland and Wales," *Twentieth Century British History* 27, no. 1 (2016): 26–50.

29. Morgan, *Modern Wales*, 14–15, 18. For the postwar consensus, see David Dutton, *British Politics Since 1945: The Rise, Fall and Rebirth of Consensus*, 2nd ed. (Oxford; Malden, MA: Wiley-Blackwell, 1997).

30. Hansard (UK), House of Commons Debate, October 17, 1944, vol. 403, cc 2312.

31. Joint ARC/MRC/DC Committee on Biological Problems of Nuclear Physics, Managing Subcommittee on Monitoring, July 19, 1957, FD 1/8461, TNA.

32. "Co-ordination of Work on Monitoring Fallout etc.," June 13, 1957, CAB 124/1658, TNA.

33. Managing Sub-Committee on Monitoring, Summary of Plans Approved by the Subcommittee for the Sampling and Analysis of Biological Materials in the UK (as of 3rd October, 1958), MAF 291/220, TNA.

34. Margaret Fremlin, "There Isn't a Snake in the Cupboard: A Review of the Life of J. H. Fremlin, Updated on June 22, 2012," Chapter 14, accessed October 1, 2019, http://margaret.fremlin.org/fourteen.html.

35. Memo to Minister, July 16, 1958, BD 24/171, TNA.

36. Fremlin, "There Isn't a Snake in the Cupboard," Chapter 14.

37. "Geiger Clicks in Sheep Country," *Times* (London), July 14, 1958, 5.

38. For the role of British popular newspapers in shaping nuclear culture, see Adrian Bingham, "'The Monster'? The British Popular Press and Nuclear Culture, 1945–Early 1960s," *The British Journal for the History of Science* 45, no. 4 (2012): 609–624.

39. Brooke to Prime Minister, October 17, 1958, FD 23/707, TNA.

40. "Strontium Warning by M.O.H.," *Liverpool Daily Post*, January 5, 1959, filed in BD 25/99, TNA; and "Sr90—A Doctor Speaks Again," *Western Mail* (Cardiff), January 14, 1959, filed in BD 24/171, TNA.

41. "Welsh M.O.H.s Urge Radio-active Substance Research Laboratory," *Liverpool Daily Post*, August 10, 1959, filed in BD 24/171, TNA.

42. Memo to Beddoe, November 3, 1960, BD 11/3050, TNA.

43. Statement by Maurice B. Visscher, February 6, 1959, Atomic Energy Development Problems in MN, 1959, Box 5, MBVP.

44. Meeting of the Subcommittee on the Biological Significance of Radiation, May 15, 1958, n.d., Atomic Energy Development Problems in MN, February–May 1958, Box 5, MBVP.

45. Visscher to Freeman, January 31, 1959, Atomic Energy Development Problems in MN, 1959, Box 5, MBVP.

46. Statement by Maurice B. Visscher, February 6, 1959.

47. Visscher to Hubert Humphrey and Eugene McCarthy, February 9, 1959, Atomic Energy Development Problems in MN, 1959, Box 5, MBVP.

48. New York to London, no. 618, October 14, 1959, CAB 124/2500, TNA; and A. R. Moore to D. S. Cape, November 3, 1959, CAB 124/2500, TNA.

49. Daniel Heidt, "'I Think That Would Be the End of Canada': Howard Green, the Nuclear Test Ban, and Interest-Based Foreign Policy, 1946–1963," *American Review of Canadian Studies* 42, no. 3 (2012): 343–369.

50. Cabinet Conclusions, October 17, 1959, quoted in Heidt, "'I Think That Would Be the End of Canada'," 352.

51. London to New York, no. 1341, November 9, 1959, CAB 124/2500, TNA; and Moore to Cape, November 3, 1959.

52. Draft, Hollister, Western, Bruner, March 5, 1959, Sunshine General 1958, Box 1, RRFS, RG 326, NACP; and Minutes of Cabinet Meeting, March 6, 1959, Staff Notes March 1–15, 1959 (2), Box 39, DDS, AWF, DDEL.

53. Minutes of Cabinet Meeting, March 6, 1959, Staff Notes Mar 1–15, 1959 (2), Box 39, DDS, AWF, DDEL.

54. "Statement on Strontium 90 in Minnesota Wheat Made Before the Joint Committee on Atomic Energy During the Authorization Hearings, February 27, 1959, by Willard F. Libby," February 27, 1959, Data Files—Radioactivity Foods, Box 28, RRFS, RG 326, NACP.

55. Libby to Visscher, March 6, 1959, Fallout—Visscher, Maurice B., Box 312, General Correspondence of the Joint Committee on Atomic Energy (GCJCAE), Records of the Joint Committees of Congress (RG 128), The National Archives Building (NAB), Washington, DC.

56. "The Milk All of Us Drink," 108–110.

57. U.S. Congress, Joint Committee on Atomic Energy, Subcommittee on Research, Development, and Radiation, *Fallout, Radiation Standards, and Countermeasures* (Washington, DC: Government Printing Office, 1963), vol. 2, 499.

58. John D. Porterfield to James T. Ramey, May 17, 1959, Fallout-Strontium90-Milk, Box 312, GCJCAE, RG 128, NAB.

59. "The Milk All of Us Drink," 103.

60. Ibid., 111.

61. E. Jerry Jessee, "Radiation Ecologies: Bombs, Bodies, and Environment During the Atmospheric Nuclear Weapons Testing Period, 1942–1965" (PhD diss., Montana State University, 2013), 330–339.

62. U.S. Congress, Joint Committee on Atomic Energy, Special Subcommittee on Radiation, *Fallout from Nuclear Weapons Tests* (Washington, DC: Government Printing Office, 1959), 763–764.

63. A. R. Luedecke to the Joint Committee on Atomic Energy, March 21, 1959, Fallout, vol. 6, January thru July 1959, Box 310, GCJCAE, RG 128, NAB.

64. Henry Wexler to Bronk, January 8, 1958, Meteorological General 1958, CBEAR, NAA.

65. Wexler to Bronk, May 19, 1958, Meteorological General 1958, CBEAR, NAA.

66. "Texts of Documents Made Public by Anderson on Fall-Out of Radioactive Materials," *New York Times*, March 22, 1959, 66.

67. Ibid.; and *Newsweek*, April 6, 1959, 35.

68. Statement by John A. McCone, March 24, 1959.

69. Luedecke to Dunham, April 22, 1959, Info & Publications 9 (1959), Box 51, CSF, DBER, RG 326, NACP. The report first appeared in September and continued until April 1960, when it was incorporated into the USPHS's monthly report *Radiological Health Data*.

70. Andrew Stark, *Conflict of Interest in American Public Life* (Cambridge, MA: Harvard University Press, 2000).

71. National Advisory Committee on Radiation, *Report to the Surgeon General on the Control of Radiation Hazards in the United States* (Washington, DC: n.p., March 1959).

72. George T. Mazuzan and J. Samuel Walker, *Controlling the Atom: The Beginning of Nuclear Regulation, 1946–1962* (Berkeley: University of California Press, 1984), 253.

73. Ibid., 257.

74. Ibid., 257–258.

75. Benjamin P. Greene, *Eisenhower, Science Advice, and the Nuclear Test Ban Debate, 1945–1963* (Stanford, CA: Stanford University Press, 2007), 170–174.

76. David Griggs and Edward Teller, *Deep Underground Test Shots* (UCRL-4659) (Livermore, CA: University of California Radiation Laboratory, 1956). For more about the origins of underground testing, see Kai-Henrik Barth, "Detecting the Cold War: Seismology and Nuclear Weapons Testing, 1945–1970" (PhD diss., University of Minnesota, 2000).

77. Edward Teller, *Memoirs: A Twentieth-Century Journey in Science and Politics* (Cambridge, MA: Perseus, 2001), 436; and Rubinson, *Redefining Science*, 101–105.

78. *Foreign Relations of the United States, 1958–1960*, vol. 3 (Washington, DC: Government Printing Office, 1996), 737.

79. Ibid., 708.

80. Ibid., 714–715.

81. "A Lead on Tests," *Manchester Guardian*, May 4, 1959, 6.

82. Susanna Schrafstetter and Stephen Twigge, *Avoiding Armageddon: Europe, the United States, and the Struggle for Nuclear Nonproliferation, 1945–1970* (Westport, CT: Praeger, 2004), 97–98.

83. *Foreign Relations of the United States, 1958–1960*, vol. 3, 723.

84. Ibid., 724.

85. U.S. Department of State, *Documents on Disarmament, 1945–1959* (Washington, DC: Government Printing Office, 1960), 1392–1393.

86. Ibid., 1396–1398.

87. Washington to London, no. 269, July 14, 1959, PREM 11/2862, TNA.

88. Gordon Chase to McGeorge Bundy, August 6, 1963, ACDA Disarmament, Subjects, Nuclear Test Ban, General, 8/63 and undated, Box 261A, Departments & Agencies, National Security Files (NSF), Presidential Papers (PP), John F. Kennedy Presidential Library (JFKL), Boston.

89. "Moratorium on Nuclear Testing," December 16, 1959, Suspension of Nuclear Testing, and Surprise Attack (2), Box 2, Briefing Notes Subseries, National Security Council Series, Records of the Office of the Special Assistant for National Security Affairs, WHO, DDEL.

90. Robert S. Norris and William M. Arkin, "Known Nuclear Tests Worldwide, 1945–98," *Bulletin of the Atomic Scientists* 54, no. 6 (1998): 66.

91. Matthew Evangelista, *Unarmed Forces: The Transnational Movement to End the Cold War* (Ithaca, NY: Cornell University Press, 1999), 95, 100–109.

92. Greene, *Eisenhower, Science Advice*, 190–193, 196–197.

93. London to Washington, no. 2123, April 29, 1959, PREM 11/2861, TNA; and Kendrick Oliver, *Kennedy, Macmillan and the Nuclear Test-Ban Debate, 1961–63* (Basingstoke, UK: Macmillan, 1998), 11–13.

94. *Foreign Relations of the United States, 1958–1960*, vol. 3, 819.

95. Macmillan to Selwyn Lloyd, August 7, 1959, PREM 11/2862, TNA.

96. Memo of Conference with the President, May 5, 1959, May 6, 1959, Staff Notes, May 1959 (3), Box 41, DDS, AWF, DDEL.

97. Memo of Conference with the President, September 22, 1959, September 24, 1959, Staff Notes September 1959 (1), Box 44, DDS, AWF, DDEL.

98. George W. Beadle to Howard L. Andrews, April 1, 1960, Effects of Strontium-90 General 1956–1960, CBEAR, NAA.

99. George B. Kistiakowsky, *A Scientist at the White House: The Private Diary of President Eisenhower's Special Assistant for Science and Technology* (Cambridge, MA: Harvard University Press, 1976), 291.

100. Memo of Conference with the President, April 22, 1960, April 28, 1960, Staff Notes, April 1960 (1), Box 49, DDS, AWF, DDEL; and Greene, *Eisenhower, Science Advice*, 220–221.

101. Greene, *Eisenhower, Science Advice*, 208–218. For the U-2 incident of 1960, see Michael R. Beschloss, *May-Day: Eisenhower, Khrushchev, and the U-2 Affair* (New York: Harper & Row, 1986).

102. *Public Papers of the President, 1960–1961* (Washington, DC: Government Printing Office, 1961), 627.

Chapter 7

1. "Soviet Statement on Resumption of Nuclear Testing," *Current Digest of the Soviet Press* 13, no. 35 (September 27, 1961): 6.
2. Ibid., 8.
3. Harold Brown Oral History Interview—JFK #5, 6/25/1964, 4, JFKL.
4. "Position on Nuclear Testing," XX JCAE re. Suspension, discontinuance of nuclear weapons testing, 86th Cong., Box 818, CPAP, LOC.
5. "Nixon Hits Critics of A-Test Ban," *New York Times*, November 13, 1959, 16.
6. "Remarks of Senator John F. Kennedy, Student Convocation, UCLA, November 2, 1959," John F. Kennedy Presidential Library and Museum, accessed October 1, 2019, https://www.jfklibrary.org/archives/other-resources/john-f-kennedy-speeches/university-of-california-los-angeles-19591102.
7. Peter J. Roman, *Eisenhower and the Missile Gap* (Ithaca, NY: Cornell University Press, 1995).
8. Shane J. Maddock, *Nuclear Apartheid: The Quest for American Atomic Supremacy from World War II to the Present* (Chapel Hill: University of North Carolina Press, 2010), 128–129.
9. Benjamin P. Greene, *Eisenhower, Science Advice, and the Nuclear Test Ban Debate, 1945–1963* (Stanford, CA: Stanford University Press, 2007), 236; and Glenn T. Seaborg, *Kennedy, Khrushchev, and the Test Ban* (Berkeley: University of California Press, 1981), 57–59.
10. A. A. Fursenko, ed., *Prezidium TsK KPSS, 1954–1964, Tom 1, Chernovye protokol'nye zapisi zasedanii, stenogrammy* [Presidium of the Central Committee of the CPSU, 1954–1964, Book 1: Draft Protocols of Meetings, Stenographic Records] (Moscow: ROSSPEN, 2003), 506–507.
11. Greene, *Eisenhower, Science Advice*, 236; Susanna Schrafstetter and Stephen Twigge, *Avoiding Armageddon: Europe, the United States, and the Struggle for Nuclear Nonproliferation, 1945–1970* (Westport, CT: Praeger, 2004), 106–107; Seaborg, *Kennedy, Khrushchev*, 66–68; and "Nuclear Tests Negotiations," June 9, 1961, enclosed with "Nuclear Tests," June 9, 1961, PREM 11/3590, TNA.
12. *Foreign Relations of the United States, 1961–1963*, vol. 7 (Washington, DC: Government Printing Office, 1995), 75.
13. London to Washington, no. 2921 Saving, "Nuclear Tests Conference," June 19, 1961, PREM 11/3591, TNA.
14. *Foreign Relations of the United States, 1961–1963*, vol. 7, 131.
15. Ibid., 139.
16. Fursenko, *Prezidium TsK KPSS, 1954–1964, Tom 1*, 500–501.

17. For the 1961 Berlin crisis, see Marc Trachtenberg, *A Constructed Peace: The Making of the European Settlement, 1945–1963* (Princeton, NJ: Princeton University Press, 1999), 322–351. For the nuclear dimensions of the Berlin crisis, see Campbell Craig, *Destroying the Village: Eisenhower and Thermonuclear War* (New York: Columbia University Press, 1998); and Francis J. Gavin, *Nuclear Statecraft: History and Strategy in America's Atomic Age* (Ithaca, NY: Cornell University Press, 2012), 70–87.

18. Andrei D. Sakharov, *Memoirs*, trans. Richard Lourie (London: Hutchinson, 1990), 215.

19. Nikita S. Khrushchev, *Memoirs of Nikita Khrushchev*, vol. 3 (University Park: Pennsylvania State University Press, 2007), 483–484.

20. Sakharov, *Memoirs*, 217. Also see Andrei D. Sakharov, *Sakharov Speaks* (New York: Alfred A. Knopf, 1974), 33.

21. Sakharov, *Memoirs*, 218. For his estimate, see Sakharov, "Radioactive Carbon in Nuclear Explosions and Nonthreshold Biological Effects," in *Soviet Scientists on the Danger of Nuclear Tests*, ed. A. V. Lebedinsky, 39–49 (Moscow: Foreign Languages Publishing House, 1960), 45.

22. Sergei Khrushchev, *Nikita Khrushchev and the Creation of a Superpower* (University Park: Pennsylvania State University Press, 2000), 446–447.

23. Sakharov, *Memoirs*, 218.

24. S. N. Voronin, et al., *To vremia ukhodit v istoriiu* [That Time Becomes a History] (Sarov, Russia: RFIaTs-VNIIEF, 2008), 89.

25. *Foreign Relations of the United States, 1961–1963*, vol. 7, 158.

26. Washington to London, no. 2242, September 3, 1961, PREM 11/3591, TNA.

27. Memo, Prime Minister, September 3, 1961, PREM 11/3591, TNA; London to Washington, no. 6156, September 3, 1961, PREM 11/3591, TNA; Washington to London, no. 2245, September 3, 1961, PREM 11/3591, TNA; and *Foreign Relations of the United States, 1961–1963*, vol. 7, 160.

28. U.S. Arms Control and Disarmament Agency, *Documents on Disarmament 1961* (Washington, DC: Government Printing Office, 1962), 389.

29. Washington to Ottawa, no. 2917, September 15, 1961, CSC 1644:7 part 4, vol. 21283, Records of the Department of National Defence (RG 24), NLAC.

30. UNGA, Special Political Committee 262nd Meeting, October 16, 1961, A/SPC/SR.262, Agenda Item 24, p. 7, Series 262, Box 17, File 3, ACC. A/706 75/7, UNA.

31. Rusk to Paris, no. TOPOL 481, October 4, 1961, Nuclear Testing 3/61-12/61, Box WH-42, Classified Subject Files, White House Files, Papers of Arthur M. Schlesinger, Jr., JFKL.

32. Kathleen Teltsch, "U.N., 87-11, Appeals to Soviet on Test," *New York Times*, October 28, 1961, 1.

33. Lawrence Freedman, *Kennedy's Wars: Berlin, Cuba, Laos, and Vietnam* (New York; Oxford: Oxford University Press, 2000), 72–91.

34. "6 Nations in U.N. Bid Soviet Cancel 50–Megaton Test," *New York Times*, October 21, 1961, 1.

35. "In Memorium," *Washington Post*, October 24, 1961, A14.

36. V. B. Adamskii and Iu. N. Smirnov, "50-Megatonnyi vzryv nad novoi zemlei" [50-Megaton Explosion over Novaya Zemlya], *Voprosy istorii estestvoznaniia i tekhniki* [Problems of the History of Science and Technology] 3 (1995): 79–99.

37. *Foreign Relations of the United States, 1961–1963*, vol. 7, 286.

38. Ibid., 241–248.

39. Macmillan to Kennedy, January 5, 1962, United Kingdom, Security 1/5/62–3/10/62, Box 127A, Countries, The President's Office Files (POF), PP, JFKL.

40. Bundy, TAB A: The President's Review of the Atmospheric Testing Problem, n.d., Nuclear Weapons Testing General 2/17/62–4/4/62 and undated, Box 300, Subjects, NSF, PP, JFKL.

41. *Foreign Relations of the United States, 1961–1963*, vol. 7, 320–321.

42. *Public Papers of the President, 1962* (Washington, DC: Government Printing Office, 1963), 186–192, quote from 187.

43. "Memo," n.d., Atomic Energy—Hearing Outline; and Memoranda (Fallout Hearings) May 1959, Box 822, CPAP, LOC.

44. Federal Radiation Council, *Report No. 2, Background Material for the Development of Radiation Protection Standards* (Washington, DC: Government Printing Office, 1961), 4–5.

45. R. C. Norton to Edith Walker, "Levels of I.131 from Fallout in Milk," October 2, 1961, MAF 298/145, TNA; Committee on Protection Against Ionizing Radiations, Minutes of the 35th Meeting, October 5, 1961, FD 7/13, TNA; and Food Science & Atomic Energy Division, Submission: Fallout: Strontium 90, November 29, 1961, MAF 298/145, TNA.

46. Federal Radiation Council, *Report No. 2*, 6.

47. NACOR Minutes, Meeting of October 6, 1961, pp. 24, 31–32, Carton 4, General Subject File (GSF) 1959–1966, Bureau of Radiological Health (BRH), Records of the Public Health Service (RG 90), NACP.

48. NACOR Minutes, Meeting of January 31, 1962, p. 125, Carton 4, GSF 1959–1966, BRH, RG 90, NACP.

49. "The Protection of Infants from Fallout in Milk: Submission to the Minister," October 23, 1961, MAF 298/145, TNA; and E. Walker, "Fallout from Weapon Testing," October 26, 1961, MAF 298/145, TNA.

50. Minutes of the Meeting, October 27, 1961, FD 7/3512, TNA.

51. P. Humphreys-Davies to Franklin et al., "Fall-out and Milk," October 25, 1961, MAF 298/145, TNA.

52. National Milk Publicity Council, Bulletin no. 2, "Radio-active Fall-out and Milk," October 26, 1961, MAF 251/648, TNA.

53. NACOR Minutes, Meeting of October 5, 1961, pp. 153–154, 159, Carton 3, GSF 1959–1966, BRH, RG 90, NACP.

54. Nate Haseltine, "Experts Say 50-Megaton A-Blast by Reds Won't Change U.S. Life: No Safe Threshold," *Washington Post*, October 28, 1961, 4.

55. Robert C. Pendleton, Ray D. Lloyd, and Charles W. Mays, "Iodine 131 in Utah During July and August 1962," *Science* 141, no. 3581 (1963): 640–642.

56. U.S. Congress, Joint Committee on Atomic Energy, Subcommittee on Research, Development, and Radiation, *Fallout, Radiation Standards, and Countermeasures* (Washington, DC: Government Printing Office, 1963), vol. 1, 337.

57. Dunning, "Meeting in Utah re High Iodine Levels in Milk," August 13, 1963, Fallout—Iodine 131, Box 21, RRFS, RG 326, NACP.

58. U.S. Congress, Joint Committee on Atomic Energy, Subcommittee on Research, Development, and Radiation, *Fallout, Radiation Standards, and Countermeasures*, vol. 2, 544.

59. Ibid., vol. 2, 558.

60. Ibid., vol. 1, 19.

61. Richard Starners, "When We Strip, We Will Glow in the Dark," *Washington Daily News*, June 10, 1963, 1.

62. Amy G. Swerdlow, *Women Strike for Peace: Traditional Motherhood and Radical Politics in the 1960s* (Chicago: University of Chicago Press, 1993), 15–26.

63. Amy G. Swerdlow, "The Politics of Motherhood: The Case of Women Strike for Peace and the Test Ban Treaty" (PhD diss., Rutgers University—New Brunswick, 1984), 151–161.

64. Elaine Tyler May, *Homeward Bound: American Families in the Cold War Era* (New York: Basic Books, 1988).

65. Amy G. Swerdlow, "Pure Milk, Not Poison: Women Strike for Peace and the Test Ban Treaty of 1963," in *Rocking the Ship of State: Toward a Feminist Peace Politics*, ed. Adrienne Harris and Ynestra King, 225–237 (Boulder, CO: Westview Press, 1989), 227–229.

66. Molly Ladd-Taylor and Lauri Umansky, "Introduction," in *"Bad" Mothers: The Politics of Blame in Twentieth-Century America*, ed. Ladd-Taylor and Umansky, 1–28 (New York: New York University Press, 1998), 3–4.

67. Rima D. Apple, *Perfect Motherhood: Science and Childrearing in America* (New Brunswick, NJ: Rutgers University Press, 2006), 107–134; and Julia

Grant, *Raising Baby by the Book: The Education of American Mothers* (New Haven, CT; London: Yale University Press, 1998), 3–6.

68. Lawrence S. Wittner, *The Struggle Against the Bomb, vol. 2: Resisting the Bomb: A History of the World Nuclear Disarmament Movement, 1954–1970* (Stanford, CA: Stanford University Press, 1997), 251. A similar grassroots monitoring campaign by women also began in Canada, where Voice of Women, launched in July 1960, collected thousands of baby teeth. See Wittner, 198–199.

69. WARI Memo, n.d., CND/2008/15/4, CNDP.

70. Sample letter, Atomic Energy Information Agency, November 24, 1962, CND/2008/15/4, CNDP.

71. Wittner, *The Struggle Against the Bomb, vol.* 2, 52–53.

72. Ibid., 246–248, 324–329.

73. Lynn Z. Bloom, *Doctor Spock: Biography of a Conservative Radical* (Indianapolis: Bobbs-Merrill, 1972), 242–243.

74. "Danger Radioactive—Committee of 100," n.d., [October 1961], CND/2008, 3/5, CNDP.

75. "What to Eat in a Nuclear Age," *Times* (London), January 16, 1962, 7.

76. "Is This What It's Coming to?" *New York Times*, July 5, 1962, 40.

77. Protest—Protest [1962], literature 1962, Box 2, Series A-2, Women Strike for Peace Records (DG 115), Swarthmore College Peace Collection, Swarthmore, Pennsylvania.

78. Swerdlow, *Women Strike for Peace*, 83.

79. U.S. Arms Control and Disarmament Agency, *Documents on Disarmament 1962* (Washington, DC: Government Printing Office, 1963), 479–484, 693–697; "Sweden Urges Partial Ban," *New York Times*, August 2, 9.

80. *Foreign Relations of the United States, 1961–1963*, vol. 7, 471–475, 502–503, 512.

81. Kennedy to Macmillan, July 27, 1962, United Kingdom, Security 7/62–9/62, Box 127A, Countries, POF, PP, JFKL.

82. U.S. Arms Control and Disarmament Agency, *Documents on Disarmament 1962*, 804–807.

83. *Foreign Relations of the United States, 1961–1963*, vol. 7, 561–562.

84. Ibid., 497.

85. S. Khrushchev, *Nikita Khrushchev*, 514, 526–527.

86. *UNSCEAR 2000 Report, Volume 1: Sources* (New York: United Nations, 2000), 203, 207.

87. Sakharov, *Memoirs*, 228.

88. Ibid., 229.

89. V. B. Adamsky, "Becoming a Citizen," in *Andrei Sakharov: Facets of a Life*, ed. B. L. Altshuler et al., 21–43 (Gif-sur-Yvette, France: Editions Frontières,

1991), 38–39; V. B. Adamskii, "Dear Mr. Khrushchev," *Bulletin of the Atomic Scientists* 51, no. 6 (1995): 28–31; Matthew Evangelista, *Unarmed Forces: The Transnational Movement to End the Cold War* (Ithaca, NY: Cornell University Press, 1999), 83–84; Sakharov, *Memoirs*, 230–231; and Sakharov, *Sakharov Speaks*, 34.

90. Sakharov, *Memoirs*, 230–231.

91. Kendrick Oliver, *Kennedy, Macmillan and the Nuclear Test-Ban Debate, 1961–63* (Basingstoke, UK: Macmillan, 1998), 150–151. For the Sino-Indian clash, see John W. Garver, *Protracted Contest: Sino-Indian Rivalry in the Twentieth Century* (Seattle: University of Washington Press, 2001), 124.

92. Moscow to London, no. 2251, "Farewell Interview with Khrushchev," November 12, 1962, PREM 11/4554, TNA.

93. Anatoly F. Dobrynin, *In Confidence: Moscow's Ambassador to America's Six Cold War Presidents* (New York: Random House, 1995), 100; and S. Khrushchev, *Nikita Khrushchev*, 692–693.

94. *Foreign Relations of the United States, 1961–1963*, vol. 7, 623–625; Oliver, *Kennedy, Macmillan*, 142, 146; Seaborg, *Kennedy, Khrushchev*, 180–181; and "Main Points of JBW Discussion with Dobrynin on Test Ban," May 16, 1963, Disarmament—Subjects Nuclear Testing, 12/62–8/63, Box 368a, Carl Kaysen, NSF, PP, JFKL.

95. Vojtech Mastny, "The 1963 Nuclear Test Ban Treaty: A Missed Opportunity for Détente?" *Journal of Cold War Studies* 10, no. 1 (2008): 14.

96. Oliver, *Kennedy, Macmillan*, 153–154, 156–158; Geneva to London, no. 54, February 26, 1963, PREM 11/4555, TNA; and *Congressional Record, 1963* (United States), 1588–1589.

97. Sean J. Savage, *JFK, LBJ, and the Democratic Party* (Albany: State University of New York Press, 2004), 108.

98. A. A. Fursenko, ed., *Prezidium TsK KPSS, 1954–1964, Tom 3, Postanovleniia, 1959–1964* [Presidium of the Central Committee of the CPSU: Book 3, Resolutions, 1959–1964] (Moscow: ROSSPEN, 2008), 483. Also see Cousins's own account. Norman Cousins, *The Improbable Triumvirate: John F. Kennedy, Pope John, Nikita Khrushchev* (New York: W.W. Norton, 1972).

99. Fursenko, *Prezidium TsK KPSS, 1954–1964, Tom 3*, 481–482. Khrushchev shared this conclusion with both U.S. and British ambassadors in Moscow. See Moscow to London, no. 751, April 25, 1963, PREM 11/4557, TNA; and *Foreign Relations of the United States, 1961–1963*, vol. 7, 685–686.

100. S. Khrushchev, *Nikita Khrushchev*, 693–694.

101. Fursenko, *Prezidium TsK KPSS, 1954–1964, Tom 1*, 701.

102. Moscow to London, May 8, 1963, PREM 11/4557, TNA.

103. *Foreign Relations of the United States, 1961–1963*, vol. 7, 710–711.

104. Ibid., 713.

105. Ibid., 731, 752–755, 757–758, 769, 785–786.

106. Hailsham to Secretary of State for Foreign Affairs, July 3, 1963, PREM 11/4558, TNA.

107. Moscow to Washington, no. 154, July 16, 1963, Test Ban Treaty Briefing Book Incoming 5, Box 540, W. Averell Harriman Papers (WAHP), LOC.

108. Moscow to Washington, no. 153, July 15, 1963, Disarmament Subjects, Harriman Mission, Outgoing Cables 7/12/63–7/18/63 + undated, Box 369, Carl Kaysen, NSF, PP, JFKL;

109. Moscow to Washington, no. 183, July 17, 1963, and no. 186, July 17, 1963, both Disarmament Subjects, Harriman Mission, Outgoing Cables 7/12/63–7/18/63 + undated, Box 369, Carl Kaysen, NSF, PP, JFKL.

110. *Public Papers of the President, 1963* (Washington, DC: Government Printing Office, 1964), 603.

111. U.S. Congress, Senate Committee on Foreign Relations, *Nuclear Test Ban Treaty* (Washington, DC: Government Printing Office, 1963), 214.

112. U.S. Congress, Senate Committee on Foreign Relations, *Nuclear Test Ban Treaty*, 227; U.S. Congress, Senate Committee on Armed Services, Preparedness Investigating Subcommittee, *Military Aspects and Implication of Nuclear Test Ban Proposals and Related Matters* (Washington, DC: Government Printing Office, 1963), 356, 471, 944–945.

113. "Test Ban Treaty," enclosed with Frederick G. Dutton to Lawrence O'Brien, August 12, 1963, August 13, 1963, #2, Box 28, White House Staff Files of Lawrence F. O'Brien, Presidential Papers, JFKL.

114. Dutton to Theodore C. Sorensen, "Public Campaign in Support of Test Ban Treaty," August 16, 1963, Disarmament—Nuclear Test Ban Negotiations, 7/63 Meeting in Moscow Part II, Box 100, Subject, POF, PP, JFLK.

115. George F. Johnson (Pennsylvania) to Harriman, August 7, 1963, Test Ban Treaty Public Reaction 16, Test Ban Treaty Public Reaction 16, Box 542, WAHP.

116. Mrs. William A. Daniel (California) to John Pillion, August 31, 1963, Test Ban Treaty Public Reaction 16, Box 542, WAHP.

117. *Public Papers of the President, 1964* (Washington, DC: Government Printing Office, 1965), 502–503; and Wittner, *The Struggle Against the Bomb*, vol. 2, 430–432.

118. Robert S. Norris and William M. Arkin, "Known Nuclear Tests Worldwide, 1945–98," *Bulletin of the Atomic Scientists* 54, no. 6 (1998): 66.

119. Maddock, *Nuclear Apartheid*, 173–175.

120. *Public Papers of the President, 1964*, 908.

121. Edward R. Murrow to Sorensen, "Notes for Inclusion in the President's Speech," July 25, 1963, ACDA Disarmament, Subjects, Nuclear Test Ban Treaty, President's Speech, 7/26/63, 7/25/63-7/26/63, Box 264A, Departments & Agencies, NSF, PP, JFKL.

Conclusion

1. Toshihiro Higuchi, "'Kankyō taikoku' Nihon no genten? 1972 nen Stokkuhorumu ningen kankyō kaigi to Nihon no kankyō gaikō" [The Origin of Japan as a "Green Great Power"? The 1972 Stockholm Human Environmental Conference and Japanese Environmental Diplomacy], in *Reisen henyōki no Nihon gaikō: "Hiyowana taikoku" no kiki to mosaku* [Japanese Diplomacy at the Turning Point in the Cold War: A "Fragile Great Power" in Crisis and Struggle], ed. Sumio Hatano, 256–278 (Kyoto: Mineruva Shobō, 2013); and Rebecca Priestley, *Mad on Radium: New Zealand in the Atomic Age* (Auckland: Auckland University Press, 2012), 140–141, 213–219.

2. Stockholm Declaration of the United Nations Conference on the Human Environment, June 16, 1972, accessed October 1, 2019, https://www.jus.uio.no/english/services/library/treaties/06/6-01/stockholm_decl.xml. For the politics of this anti-fallout clause at the Stockholm Conference, see Higuchi, "'Kankyō taikoku' Nihon no genten?"

3. Bernard Weinraub, "India Becomes 6th Nation to Set Off Nuclear Device," *New York Times*, May 19, 1974, 1; and "India Joins Select Nuclear Club—With an Underground Blast," *Chicago Tribune*, May 19, 1974, 3.

4. "Pakistani Government Statement on Nuclear Tests, May 29, 1998," atomicarchive.com, accessed October 1, 2019, http://www.atomicarchive.com/Docs/Deterrence/PakistanStatement2.shtml.

5. "North Korea Hydrogen Bomb: Read the Full Announcement from Pyongyang," *CNBC*, September 3, 2017, accessed October 1, 2019, https://www.cnbc.com/2017/09/03/north-korea-hydrogen-bomb-read-the-full-announcement-from-pyongyang.html.

6. Kate Brown, *Plutopia: Nuclear Families, Atomic Cities, and the Great Soviet and American Plutonium Disasters* (Oxford: Oxford University Press, 2013).

7. Gabrielle Hecht, *Being Nuclear: Africans and the Global Uranium Trade* (Cambridge, MA: MIT Press, 2012); and Robynne N. Mellor, "The Cold War Underground: An Environmental History of Uranium Mining in the United States, Canada, and the Soviet Union, 1945–1991" (PhD diss., Georgetown University, 2018).

8. Michelle Murphy, *Sick Building Syndrome and the Problem of Uncertainty: Environmental Politics, Technoscience, and Women Workers* (Durham, NC: Duke University Press, 2006).

9. Paul Boyer, *By the Bomb's Early Light: American Thought and Culture at the Dawn of the Atomic Age* (New York: Pantheon, 1985); Angela N. H. Creager, *Life Atomic: A History of Radioisotopes in Science and Medicine* (Chicago: University of Chicago Press, 2013); and Spencer R. Weart, *Nuclear Fear: A History of Images* (Cambridge, MA: Harvard University Press, 1988).

10. *Report of the United Nations Scientific Committee on the Effects of Atomic Radiation* (New York: United Nations, 1958), 41.

11. For U.S. debates over the safety of nuclear power, see Gary L. Downey, "Reproducing Cultural Identity in Negotiating Nuclear Power: The Union of Concerned Scientists and Emergency Core Cooling," *Social Studies of Science* 18, no. 2 (1988): 231–264; J. Samuel Walker, *Three Mile Island: A Nuclear Crisis in Historical Perspective* (Berkeley: University of California Press, 2004); J. Samuel Walker and George T. Mazuzan, *Containing the Atom: Nuclear Regulation in a Changing Environment, 1963–1971* (Berkeley: University of California Press, 1992); and Thomas R. Wellock, "Engineering Uncertainty and Bureaucratic Crisis at the Atomic Energy Commission, 1964–1973," *Technology and Culture* 53, no. 4 (2012): 846–884. For the Soviet perspective, see Sonja D. Schmid, *Producing Power: The Pre-Chernobyl History of the Soviet Nuclear Industry* (Cambridge, MA: MIT Press, 2015).

12. For the radioiodine controversy, see Scott Kirsch, "Harold Knapp and the Geography of Normal Controversy: Radioiodine in the Historical Environment," *Osiris* 19 (2004): 167–181.

13. For the struggle of those affected by the U.S. tests, see Laura J. Harkewicz, "'The Ghost of the Bomb': The Bravo Medical Program, Scientific Uncertainty, and the Legacy of U.S. Cold War Science, 1954–2005" (PhD diss., University of California-San Diego, 2010); and Holly M. Barker, *Bravo for the Marshallese: Regaining Control in a Post-Nuclear, Post-Colonial World*, 2nd ed. (Belmont, CA: Wadsworth, 2013). For the atomic veterans and downwinders in the United States, see Howard Ball, *Justice Downwind: America's Atomic Testing Program in the 1950s* (New York: Oxford University Press, 1986); and Barton C. Hacker, *Elements of Controversy: The Atomic Energy Commission and Radiation Safety in Nuclear Weapons Testing, 1947–1974* (Berkeley: University of California Press, 1994). For the Soviet case, see Susanne Bauer, "Radiation Science After the Cold War: The Politics of Measurement, Risk, and Compensation in Kazakhstan," in *Health, Technologies, and Politics in Post-Soviet Settings: Navigating Uncertainties*, ed. Olga Zvonareva, Evgeniya Popova, and Klasien Horstman, 225–249 (Basingstoke, UK: Palgrave Macmillan, 2017); Magdalena E. Stawkowski, "Radioactive Knowledge: State Control of Scientific Information in Post-Soviet Kazakhstan" (PhD diss., University of Colorado-Boulder, 2014); and Cynthia Werner and Kathleen Purvis-Roberts, "After the Cold War: International Politics, Domestic Policy and the Nuclear Legacy in Kazakhstan," *Central Asian Survey* 25, no. 4 (2006): 461–480. For the British case, see Roger Cross, "British Nuclear Tests and the Indigenous People of Australia," in *The British Nuclear Weapons Programme, 1952–2002*, ed. Douglas Holdstock and Frank

Barnaby, 75–88 (London: Frank Cass, 2003); and Sue Rabbitt Roff, "Long-Term Health Effects in UK Test Veterans," in *The British Nuclear Weapons Programme, 1952–2002*, 99–112.

14. Subcommission on Quaternary Stratigraphy, "Working Group on the 'Anthropocene'," accessed October 1, 2019, http://quaternary.stratigraphy.org/working-groups/anthropocene; and Jan Zalasiewicz et al., "When Did the Anthropocene Begin? A Mid-Twentieth Century Boundary Level Is Stratigraphically Optimal," *Quaternary International* 383 (2015): 196–203.

15. Treaty on the Prohibition of Nuclear Weapons, July 7, 2017, accessed October 1, 2019, https://undocs.org/A/CONF.229/2017/8.

GLOSSARY

The entries with asterisks are provided through the courtesy of Timothy J. Jorgensen.

Alpha particles: The composite of two protons and two neutrons, with the lowest penetration depth of all types of ionizing radiation. An alpha particle can be stopped by a few centimeters of air or a thin sheet of paper.

Beta particles: High-energy, high-speed electrons or positrons emitted by certain types of radioactive nuclei. They have the medium penetrating power and the medium ionizing power.

Counts per minute (cpm)*: Refers to the number of interactions (that is, "counts") of photons or particulate radiation within the detector of a radiation survey meter—typically a Geiger counter—over a one-minute time period. A cpm reading from a survey meter has meaning only in the context of that specified type of meter used in a highly defined way. Geiger counters have very low counting efficiencies, which are highly dependent upon the exact geometry of the detector with the particular sample. The cpm values reflect but a fraction of the actual number of radioactive atoms that decay in the sample.

curie (Ci)*: A unit of radioactivity. The unit was named after Pierre Curie, who was codiscoverer, along with his wife Marie Sklodowska Curie, of the radioactive element radium. A Ci was originally defined as a quantity of radioactivity equal to the amount of radioactivity in one gram of radium (37,000,000,000 disintegrations per second). Since one Ci represents a large quantity of radioactivity, it is more common to see the smaller subunits: millicurie (one thousandth of a Ci) or microcurie (one millionth of a Ci). The Ci has been replaced by the SI (standardized international) unit, becquerel. A becquerel is simply defined as one disintegration per second. One Ci equals 3.7×10^{10} becquerel.

Gamma ray: Electromagnetic radiation of high frequency emitted by the nucleus. It has the most penetrating power, which can irradiate the whole body from outside.

Half-life: The rate of nuclear decay. The time required for one half of the atoms of that material to decay to some other material.

rad*: A unit of radiation dose. This unit's name is an acronym of "radiation absorbed dose." It is a measure of the amount of energy deposited in a defined amount of matter by exposure to ionizing radiation. One rad was originally defined as the ionizing radiation dose resulting in 100 ergs of energy being deposited in one gram of matter. The rad has been replaced by the SI unit, gray (Gy), which was named after the radiation scientist Louis Harold Gray. A Gy represents one joule of ionizing radiation energy deposited in one kilogram of matter. One rad equals 0.01 Gy.

rem*: A unit of ionizing radiation dose equivalency. This is an acronym of "roentgen equivalent man." This unit was originally developed to account for the fact that all types of ionizing radiation are not the same in their ability to produce biological effects, so that knowledge of dose level alone was not a sufficient predicator of the magnitude of any radiation-induced health outcome; information about the type of ionizing radiation was also needed. To deal with this complication, equality factors (QFs) were developed for all the different types of ionizing radiation. QFs are radiation-type-specific multiplication factors that are normalized to a defined biological radiation effect from a defined dose of x-rays. The dose in rads multiplied by the QF for the particular type of radiation thus results in a dose equivalency, with the unit of rem. By using rems rather than rads to measure "dose," biological effects such as cancer risk can be predicted regardless of radiation type. The rem has been replaced by the SI unit sievert (Sv), named after the radiation scientist Rolf Maximilian Sievert. One Sv equals 100 rem.

roentgen (R)*: A unit of ionizing radiation *exposure*. Named after Wilhelm Roentgen, the discoverer of x-rays, this unit is the measure of the amount of ionizations that are produced within a specified volume of gas due to exposure to ionizing radiation. Since the average energy required to produce a gas molecule ionization is well established, this unit is an indirect measure of the amount of energy being emitted from a radiation source. In practice, subunits of the roentgen are often reported—milliroentgen (one thousandth of a R). The roentgen has been replaced by the SI unit coulomb/kilogram (C/kg). One roentgen equals 0.000258 C/kg.

Strontium unit (SU): The unit to measure the amount of radioactivity from strontium-90 (Sr-90) in relation to calcium. One SU is equal to one picocurie of Sr-90 per gram of calcium.

BIBLIOGRAPHY

Archival Sources

CANADA

National Library and Archives of Canada, Ottawa
    RG 24, Records of the Department of National Defence
    RG 25, Records of the Department of External Affairs

JAPAN

Daigo Fukuryū Maru Kinenkan, Tokyo
    FOIA 2004-00197

Gaikō Shiryōkan, Tokyo
    B'2.1.0.3-2 Kokusai rengō sōkai hojo kikan kankei zakken, kagaku iinkai kankei (UNSCEAR) [Miscellaneous matters related to the auxiliary organizations of the United Nations General Assembly, On the Scientific Committee (UNSCEAR)]
    C'4.2.1.1-2 Gensuibaku jikken kankei, Eikoku kankei, Matsushita tokushi haken kankei [Matters related to A- and H-bomb tests, United Kingdom, on the dispatch of special emissary Matsushita]
    C'4.2.1.2 Kakubakuhatsu jikken ni taisuru honpō no taido [Our country's position regarding nuclear explosions tests]
    C'4.2.1.5 Dai go Fukuryū Maru sonota Bikini genbaku hisai jiken kankei ikken [Matters related to the A-bomb incident of the Lucky Dragon No. 5 and others]

RUSSIA

Arkhiv Rossiiskoi akademii nauk, Moscow
    f.2, Sekretariat Prezidiuma Rossiiskoi akademii nauk [Secretariat of the Presidium of the Russian Academy of Sciences]
    f.534, Otdelenie biologicheskikh nauk Akademii nauk SSSR [Department of Biological Sciences of the USSR Academy of Sciences]
    f.1522, Otdel rabot po atomnoi energii Akademii nauk SSSR [Department of Work on Atomic Energy of the USSR Academy of Sciences]

Arkhiv Vneshnei Politiki Rossiiskoi Federatsii, Moscow
    f. Otdela mezhdunarodnykh orzanizatsii [Department of International Organizations]

Gosudarstvennyi Arkhiv Rossiiskoi Federatsii, Moscow
    f. R-8009, Ministerstvo zdravookhraneniia SSSR [The USSR Ministry of Welfare]

UNITED KINGDOM

Cambridge University Library Special Collections, Cambridge
    Joseph S. Mitchell Papers

LSE Archives and Special Collections, London
    Campaign for Nuclear Disarmament Papers

The National Archives of the United Kingdom, Kew
    AB, United Kingdom Atomic Energy Authority and its predecessors
    BD, Welsh Office and the Wales Office
    CAB, Cabinet Office
    DEFE, Ministry of Defence
    FCO, Foreign and Commonwealth Office and predecessors
    FD, Medical Research Council
    FO, Foreign Office
    MAF, Agriculture, Fisheries and Food Departments, and related bodies
    PREM, Prime Minister's Office

Wellcome Library, London
    Alexander Haddow Papers

UNITED STATES

The Caltech Archives, Pasadena, California
    Alfred H. Sturtevant Papers

The Countway Library of Medicine, Boston
　　Lauriston S. Taylor Papers

The Department of Energy OpenNet System
　　https://www.osti.gov/opennet/

Dwight D. Eisenhower Presidential Library, Abilene, Kansas
　　Ann Whitman File
　　　　DDE Diary Series
　　John Foster Dulles Papers
　　　　JFD Chronological Series
　　White House Office
　　　　National Security Council Staff Papers
　　　　Records of the Office of the Special Assistant for Disarmament
　　　　Records of the Office of the Special Assistant for National Security Affairs
　　　　Records of the Office of the Special Assistant for Science and Technology

John F. Kennedy Presidential Library, Boston
　　Harold Brown Oral History Interview—JFK #5, 6/25/1964
　　Papers of Arthur M. Schlesinger, Jr.
　　Presidential Papers
　　　　National Security Files
　　The President's Office Files
　　　　White House Staff Files of Lawrence F. O'Brien

Manuscript Division, Library of Congress, Washington, DC
　　Clinton Presba Anderson Papers
　　W. Averell Harriman Papers

Lilly Library, Indiana University, Bloomington
　　Papers of Hermann Joseph Muller

National Academies Archives, Washington, DC
　　Committees on Biological Effects of Atomic Radiation

The National Archives Building, Washington, DC
　　RG 128, Records of the Joint Committees of Congress
　　　　General Correspondence of the Joint Committee on Atomic Energy

The National Archives at College Park, Maryland
　　RG 59, Records of the Department of State
　　　　Central Decimal Files
　　　　Executive Secretariat—Conference Files

Special Assistant to the Secretary of State for Atomic Energy and Outer Space, Records Relating to Atomic Energy Matters, 1948–1962
RG 88, Records of the Food and Drug Administration
Subject Files
RG 90, Records of the Public Health Service
Bureau of Radiological Health, General Subject File 1959–1966
RG 326, Records of the Atomic Energy Commission
Division of Biomedical and Environment Research, Central Subject Files
Office Files of Commissioner Lewis L. Strauss
Office Files of Commissioner Willard F. Libby
Office of the Secretary General Correspondence, 1951–1958
Records Relating to Fallout Studies

Oregon State University Libraries Special Collections, Corvallis
Ava Helen and Linus Pauling Papers

Rockefeller Archive Center, Sleepy Hollow, New York
John C. Bugher Papers

Seeley G. Mudd Manuscript Library, Princeton University, Princeton, New Jersey
Adlai E. Stevenson Papers

Swarthmore College Peace Collection, Swarthmore, Pennsylvania
DG 058, SANE, Inc. Records
DG 115, Women Strike for Peace Records

University Archives, University of Minnesota, Minneapolis
Maurice B. Visscher Papers

University of Washington Special Collections, Seattle
University of Washington Laboratory of Radiation Biology Records

OTHERS
United Nations Archives, New York
UNSCEAR Library, Vienna

Periodicals

Asahi Shimbun
Bulletin of the Atomic Scientists
Chicago Tribune
CNBC (U.S.)
Consumer Reports
Current Digest of the Soviet Press
Gyoson [Fishing Village]
Hansard (Australia)
Hansard (United Kingdom)
Manchester Guardian
New York Times
The News (Adelaide)
Newsweek
Pravda
San Francisco Chronicle
Sydney Morning Herald
Times (London)
U.S. Congressional Record
Washington Daily News
Washington Post
Yomiuri Shimbun

Published Sources

Adams, Mark B. "The Founding of Population Genetics: Contributions of the Chetverikov School 1924–1934." *Journal of the History of Biology* 1, no. 1 (1968): 23–39.

———. "The Soviet Nature-Nurture Debate." In *Science and the Soviet Social Order*, edited by Loren R. Graham, 94–138. Cambridge, MA: Harvard University Press, 1990.

Adamskii, V. B. "Dear Mr. Khrushchev." *Bulletin of the Atomic Scientists* 51, no. 6 (1995): 28–31.

———, and Iu. N. Smirnov. "50-Megatonnyi vzryv nad novoi zemlei." [50-Megaton Explosion Over Novaya Zemlya] *Voprosy istorii estestvoznaniia i tekhniki* [Problems of the History of Science and Technology] 3 (1995): 79–99.

Adamsky, V. B. "Becoming a Citizen." In *Andrei Sakharov: Facets of a Life*, edited by B. L. Altshuler et al., 21–43. Gif-sur-Yvette, France: Editions Frontières, 1991.

Afiani, V. Iu., and V. D. Esakov, eds. *Akademiia nauk v resheniiakh TsK KPSS: Biuro prezidiuma, prezidium, sekretariat TsK KPSS, 1952–1958* [Academy of

Sciences in Resolutions of the Central Committee of the Communist Party of the Soviet Union: Bureau of the Presidium, Presidium, and Secretariat, 1952–1958]. Moscow: ROSSPEN, 2010.

Agar, Jon, and Brian Balmer. "British Scientists and the Cold War: The Defence Research Policy Committee and Information Networks, 1947–1963." *Historical Studies in the Physical and Biological Sciences* 28, no. 2 (1998): 209–252.

Aldrich, Richard J. *The Hidden Hand: Britain, America and Cold War Secret Intelligence.* London: John Murray, 2001.

Aleksandrov, V. Ia. *Trudnye godyi sovetskoi biologii: Zapiski sovremennika* [The Tragic Years of Soviet Biology: Memoirs of a Contemporary]. St. Petersburg: Nauka, 1993.

Andreev, A. F., ed. *Kapitsa, Tamm, Semenov: V ocherkakh i pis'makh* [Kapitsa, Tamm, Semenov: In Surveys and Letters]. Moscow: Vagris Priroda, 1998.

ANS Nuclear Café. "Elk River—Rural America's First Atomic Power Plant." July 28, 2016. Accessed October 1, 2019. http://ansnuclearcafe.org/2016/07/28/elk-river-rural-americas-first-atomic-power-plant/#sthash.cpxGm4UI.dpbs.

Apple, Rima D. *Perfect Motherhood: Science and Childrearing in America.* New Brunswick, NJ: Rutgers University Press, 2006.

Arnold, Lorna. *Britain and the H-Bomb.* Basingstoke, UK; New York: Palgrave, 2001.

———. *A Very Special Relationship: British Atomic Weapon Trials in Australia.* London: H.M.S.O., 1987.

———, and Mark Smith. *Britain, Australia and the Bomb: The Nuclear Tests and Their Aftermath.* Basingstoke, UK; New York: Palgrave Macmillan, 2006.

Babkov, V. V., and E. S. Sakanian. *Nikolai Vladimirovich Timofeev-Resovskii, 1900-1981.* Moscow: Pamiatniki Istoricheskoi Mysli, 2002.

Bailes, Kendall E. *Technology and Society Under Lenin and Stalin: Origins of the Soviet Technical Intelligentsia, 1917–1941.* Princeton, NJ: Princeton University Press, 1978.

Ball, Howard. *Justice Downwind: America's Atomic Testing Program in the 1950s.* New York: Oxford University Press, 1986.

Barker, Holly M. *Bravo for the Marshallese: Regaining Control in a Post-Nuclear, Post-Colonial World.* 2nd ed. Belmont, CA: Wadsworth, 2013.

Barth, Kai-Henrik. "Detecting the Cold War: Seismology and Nuclear Weapons Testing, 1945–1970." PhD diss., University of Minnesota, 2000.

Bauer, Susanne. "Radiation Science After the Cold War: The Politics of Measurement, Risk, and Compensation in Kazakhstan." In *Health, Technologies, and Politics in Post-Soviet Settings: Navigating Uncertainties,* edited by Olga Zvonareva, Evgeniya Popova, and Klasien Horstman, 225–249. Basingstoke, UK: Palgrave Macmillan, 2017.

Beatty, John. "Weighing the Risks: Stalemate in the Classical/Balance Controversy." *Journal of the History of Biology* 20, no. 3 (1987): 289–319.

Beck, Ulrich. *Risk Society: Towards a New Modernity*. Translated by Mark Ritter. London: Sage, 1992.
———. *World at Risk*. Translated by Ciaran Cronin. Cambridge; Malden, MA: Polity Press, 2009.
Beschloss, Michael R. *May-Day: Eisenhower, Khrushchev, and the U-2 Affair*. New York: Harper & Row, 1986.
Bingham, Adrian. "'The Monster'? The British Popular Press and Nuclear Culture, 1945–Early 1960s." *The British Journal for the History of Science* 45, no. 4 (2012): 609–624.
Blair, J. M., D. H. Frisch, and S. Katcoff. "Detection of Nuclear-Explosion Dust in the Atmosphere," October 2, 1945, Los Alamos National Laboratory.
Bloom, Lynn Z. *Doctor Spock: Biography of a Conservative Radical*. Indianapolis: Bobbs-Merrill, 1972.
Boudia, Soraya. "Global Regulation: Controlling and Accepting Radioactivity Risks." *History and Technology* 23, no. 4 (2007): 389–406.
Boyer, Paul S. *By the Bomb's Early Light: American Thought and Culture at the Dawn of the Atomic Age*. New York: Pantheon, 1985.
Brady, Lisa M. "Life in the DMZ: Turning a Diplomatic Failure into an Environmental Success." *Diplomatic History* 32, no. 4 (2008): 585–611.
Brock, Peggy. "South Australia." In *Contested Ground: Australian Aborigines Under the British Crown*, edited by Ann McGrath, 208–239. St. Leonards, Australia: Allen & Unwin, 1995.
Brown, Harrison. "What Is a 'Small' Risk?" *The Saturday Review* (May 25, 1957): 9–10.
Brown, Kate. *Plutopia: Nuclear Families, Atomic Cities, and the Great Soviet and American Plutonium Disasters*. Oxford: Oxford University Press, 2013.
Bryant, F. J., A. C. Chamberlain, A. Morgan, and G. S. Spicer. *Radiostrontium Fallout in Biological Materials in Britain*. Harwell, UK: Atomic Energy Research Establishment, 1956.
———. "Radiostrontium in Soil, Grass, Milk and Bone in U.K.: 1956 Results." *Journal of Nuclear Energy* 6, no. 1–2 (1957): 22–40.
Burnazian, A. I. "O radiatsionnoi bezopasnosti" [On Radiation Safety]. In *Vospominaniia ob Igore Vasil'eviche Kurchatove* [Reminiscences on Igor Vasil'vich Kurchatov], edited by A. P. Aleksandrov, 305–311. Moscow: Nauka, 1988.
Carlson, Elof Axel. *Genes, Radiation, and Society: The Life and Work of H. J. Muller*. Ithaca, NY: Cornell University Press, 1981.
Carlton, David. *Britain and the Suez Crisis*. Oxford: Basil Blackwell, 1988.
Carter, T. C. "The Genetic Problem of Irradiated Human Populations." *Bulletin of the Atomic Scientists* 11, no. 10 (1955): 362–363, 366.
Clark, Claudia. *Radium Girls: Women and Industrial Health Reform, 1910–1935*. Chapel Hill: University of North Carolina Press, 1997.
Clark, Herbert M. "The Occurrence of an Unusually High-Level Radioactive Rainout in the Area of Troy, N.Y." *Science* 119, no. 3097 (1954): 619–622.

Clarke, Roger, and Jack Valentin. "A History of the International Commission on Radiological Protection." *Health Physics* 88, no. 6 (2005): 717–732.

Cochrane, Rexmond C. *The National Academy of Sciences: The First Hundred Years, 1863–1963*. Washington, DC: The National Academy of Sciences, 1978.

Cohen, Lizabeth. *A Consumers' Republic: The Politics of Mass Consumption in Postwar America*. New York: Alfred A. Knopf, 2003.

Cousins, Norman. *Dr. Schweitzer of Lambaréné*. Westport, CT: Greenwood Press, 1973.

———. *The Improbable Triumvirate: John F. Kennedy, Pope John, Nikita Khrushchev*. New York: W.W. Norton, 1972.

Craig, Campbell. *Destroying the Village: Eisenhower and Thermonuclear War*. New York: Columbia University Press, 1998.

Creager, Angela N. H. *Life Atomic: A History of Radioisotopes in Science and Medicine*. Chicago: University of Chicago Press, 2013.

Cross, Roger. "British Nuclear Tests and the Indigenous People of Australia." In *The British Nuclear Weapons Programme, 1952–2002*, edited by Douglas Holdstock and Frank Barnaby, 75–88. London: Frank Cass, 2003.

Crutzen, Paul J. "Geology of Mankind." *Nature* 415, no. 6867 (2002): 23.

———, and Eugene F. Stoermer. "The 'Anthropocene'." *IGBP Global Change Newsletter* 41 (2000): 17–18.

Cullather, Nick. *The Hungry World: America's Cold War Battle Against Poverty in Asia*. Cambridge, MA: Harvard University Press, 2010.

Dalton, Hugh. *The Political Diary of Hugh Dalton, 1918–40, 1945–60*. London: Cape, 1986.

David-Fox, Michael. *Showcasing the Great Experiment: Cultural Diplomacy and Western Visitors to the Soviet Union, 1921–1941*. New York; Oxford: Oxford University Press, 2012.

deJong-Lambert, William. "Hermann J. Muller, Theodosius Dobzhansky, Leslie Clarence Dunn, and the Reaction to Lysenkoism in the United States." *Journal of Cold War Studies* 15, no. 1 (2013): 78–118.

———, and Nikolai Krementsov. "On Labels and Issues: The Lysenko Controversy and the Cold War." *Journal of the History of Biology* 45, no. 3 (2012): 373–388.

———, eds. *The Lysenko Controversy as a Global Phenomenon*, 2 vols. New York: Palgrave Macmillan, 2017.

Department of Fisheries and Wildlife, Western Australia. *The Pearling Industry of Western Australia, 1850–1979*. Perth: Extension and Publicity Service, Department of Fisheries & Wildlife, Western Australia, 1979.

DiMoia, John P. *Reconstructing Bodies: Biomedicine, Health, and Nation-Building in South Korea Since 1945*. Stanford, CA: Stanford University Press, 2013.

Dingman, Roger. "Alliance in Crisis: The Lucky Dragon Incident and Japanese-

American Relations." In *The Great Powers in East Asia, 1953–1960*, edited by Warren I. Cohen and Akira Iriye, 187–214. New York: Columbia University Press, 1990.

Divine, Robert A. *Blowing on the Wind: The Nuclear Test Ban Debate, 1954–1960*. New York: Oxford University Press, 1978.

Dobrynin, Anatoly. *In Confidence: Moscow's Ambassador to America's Six Cold War Presidents*. New York: Random House, 1995.

Documents on Canadian External Relations, vol. 21. Ottawa: Department of Foreign Affairs and International Trade, 1999.

Documents on Canadian External Relations, vol. 22. Ottawa: Department of Foreign Affairs and International Trade, 2001.

Downey, Gary L. "Reproducing Cultural Identity in Negotiating Nuclear Power: The Union of Concerned Scientists and Emergency Core Cooling." *Social Studies of Science* 18, no. 2 (1988): 231–264.

Dubinin, Nikolai P. *Vechnoe dvizhenie* [Perpetual Motion]. Moscow: Politizdat, 1989.

———, "Work of Soviet Biologists: Theoretical Genetics." *Science* 105, no. 2718 (1947): 109–112.

Dupree, A. Hunter. "The Founding of the National Academy of Sciences: A Reinterpretation." *Proceedings of the American Philosophical Society* 101, no. 5 (1957): 434–440.

Dutton, David. *British Politics Since 1945: The Rise, Fall, and Rebirth of Consensus*, 2nd ed. Oxford; Malden, MA: Wiley-Blackwell, 1997.

Edgerton, David. *Warfare State: Britain, 1920–1970*. Cambridge: Cambridge University Press, 2006.

Egan, Michael. *Barry Commoner and the Science of Survival: The Remaking of American Environmentalism*. Cambridge, MA: MIT Press, 2007.

Eisenbud, Merril. *An Environmental Odyssey: People, Pollution, and Politics in the Life of a Practical Scientist*. Seattle: University of Washington Press, 1990.

———, and John H. Harley. "Radioactive Dust from Nuclear Detonations." *Science* 117, no. 3033 (1953): 141–147.

Elagin, A. S., ed. *Semipalatinsk*. Alma-Ata, Kazakh SSR: Nauka, 1984.

Elazar, Daniel J., Virginia Gray, and Wyman Spano. *Minnesota Politics & Government*. Lincoln: University of Nebraska Press, 1999.

Emel'ianov, B. M., and V. S. Gavril'chenko. *Laboratoriia "B": Sungul'skii fenomen*. Snezhinsk, Russia: RFIaTs—VNIITF, 2000.

Evangelista, Matthew. *Unarmed Forces: The Transnational Movement to End the Cold War*. Ithaca, NY: Cornell University Press, 1999.

Evans, Robley D. "Acceptance of the Coolidge Award." *Medical Physics* 11, no. 5 (1984): 579–581.

Falconer, Donald. "Quantitative Genetics in Edinburgh: 1947–1980." *Genetics* 133, no. 2 (1993): 137–142.

Federal Radiation Council. *Report No. 2: Background Material for the Devel-

opment of Radiation Protection Standards. Washington, DC: Government Printing Office, 1961.

Foreign Relations of the United States, 1955–1957, vol. 20. Washington, DC: Government Printing Office, 1990.

Foreign Relations of the United States, 1955–1957, vol. 27. Washington, DC: Government Printing Office, 1992.

Foreign Relations of the United States, 1958–1960, vol. 3. Washington, DC: Government Printing Office, 1996.

Foreign Relations of the United States, 1961–1963, vol. 7. Washington, DC: Government Printing Office, 1995.

Forman, Paul. "Behind Quantum Electronics: National Security as Basis for Physical Research in the United States, 1940–1960." *Historical Studies in the Physical and Biological Sciences* 18, no. 1 (1987): 149–229.

Freedman, Lawrence. *Kennedy's Wars: Berlin, Cuba, Laos, and Vietnam*. New York; Oxford: Oxford University Press, 2000.

Fremlin, Margaret. "There Isn't a Snake in the Cupboard: A Review of the Life of J. H. Fremlin, Updated on June 22, 2012." Accessed October 1, 2019. http://margaret.fremlin.org/fourteen.html.

Funtowicz, Silvio O., and Jerome R. Ravetz. "Science for the Post-Normal Age." *Futures* 25, no. 7 (1993): 739–755.

Fursenko, A. A., ed. *Prezidium TsK KPSS, 1954–1964, Tom 1, Chernovye protokol'nye zapisi zasedanii, stenogrammy* [Presidium of the Central Committee of the CPSU, 1954–1964, Book 1: Draft Protocols of Meetings and Stenographic Records]. Moscow: ROSSPEN, 2003.

———. ed. *Prezidium TsK KPSS, 1954–1964, Tom 2, Postanovleniia, 1954–1958* [Presidium of the Central Committee of the CPSU, Book 2, Resolutions, 1954–1958]. Moscow: ROSSPEN, 2006.

———, ed. *Prezidium TsK KPSS, 1954–1964, Tom 3, Postanovleniia, 1959–1964* [Presidium of the Central Committee of the CPSU: Book 3, Resolutions, 1959–1964]. Moscow: ROSSPEN, 2008.

Garver, John W. *Protracted Contest: Sino-Indian Rivalry in the Twentieth Century*. Seattle: University of Washington Press, 2001.

Gavin, Francis J. *Nuclear Statecraft: History and Strategy in America's Atomic Age*. Ithaca, NY: Cornell University Press, 2012.

Gilman, Rhoda R. *Stand Up! The Story of Minnesota's Protest Tradition*. St. Paul: Minnesota Historical Society Press, 2012.

Glass, H. Bentley. "The Hazards of Atomic Radiations to Man: British and American Reports." *Bulletin of the Atomic Scientists* 12, no. 8 (1956): 312–317.

———. "Timofeeff-Ressovsky, Nikolai Vladimirovich." *Complete Dictionary of Scientific Biography*. Encyclopedia.com. Accessed October 1, 2019. https://www.encyclopedia.com/science/dictionaries-thesauruses-pictures-and-press-releases/timofeeff-ressovsky-nikolai-vladimirovich.

Glasstone, Samuel, ed. *The Effects of Atomic Weapons*. Washington, DC: Government Printing Office, 1950.

Glickman, Lawrence B. *Buying Power: A History of Consumer Activism in America*. Chicago: University of Chicago Press, 2009.

Goikhman, Izabella. "Soviet-Chinese Academic Interactions in the 1950s: Questioning the 'Impact-Response' Approach." In *China Learns from the Soviet Union, 1949–Present*, edited by Thomas P. Bernstein and Hua-Yu Li, 275–302. Lanham, MD: Lexington, 2010.

Goodchild, Peter. *Edward Teller: The Real Dr. Strangelove*. Cambridge, MA: Harvard University Press, 2004.

Goodwyn, Lawrence. *The Populist Moment: A Short History of the Agrarian Revolt in America*. New York: Oxford University Press, 1978.

Gordin, Michael D. "How Lysenkoism Became Pseudoscience: Dobzhansky to Velikovsky." *Journal of the History of Biology* 45, no. 3 (2012): 443–468.

———. "Lysenko Unemployed: Soviet Genetics After the Aftermath." *Isis* 109, no. 1 (2018): 56–78.

———. *Red Cloud at Dawn: Truman, Stalin, and the End of the Atomic Monopoly*. New York: Farrar, Straus and Giroux, 2009.

———, and Karl Hall. "Introduction: Intelligentsia Science Inside and Outside Russia." *Osiris* 23 (2008): 1–19.

Gowing, Margaret. *Independence and Deterrence: Britain and Atomic Energy, 1945–1952*, 2 vols. New York: St. Martin's Press, 1974.

Graham, Loren R. *Science in Russia and the Soviet Union: A Short History*. New York: Cambridge University Press, 1993.

———. *What Have We Learned About Science and Technology from the Russian Experience?* Stanford, CA: Stanford University Press, 1998.

Grant, Julia. *Raising Baby by the Book: The Education of American Mothers*. New Haven, CT; London: Yale University Press, 1998.

Greene, Benjamin P. *Eisenhower, Science Advice, and the Nuclear Test Ban Debate, 1945–1963*. Stanford, CA: Stanford University Press, 2007.

Griggs, David, and Edward Teller. *Deep Underground Test Shots* (UCRL-4659). Livermore, CA: University of California Radiation Laboratory, 1956.

Hacker, Barton C. *The Dragon's Tail: Radiation Safety in the Manhattan Project, 1942–1946*. Berkeley: University of California Press, 1987.

———. *Elements of Controversy: The Atomic Energy Commission and Radiation Safety in Nuclear Weapons Testing, 1947–1974*. Berkeley: University of California Press, 1994.

Hamblin, Jacob Darwin. *Arming Mother Nature: The Birth of Catastrophic Environmentalism*. New York: Oxford University Press, 2013.

———. "'A Dispassionate and Objective Effort': Negotiating the First Study on the Biological Effects of Atomic Radiation." *Journal of the History of Biology* 40, no. 1 (2007): 147–177.

———. *Poison in the Well: Radioactive Waste in the Oceans at the Dawn of the Nuclear Age*. New Brunswick, NJ: Rutgers University Press, 2008.

Hansson, Sven Ove. "Risk and Ethics: Three Approaches." In *Risk: Philosophical Perspectives*, edited by Tim Lewens, 21–35. London: Routledge, 2007.

Harkewicz, Laura J. "'The Ghost of the Bomb': The Bravo Medical Program, Scientific Uncertainty, and the Legacy of U.S. Cold War Science, 1954–2005." PhD diss., University of California-San Diego, 2010.

Hecht, Gabrielle. *Being Nuclear: Africans and the Global Uranium Trade*. Cambridge, MA: MIT Press, 2012.

Heidt, Daniel. "'I Think That Would Be the End of Canada': Howard Green, the Nuclear Test Ban, and Interest-Based Foreign Policy, 1946–1963." *American Review of Canadian Studies* 42, no. 3 (2012): 343–369.

Herken, Gregg. *Cardinal Choices: Presidential Science Advising from the Atomic Bomb to SDI*. New York: Oxford University Press, 1992.

Herran, Néstor. "Spreading Nucleonics: The Isotope School at the Atomic Energy Research Establishment, 1951–67." *The British Journal for the History of Science* 39, no. 4 (2006): 569–586.

———. "'Unscare' and Conceal: The United Nations Scientific Committee on the Effects of Atomic Radiation and the Origin of International Radiation Monitoring." In *The Surveillance Imperative: Geosciences During the Cold War and Beyond*, edited by Simone Turchetti and Peder Roberts, 69–84. New York: Palgrave Macmillan, 2014.

Higuchi, Toshihiro. "'Clean' Bombs: Nuclear Technology and Nuclear Strategy in the 1950s." *Journal of Strategic Studies* 29, no. 1 (2006): 83–116.

———. "An Environmental Origin of Antinuclear Activism in Japan, 1954–1963: The Government, the Grassroots Movement, and the Politics of Risk." *Peace & Change* 33, no. 3 (2008): 333–367.

———. "'Kankyō taikoku' Nihon no genten? 1972 nen Stokkuhorumu ningen kankyō kaigi to Nihon no kankyō gaikō" [The Origin of Japan as a "Green Great Power"? The 1972 Stockholm Human Environmental Conference and Japanese Environmental Diplomacy]. In *Reisen henyōki no Nihon gaikō: "Hiyowana taikoku" no kiki to mosaku* [Japanese Diplomacy at the Turning Point in the Cold War: A "Fragile Great Power" in Crisis and Struggle], edited by Sumio Hatano, 256–278. Kyoto: Mineruva Shobō, 2013.

———. "The Strange Career of Dr. Fish: Yoshio Hiyama, Radioactive Fallout, and Nuclear Fear Management in Japan, 1954–1958." *Historia Scientiarum* 25, no. 1 (2015): 57–77.

Hill, Christopher R. "Nations of Peace: Nuclear Disarmament and the Making of National Identity in Scotland and Wales." *Twentieth Century British History* 27, no. 1 (2016): 26–50.

Hines, Neal O. *Proving Ground: An Account of the Radiobiological Studies in the Pacific, 1946–1961*. Seattle: University of Washington Press, 1963.

Hirschfelder, Joseph O. "The Scientific-Technological Miracle at Los Alamos."

In *Reminiscences of Los Alamos, 1943–1945*, edited by Lawrence Badash, Joseph O. Hirschfelder, and Herbert P. Broida, 67–88. Dordrecht, the Netherlands: D. Reidel, 1980.
Holloway, David. "Nuclear Weapons and the Escalation of the Cold War, 1945–1962." In *The Cambridge History of the Cold War, vol. 1: Origins*, edited by Melvyn P. Leffler and Odd Arne Westad, 376–397. Cambridge: Cambridge University Press, 2010.
———. "Physics, the State, and Civil Society in the Soviet Union." *Historical Studies in the Physical and Biological Sciences* 30, no. 1 (1999): 173–192.
———. *Stalin and the Bomb: The Soviet Union and Atomic Energy, 1939–1956*. New Haven, CT: Yale University Press, 1994.
Hünemörder, Kai. "Environmental Crisis and Soft Politics: Détente and the Global Environment, 1968–1975." In *Environmental Histories of the Cold War*, edited by John R. McNeill and Corinna R. Unger, 257–276. Cambridge: Cambridge University Press, 2010.
Iakubovskaia, E. L., V. I. Nagibin, and V. P. Sislin. *Semipalatinskii ispytatel'nyi poligon: Nezavisimyi analiz problemy* [Semipalatinsk Proving Ground: An Independent Problem Analysis]. Novosibirsk: Novosibirskii Poligrafkombinat, 2003.
Ichikawa, Hiroshi. *Soviet Science and Engineering in the Shadow of the Cold War*. London: Routledge, 2018.
Iida, Kaori. "A Controversial Idea as a Cultural Resource: The Lysenko Controversy and Discussions of Genetics as a 'Democratic' Science in Postwar Japan." *Social Studies of Science* 45, no. 4 (2015): 546–569.
Jessee, E. Jerry. "Radiation Ecologies: Bombs, Bodies, and Environment During the Atmospheric Nuclear Weapons Testing Period, 1942–1965." PhD diss., Montana State University, 2013.
Johnson, Walter, ed. *The Papers of Adlai E. Stevenson*, vol. 6. Boston: Little, Brown, 1976.
Jolly, J. Christopher. "Thresholds of Uncertainty: Radiation and Responsibility in the Fallout Controversy." PhD diss., Oregon State University, 2003.
Jones, Greta. "British Scientists, Lysenko and the Cold War." *Economy and Society* 8, no. 1 (1979): 26–58.
Jones, Matthew. *After Hiroshima: The United States, Race and Nuclear Weapons in Asia, 1945–1965*. Cambridge; New York: Cambridge University Press, 2010.
Jonter, Thomas. *The Key to Nuclear Restraint: The Swedish Plans to Acquire Nuclear Weapons During the Cold War*. London: Palgrave Macmillan, 2016.
Joravsky, David. *The Lysenko Affair*. Chicago: University of Chicago Press, 1986 [1970].
Josephson, Paul R. *Red Atom: Russia's Nuclear Power Program from Stalin to Today*. New York: W.H. Freeman, 2000.
Jundt, Thomas. *Greening the Red, White, and Blue: The Bomb, Big Business,*

and Consumer Resistance in Postwar America. New York: Oxford University Press, 2014.

Kataoka, Chikashi. *Nanyō no Nihonjin gyogyō* [Japanese Fisheries in South Sea]. Tokyo: Dōbunkan Shuppan, 1991.

Khrushchev, Nikita S. *Memoirs of Nikita Khrushchev*, vol. 3. University Park: Pennsylvania State University Press, 2007.

Khrushchev, Sergei. *Nikita Khrushchev and the Creation of a Superpower*. University Park: Pennsylvania State University Press, 2000.

Kirsch, Scott. "Harold Knapp and the Geography of Normal Controversy: Radioiodine in the Historical Environment." *Osiris* 19 (2004): 167–181.

Kistiakowsky, George B. *A Scientist at the White House: The Private Diary of President Eisenhower's Special Assistant for Science and Technology*. Cambridge, MA: Harvard University Press, 1976.

Kobayashi, Tōru, ed. *Gensuibaku kinshi undō shiryōshū* [The Collection of Materials on the Ban-the-Bomb Movement]. 7 vols. Tokyo: Ryokuin Shobō, 1995–1996.

Kōchiken Bikini Suibaku Jikken Hisai Chōsadan. *Mōhitotsu no Bikini jiken: 1000 seki o koeru hisaisen o ou* [Another Bikini Test: Searching for Over 1,000 Ships Affected by the Disaster]. Tokyo: Heiwabunka, 2004.

Kondō, Yasuo, ed. *Suibaku jikken to Nihon gyogyō* [H-Bomb Tests and Japan's Fisheries]. Tokyo: Tokyo Daigaku Shuppankai, 1958.

Kosenko, Mira M. *Where Radiobiology Began in Russia: A Physician's Perspective*. Fort Belvoir, VA: Defense Threat Reduction Agency, 2010.

Krache-Morris, Evelyn Frances. "Into the Wind: The Kennedy Administration and the Use of Herbicides in South Vietnam." PhD diss., Georgetown University, 2012.

Krementsov, Nikolai. *Stalinist Science*. Princeton, NJ: Princeton University Press, 1997.

Krige, John. *American Hegemony and the Postwar Reconstruction of Science in Europe*. Cambridge, MA: MIT Press, 2006.

Kuchinskaya, Olga. *The Politics of Invisibility: Public Knowledge About Radiation Health Effects After Chernobyl*. Cambridge, MA: MIT Press, 2014.

Kuletz, Valerie L. *The Tainted Desert: Environmental Ruin in the American West*. New York: Routledge, 1998.

Kulp, J. Laurence. "World-wide Distribution of Strontium-90 and Its Uptake in Man." *Bulletin der Schweizerischen Akademie der Medizinischen Wissenschaften* 14, no. 5–6 (1958): 420–433.

Kuzin, A. M. "Some Current Problems in Radiobiology." Translated by A. Pirscenok. *Bulletin of the Atomic Scientists* 14, no. 1 (1958): 48–51.

Laakkonen, Simo, Viktor Pál, and Richard Tucker. "The Cold War and Environmental History: Complementary Fields." *Cold War History* 16, no. 4 (2016): 377–394.

Ladd-Taylor, Molly, and Lauri Umansky. "Introduction." In *"Bad" Mothers:*

The Politics of Blame in Twentieth-Century America, edited by Ladd-Taylor and Umansky, 1–28. New York: New York University Press, 1998.

Lapp, Ralph E. "The 'Humanitarian' H-Bomb." *Bulletin of the Atomic Scientists* 12, no. 7 (1956): 261–264.

———. *The Voyage of the Lucky Dragon*. New York: Harper and Brothers, 1958.

Laucht, Christoph. "Scientists, the Public, the State, and the Debate Over the Environmental and Human Health Effects of Nuclear Testing in Britain, 1950–1958." *The Historical Journal* 59, no. 1 (2016): 221–251.

Lavrent'eva, A. "Stroiteli novogo mira" [Builders of the New World]. *V mire knig* [In the World of Books] 9 (1970): 4–5.

Lea, Douglas E. *Actions of Radiations on Living Cells*, 2nd ed. Cambridge: Cambridge University Press, 1955.

Leffler, Melvyn P. *A Preponderance of Power: National Security, the Truman Administration, and the Cold War*. Stanford, CA: Stanford University Press, 1992.

Lehman, Michael R. "Nuisance to Nemesis: Nuclear Fallout and Intelligence as Secrets, Problems, and Limitations on the Arms Race, 1940–1964." PhD diss., University of Illinois at Urbana-Champaign, 2016.

Leipunskii, Ovsei I. "Radioaktivnaia opasnost' vzryvov chistovodorodnoi bomby i obychnoi atomnoi bomby" [The Radiological Danger of the Explosions of a Clean Hydrogen Bomb and an Ordinary Atomic Bomb]. *Atomnaia energiia* [Atomic Energy] 3, no. 12 (1957): 530–539.

Levitt, Joseph. *Pearson and Canada's Role in Nuclear Disarmament and Arms Control Negotiations, 1945–1957*. Montreal: McGill-Queen's University Press, 1993.

Lewis, Edward B. "Remembering Sturtevant." In *Genes, Development and Cancer: The Life and Work of Edward B. Lewis*, edited by Howard D. Lipshitz, 511–516. Dordrecht, the Netherlands: Springer, 2007.

Libby, Willard F. "Current Research Findings on Radioactive Fallout." *Proceedings of the National Academy of Sciences* 42, no. 12 (1956): 945–962.

———. "Dosages from Natural Radioactivity and Cosmic Rays." *Science* 122, no. 3158 (1955): 57–58.

———. Interview by Greg Marlowe. Session II. April 16, 1979. Niels Bohr Library & Archives, American Institute of Physics, College Park, MD. Accessed October 1, 2019. www.aip.org/history-programs/niels-bohr-library/oral-histories/4743-1.

———. "Radioactive Fallout and Radioactive Strontium." *Science* 123, no. 3199 (1956): 657–660.

Lindee, M. Susan. "The Repatriation of Atomic Bomb Victim Body Parts to Japan: Natural Objects and Diplomacy." *Osiris* 13 (1998): 376–409.

———. *Suffering Made Real: American Science and the Survivors at Hiroshima*. Chicago: University of Chicago Press, 1994.

Loeffler, Jane C. *The Architecture of Diplomacy: Building America's Embassies.* New York: Princeton Architectural Press, 1998.

Logachev, V. A. *Semipalatinskii poligon: Obespechenie obschchei i radiatsionnoi bezopasnosti iadernykh ispytanii* [Semipalatinsk Proving Ground: Securing General and Radiation Safety of Nuclear Tests]. Moscow: Medbioekstrem RF, 1997.

Lysenko, Trofim D. *Agrobiology: Essays on Problems of Genetics, Plant Breeding and Seed Growing.* Moscow: Foreign Language Publishing House, 1954.

Lytle, Mark H. *The Gentle Subversive: Rachel Carson, Silent Spring, and the Rise of the Environmental Movement.* New York: Oxford University Press, 2007.

Maddock, Shane J. *Nuclear Apartheid: The Quest for American Atomic Supremacy from World War II to the Present.* Chapel Hill: University of North Carolina Press, 2010.

Malm, Andreas, and Alf Hornborg. "The Geology of Mankind? A Critique of the Anthropocene Narrative." *The Anthropocene Review* 1, no. 1 (2014): 62–69.

Maruhama, Eriko. *Gensuikin shomei undō no tanjō: Tokyo Suginami no jūmin pawā to suimyaku* [Birth of the Petition Campaign for a Ban on A- and H-Bombs: Power of Citizens in Suginami, Tokyo, and Its Context]. Tokyo: Gaifūsha, 2011.

Mastny, Vojtech. "The 1963 Nuclear Test Ban Treaty: A Missed Opportunity for Détente?" *Journal of Cold War Studies* 10, no. 1 (2008): 3–25.

May, Elaine Tyler. *Homeward Bound: American Families in the Cold War Era.* New York: Basic Books, 1988.

Mazuzan, George T., and J. Samuel Walker. *Controlling the Atom: The Beginnings of Nuclear Regulation, 1946–1962.* Berkeley: University of California Press, 1984.

McNeill, John R. "The Environment, Environmentalism, and International Society in the Long 1970s." In *The Shock of the Global: The 1970s in Perspective*, edited by Niall Ferguson, Charles S. Maier, Erez Manela, and Daniel J. Sargent, 263–278. Cambridge, MA: The Belknap Press of Harvard University Press, 2010.

———, and Peter Engelke. *The Great Acceleration: An Environmental History of the Anthropocene Since 1945.* Cambridge, MA: The Belknap Press of Harvard University Press, 2014.

———, and Corinna R. Unger. "Introduction: The Big Picture." In *Environmental Histories of the Cold War*, edited by McNeill and Unger, 1–18. Cambridge: Cambridge University Press, 2010.

Medical Research Council. *The Hazards to Man of Nuclear and Allied Radiations.* London: H.M.S.O., 1956.

Medvedev, Zhores A. *The Rise and Fall of T. D. Lysenko.* Translated by I. Michael Lerner. New York: Columbia University Press, 1969.

Mellor, Robynne N. "The Cold War Underground: An Environmental History of Uranium Mining in the United States, Canada, and the Soviet Union, 1945–1991." PhD diss., Georgetown University, 2018.

Mezelev, L. M. *Oni byli pervymi: Iz istorii iadernykh ispytanii* [They Were the First: From the History of Nuclear Tests], vol. 1. Moscow: n.p., 2001.

Mikhailov, V. N., ed. *Iadernye ispytaniia SSSR* [Nuclear Tests of the USSR], vol. 1. Sarov, Russia: RFIaTs-VNIIEF, 1997.

Milliken, Robert. *No Conceivable Injury*. Ringwood, Australia: Penguin, 1986.

Ministerstvo Oborony Rossiiskoi Federatsii. *Rozhdennye atomnoi eroi: Istoriia sozdaniia i razvitiia 12-e Glavnogo Upravleniia Ministerstva Oborony Rossiiskoi Federatsii* [Birth of the Atomic Era: History of the Creation and Development of the 12th Chief Directorate of the Ministry of Defense of Russian Federation], vol. 1. Moscow: Nauka, 2007.

Miura-shi, ed. *Bikini jiken Miura no kiroku* [The Bikini Incident: The Records of Miura]. Miura, Japan: Miura-shi, 1996.

Miyake, Yasuo. *Kaere Bikini e: Gensuibaku kinshi undō no genten o kangaeru* [Return to Bikini: A Reflection on the Origins of the A- and H-Bomb Ban Movement]. Tokyo: Suiyōsha, 1984.

———. *Nihon no ame* [Rain in Japan]. Tokyo: Hōsei Daigaku Shuppankai, 1956.

———. *Shi no hai to tatakau kagakusha* [Scientists Who Fight the Ashes of Death]. Tokyo: Iwanami Shoten, 1972.

———, Yoshio Hiyama, and Nobuo Kusano, eds. *Bikini suibaku hisai shiryōshū* [Collected Data from the Bikini Hydrogen Bomb Disaster], rev. ed. Tokyo: Tokyo Daigaku Shuppankai, 2014.

Mohun, Arwen P. *Risk: Negotiating Safety in American Society*. Baltimore: Johns Hopkins University Press, 2013.

Montrie, Chad. *The Myth of Silent Spring: Rethinking the Origins of American Environmentalism*. Oakland: University of California Press, 2018.

Moore, Kelly. *Disrupting Science: Social Movements, American Scientists, and the Politics of the Military, 1945–1975*. Princeton, NJ: Princeton University Press, 2008.

Morgan, Kenneth O. *Modern Wales: Politics, Places and People*. Cardiff: University of Wales Press, 1995.

Muller, Hermann J. "The Manner of Dependence of the 'Permissible Dose' of Radiation on the Amount of Genetic Damage." *Acta Radiologica* 41, no. 1 (1954): 5–20.

Murphy, Michelle. *Sick Building Syndrome and the Problem of Uncertainty: Environmental Politics, Technoscience, and Women Workers*. Durham, NC: Duke University Press, 2006.

Mutscheller, Arthur. "Physical Standards of Protection Against Roentgen-Ray Dangers." *American Journal of Roentgenology and Radiation Therapy* 13 (1925): 65–70.

Natal'ina, V. N. "Aerozol'nyi (radionuklidnyi) metod kontrolia ispytanii" [Aerosol (Radionuclide) Method of Control of Tests]. In *Rozhdennaia atomnym vekom: Sbornik istoricheskikh ocherkov, dokumentov i vospominanii veteranov k 40–letiiu sozdaniia v SSSR sluzhby spetsial'nogo kontrolia Ministerstva oborony* [Birth of the Atomic Age: Collection of Historical Assessments, Documents, and Reminiscences of Veterans Toward 40 Years of the Establishment of the USSR Service of Special Control of the Ministry of Defense], edited by A. P. Vasil'ev, vol. 1, 183–188. Moscow: Sluzhba Spetsial'nogo Kontrolia, 1998.

National Academy of Sciences. *The Biological Effects of Atomic Radiation: Summary Reports*. Washington, DC: National Academy of Sciences-National Research Council, 1956.

National Advisory Committee on Radiation. *Report to the Surgeon General, U.S. Public Health Service, on the Control of Radiation Hazards in the United States*. Washington, DC: n.p., March 1959.

National Bureau of Standards. *Maximum Permissible Amounts of Radioisotopes in the Human Body and Maximum Permissible Concentrations in Air and Water* (National Bureau of Standards Handbook 52). Washington, DC: Government Printing Office, 1953.

———. *Permissible Dose from External Sources of Ionizing Radiation: Recommendations of the National Committee on Radiation Protection* (National Bureau of Standards Handbook 59). Washington, DC: Government Printing Office, 1954.

Nihon Katsuo-Maguro Gyogyō Kyōdō Kumiai Rengōkai, ed. *Nikkatsuren shi* [The History of Nikkatsuren], 2 vols. Tokyo: Nihon Katsuo-Maguro Gyogyō Kyōdō Kumiai Rengōkai, 1966–1967.

Nisseikyō Sōritsu 50 Shūnen Kinen Rekishi Hensan Iinkai, ed. *Gendai Nihon seikyō undō shi: Shiryōshū*. [The History of the Contemporary Japanese Cooperative Movement: The Collection of Materials]. Tokyo: Nihon Seikatsu Kyōdō Kumiai Rengōkai, 2001.

Nisseikyō Sōritsu 50 Shūnen Kinen Rekishi Hensan Iinkai, ed. *Gendai Nihon seikyō undō shi*, 2 vols. [The History of the Contemporary Japanese Co-operative Movement]. Tokyo: Nihon Seikatsu Kyōdō Kumiai Rengōkai, 2002.

Norris, Robert S., and William M. Arkin. "Known Nuclear Tests Worldwide, 1945–98." *Bulletin of the Atomic Scientists* 54, no. 6 (1998): 65–67.

Nuzhdin, N. I., N. I. Shapiro, O. N. Petrova, and O. N. Kitaeva. "Effect of Ionizing Radiations on the Fertility of Mice and the Viability of Their Progeny." In *Conference of the Academy of Sciences of the USSR on the Peaceful Uses of Atomic Energy, July 1–5, 1955, Session of the Division of Biological Science*, English trans. 21–31. Washington, DC: U.S. Atomic Energy Commission, 1956.

Nye, David E. *Electrifying America: Social Meanings of a New Technology, 1880–1940*. Cambridge, MA: MIT Press, 1990.

Nye, Mary Jo. *Michael Polanyi and His Generation: Origins of the Social Construction of Science.* Chicago: University of Chicago Press, 2011.

Oikonomou, Alexandros-Panagiotis. "The Hidden Persuaders: Government Scientists and Defence in Post-War Britain." PhD diss., Imperial College London, 2011.

Oliver, Kendrick. *Kennedy, Macmillan and the Nuclear Test-Ban Debate, 1961–63.* Basingstoke, UK: Macmillan, 1998.

Orbeli, L. A. "Action of Ionizing Radiations on the Animal Organism." In *Conference of the Academy of Sciences of the USSR on the Peaceful Uses of Atomic Energy, July 1–5, 1955, Session of the Division of Biological Science*, English trans. 15–19. Washington, DC: U.S. Atomic Energy Commission, 1956.

Oreskes, Naomi. "Introduction." In *Science and Technology in the Global Cold War*, edited by Oreskes and John Krige, 1–9. Cambridge, MA; London: MIT Press, 2014.

Orr, James J. *The Victim as Hero: Ideologies of Peace and National Identity in Postwar Japan.* Honolulu: University of Hawai'i Press, 2001.

Osgood, Kenneth. *Total Cold War: Eisenhower's Secret Propaganda Battle at Home and Abroad.* Lawrence: University Press of Kansas, 2006.

"Pakistani Government Statement on Nuclear Tests, May 29, 1998." atomicarchive.com. Accessed October 1, 2019. http://www.atomicarchive.com/Docs/Deterrence/PakistanStatement2.shtml.

Paul, Diane B. "A War on Two Fronts: J. B. S. Haldane and the Response to Lysenkoism in Britain." *Journal of the History of Biology* 16, no. 1 (1983): 1–37.

Paul, Septimus H. *Nuclear Rivals: Anglo-American Atomic Relations, 1941–1952.* Columbus: Ohio State University Press, 2000.

Pauling, Linus. *No More War!* New York: Dodd, Mead, 1958.

Pendleton, Robert C., Ray D. Lloyd, and Charles W. Mays. "Iodine 131 in Utah During July and August 1962." *Science* 141, no. 3581 (1963): 640–642.

Phillips, Sarah T. *This Land, This Nation: Conservation, Rural America, and the New Deal.* Cambridge; New York: Cambridge University Press, 2007.

Pollock, Ethan. "From *Partiinost'* to *Nauchnost'* and Not Quite Back Again: Revisiting the Lessons of the Lysenko Affair." *Slavic Review* 68, no. 1 (2009): 95–115.

———. *Stalin and the Soviet Science Wars.* Princeton, NJ: Princeton University Press, 2006.

Priestley, Rebecca. *Mad on Radium: New Zealand in the Atomic Age.* Auckland: Auckland University Press, 2012.

Public Papers of the President, 1960–1961. Washington, DC: Government Printing Office, 1961.

Public Papers of the President, 1962. Washington, DC: Government Printing Office, 1963.

Public Papers of the President, 1963. Washington, DC: Government Printing Office, 1964.

Public Papers of the President, 1964. Washington, DC: Government Printing Office, 1965.

"The Pugwash Conference: Hazards Arising from the Use of Atomic Energy in Peace and War, Committee I." *Bulletin of the Atomic Scientists* 13, no. 7 (1957): 251.

Rabinowitch, Eugene. "Pugwash—History and Outlook." *Bulletin of the Atomic Scientists* 13, no. 7 (1957): 243–248.

Rader, Karen A. *Making Mice: Standardizing Animals for American Biomedical Research, 1900–1955*. Princeton, NJ: Princeton University Press, 2004.

"Radiation Effects on the Nervous System." *IAEA Bulletin* 3, no. 4 (1961): 6–7.

RAND Corporation. "Worldwide Effects of Atomic Weapons: Project Sunshine," R-251-AEC. August 6. Oak Ridge, TN: Technical Information Service Extension, 1953.

Räsänen, Tuomas, and Simo Laakkonen. "Cold War and the Environment: The Role of Finland in International Environmental Politics in the Baltic Sea Region." *Ambio* 36, no. 2–3 (2007): 229–236.

Reinhardt, Bob H. *The End of a Global Pox: America and the Eradication of Smallpox in the Cold War Era*. Chapel Hill: University of North Carolina Press, 2015.

"Remarks of Senator John F. Kennedy, Student Convocation, UCLA, November 2, 1959." John F. Kennedy Presidential Library and Museum. Accessed October 1, 2019. https://www.jfklibrary.org/archives/other-resources/john-f-kennedy-speeches/university-of-california-los-angeles-19591102.

The Report of the Royal Commission into British Nuclear Tests in Australia, vol. 1. Canberra: Australian Government Publishing Service, 1985.

Report of the United Nations Scientific Committee on the Effects of Atomic Radiation. New York: United Nations, 1958.

Riabev, L. D., ed. *Atomnyi proekt SSSR: Dokumenty i materialy* [Atomic Project of the USSR: Documents and Materials], t. 2, kn. 1. Moscow: Nauka, 1999.

———, ed. *Atomnyi proekt SSSR: Dokumenty i materialy* [Atomic Project of the USSR: Documents and Materials], t. 3, kn. 1. Moscow: Nauka, 2008.

———, ed., *Atomnyi proekt SSSR: Dokumenty i materialy* [Atomic Project of the USSR: Documents and Materials], t. 3, kn. 2. Moscow: Nauka, 2009.

Richelson, Jeffrey T. *Spying on the Bomb: American Nuclear Intelligence from Nazi Germany to Iran and North Korea*. New York: W.W. Norton, 2006.

Richter, James G. *Khrushchev's Double Bind: International Pressures and Domestic Coalition Politics*. Baltimore: Johns Hopkins University Press, 1994.

Riehl, Nikolaus, and Frederick Seitz. *Stalin's Captive: Nikolaus Riehl and the Soviet Race for the Bomb*. Washington, DC: American Chemical Society, 1996.

Roff, Sue Rabbitt. "Long-Term Health Effects in UK Test Veterans." In *The Brit-

ish Nuclear Weapons Programme, 1952–2002, edited by Douglas Holdstock and Frank Barnaby, 99–112. London: Frank Cass, 2003.

Roll-Hansen, Nils. *The Lysenko Effect: The Politics of Science*. Amherst, NY: Humanity Books, 2005.

Roman, Peter J. *Eisenhower and the Missile Gap*. Ithaca, NY: Cornell University Press, 1995.

Rubinson, Paul. *Redefining Science: Scientists, the National Security State, and Nuclear Weapons in Cold War America*. Amherst; Boston: University of Massachusetts Press, 2016.

Sabol, Steven. *Russian Colonization and the Genesis of Kazak National Consciousness*. Basingstoke, UK; New York: Palgrave Macmillan, 2003.

Sakamoto, Kazuya. "Kakuheiki to Nichi-Bei kankei: Bikini jiken no gaikō shori." [Nuclear Weapons and Japan-U.S. Relations: Diplomatic Settlement of the Bikini Incident] *Nenpō kindai Nihon kenkyū* 16 (1994): 243–271.

Sakharov, Andrei D. *Memoirs*. Translated by Richard Lourie. London: Hutchinson, 1990.

———. "Radioactive Carbon in Nuclear Explosions and Nonthreshold Biological Effects." In *Soviet Scientists on the Danger of Nuclear Tests*, edited by A. V. Lebedinsky, 39–49. Moscow: Foreign Languages Publishing House, 1960.

———. *Sakharov Speaks*. New York: Alfred A. Knopf, 1974.

Savage, Sean J. *JFK, LBJ, and the Democratic Party*. Albany: State University of New York Press, 2004.

Schmid, Sonja D. "Nuclear Colonization? Soviet Technopolitics in the Second World." In *Entangled Geographies: Empire and Technopolitics in the Global Cold War*, edited by Gabrielle Hecht, 125–154. Cambridge, MA; London: MIT Press, 2011.

———. *Producing Power: The Pre-Chernobyl History of the Soviet Nuclear Industry*. Cambridge, MA: MIT Press, 2015.

Schrafstetter, Susanna, and Stephen Twigge. *Avoiding Armageddon: Europe, the United States, and the Struggle for Nuclear Nonproliferation, 1945–1970*. Westport, CT: Praeger, 2004.

Schweitzer, Albert. "A Declaration of Conscience." *The Saturday Review* (May 18, 1957): 17–20.

Seaborg, Glenn T. *Kennedy, Khrushchev, and the Test Ban*. Berkeley: University of California Press, 1981.

Sedov, V. V. "Pervaia vsesoiuznaia konferentsiia po meditsinskoi radiologii" [The First All-Union Conference on Medical Radiology]. *Atomnaia energiia* [Atomic Energy] 1, no. 2 (1956): 98–100.

Sembritzki, Laura. "Maiak 1957 and Its Aftermath: Radiation Knowledge and Ignorance in the Soviet Union." *Jahrbücher für Geschichte Osteuropas* 66, no. 1 (2018): 45–64.

Serwer, Daniel Paul. "The Rise of Radiation Protection: Science, Medicine and Technology in Society, 1896–1935." PhD diss., Princeton University, 1976.

Sherwin, Martin J. *A World Destroyed: Hiroshima and the Origins of the Arms Race*. New York: Vintage Books, 1987.

Shimizu, Sayuri. *Creating People of Plenty: The United States and Japan's Economic Alternatives, 1950–1960*. Kent, OH: Kent State University Press, 2001.

Shōwa 29 nen ni okeru Bikini kaiiki no hōshanō eikyō chōsa hōkoku [Report of the Study of the Effects of Radioactivity in Waters in the Vicinity of Bikini in Showa 29], vol. 2. Tokyo: Suisanchō Chōsa Kenkyū Bu, 1955.

Siegelbaum, Lewis H. *Stakhanovism and the Politics of Productivity in the USSR, 1935–1941*. Cambridge; New York: Cambridge University Press, 1988.

The Situation in Biological Science: Proceedings of the Lenin Academy of Agricultural Sciences of the U.S.S.R. Session, July 31–August 7, 1948, Verbatim Report. Moscow: Foreign Languages Publishing House, 1949.

Sloan, Phillip R., and Brandon Fogel, eds. *Creating a Physical Biology: The Three-Man Paper and Early Molecular Biology*. Chicago: University of Chicago Press, 2011.

Smith, Andrew F. *American Tuna: The Rise and Fall of an Improbable Food*. Berkeley: University of California Press, 2012.

Smith, Martha J. "The Nuclear Testing Policies of the Eisenhower Administration, 1953–60." PhD diss., University of Toronto, 1997.

Smith-Howard, Kendra. *Pure and Modern Milk: An Environmental History Since 1900*. Oxford; New York: Oxford University Press, 2014.

Smith-Norris, Martha. "The Eisenhower Administration and the Nuclear Test Ban Talks, 1958–1960: Another Challenge to 'Revisionism'." *Diplomatic History* 27, no. 4 (2003): 503–541.

Solovey, Mark, ed. "Science in the Cold War," special issue. *Social Studies of Science* 31, no. 2 (2001): 165–297.

"Soviet Statement on Resumption of Nuclear Testing." *Current Digest of the Soviet Press* 13, no. 35 (1961): 3, 6–8.

Soyfer, Valery N. *Lysenko and the Tragedy of Soviet Science*. New Brunswick, NJ: Rutgers University Press, 1994.

Spoor, Norman. "Dr W. G. Marley." *Annals of Occupational Hygiene* 24, no. 2 (1981): 258–259.

Stark, Andrew. *Conflict of Interest in American Public Life*. Cambridge, MA: Harvard University Press, 2000.

Stawkowski, Magdalena E. "Radioactive Knowledge: State Control of Scientific Information in Post-Soviet Kazakhstan." PhD diss., University of Colorado-Boulder, 2014.

Stern, Curt. "One Scientist Speaks Up." *Science* 120, no. 3131 (1954): 5A.

Stewart, John. "Summerskill, Edith Clara, Baroness Summerskill (1901–1980)." In *Oxford Dictionary of National Biography*, edited by H. C. G. Matthew and Brian Harrison, vol. 53, 321–323. Oxford: Oxford University Press, 2004.

Sturtevant, Alfred H. "The Genetic Effects of High-Energy Irradiation of Human Populations." *Engineering and Science* 18 (1955): 9–12.

———. "Social Implications of the Genetics of Man." *Science* 120, no. 3115 (1954): 405–407.

Subcommission on Quaternary Stratigraphy. "Working Group on the 'Anthropocene'." Accessed October 1, 2019. http://quaternary.stratigraphy.org/working-groups/anthropocene.

Summerfield, Penny. "'Our Amazonian Colleague': Edith Summerskill's Problematic Reputation." In *Making Reputations: Power, Persuasion and the Individual in Modern British Politics*, edited by Richard Toye and Julie Gottlieb, 135–150. London: I.B. Tauris, 2005.

Swenson-Wright, John. *Unequal Allies? United States Security and Alliance Policy Toward Japan, 1945–1960*. Stanford, CA: Stanford University Press, 2005.

Swerdlow, Amy. "The Politics of Motherhood: The Case of Women Strike for Peace and the Test Ban Treaty." PhD diss., Rutgers University—New Brunswick, 1984.

———. "Pure Milk, Not Poison: Women Strike for Peace and the Test Ban Treaty of 1963." In *Rocking the Ship of State: Toward a Feminist Peace Politics*, edited by Adrienne Harris and Ynestra King, 225–237. Boulder, CO: Westview Press, 1989.

———. *Women Strike for Peace: Traditional Motherhood and Radical Politics in the 1960s*. Chicago: University of Chicago Press, 1993.

Szasz, Ferenc Morton. *The Day the Sun Rose Twice: The Story of the Trinity Site Nuclear Explosion, July 16, 1945*. Albuquerque: University of New Mexico Press, 1984.

Taylor, Lauriston S. *Organization for Radiation Protection: The Operations of the ICRP and NCRP, 1928–1974*. Springfield, VA: National Technical Information Service, 1979.

———. *The Tripartite Conferences on Radiation Protection: Canada, United Kingdom, United States (1949–1953)*. Washington, DC: Office of Scientific and Technical Information, U.S. Department of Energy, 1984.

Teller, Edward. *Memoirs: A Twentieth-Century Journey in Science and Politics*. Cambridge, MA: Perseus, 2001.

Thomson, Arthur Landsborough. *Half a Century of Medical Research, vol. 1: Origins and Policy of the Medical Research Council (U.K.)*. London: H.M.S.O., 1973.

Thorsheim, Peter. "Interpreting the London Fog Disaster of 1952." In *Smoke and Mirrors: The Politics and Culture of Air Pollution*, edited by E. Melanie DuPuis, 154–169. New York: New York University Press, 2004.

Timofeev-Resovskii, Nikolai V. "Iz istorii dialoga biologov i fizikov" [From the History of the Dialogue of Biologists and Physicists]. In *Vospominaniia o I.*

E. Tamme [Reminiscences of I. E. Tamm], edited by E. L. Feinberg, 282–285. Moscow: IZDAT, 1995.

Todes, Daniel P. *Pavlov's Physiology Factory: Experiment, Interpretation, Laboratory Enterprise*. Baltimore: Johns Hopkins University Press, 2002.

Tomonaga, Masanobu. "Leukemia in Nagasaki Atomic Bomb Survivors from 1945 Through 1959." *Bulletin of the World Health Organization* 26, no. 5 (1962): 619–631.

Tomotsugu, Shinsuke. "Decolonization and the United Kingdom's Atoms for Peace Program in the Middle East." Presented at the Annual Meeting of the Society for Historians of American Foreign Relations, Lexington, Kentucky, June 2014.

Trachtenberg, Marc. *A Constructed Peace: The Making of the European Settlement, 1945–1963*. Princeton, NJ: Princeton University Press, 1999.

Tromly, Benjamin. *Making the Soviet Intelligentsia: Universities and Intellectual Life Under Stalin and Khrushchev*. Cambridge; New York: Cambridge University Press, 2013.

Trubowitz, Peter. *Defining the National Interest: Conflict and Change in American Foreign Policy*. Chicago: University of Chicago Press, 1998.

Tsuchiya, Yuka. "Kōhō bunka gaikō to shiteno genshiryoku riyō kyanpe'en to 1950 nendai no Nichi-Bei kankei" [The Campaign for Atomic Energy Use as Public-Relations Cultural Diplomacy and U.S.-Japan Relations in the 1950s]. In *Nichi-Bei dōmei ron: Rekishi, kinō, shūhen shokoku no shiten* [On the U.S.-Japan Alliance: Perspective of History, Function, and Neighboring Countries], edited by Toshitaka Takeuchi, 180–209. Kyoto: Mineruva Shobō, 2011.

Tsukerman, V. A., and Z. M. Azarkh. *Liudi i vzryvy* [People and Explosions]. Arzamas-16, Russia: VNIIEF, 1994.

Tucker, Richard. "Containing Communism by Impounding Rivers: American Strategic Interests and the Global Spread of High Dams in the Early Cold War." In *Environmental Histories of the Cold War*, edited by John R. McNeill and Corinna R. Unger, 139–163. Cambridge: Cambridge University Press, 2010.

U.S. Arms Control and Disarmament Agency. *Documents on Disarmament 1961*. Washington, DC: Government Printing Office, 1962.

———. *Documents on Disarmament 1962*. Washington, DC: Government Printing Office, 1963.

U.S. Atomic Energy Commission. *Thirteenth Semiannual Report of the Atomic Energy Commission*. Washington, DC: Government Printing Office, 1953.

———. Health and Safety Laboratory, New York Operations Office. *Environmental Contamination from Weapon Tests* (HASL-42). Oak Ridge, TN: Technical Information Service Extension, 1958.

U.S. Congress, House Committee on Appropriations, Subcommittee on Public

Works Appropriations. *Atomic Energy Appropriations for 1958.* Washington, DC: Government Printing Office, 1957.
U.S. Congress, House Committee on Ways and Means, Subcommittee on Tuna Imports. *Tuna Imports.* Washington, DC: Government Printing Office, 1951.
U.S. Congress. Joint Committee on Atomic Energy. Special Subcommittee on Radiation. *Fallout from Nuclear Weapons Tests.* Washington, DC: Government Printing Office, 1959.
———. *The Nature of Radioactive Fallout and Its Effects on Man.* Washington, DC: Government Printing Office, 1957.
———. *Radiation Protection Criteria and Standards: Their Basis and Use.* Washington, DC: Government Printing Office, 1960.
U.S. Congress. Joint Committee on Atomic Energy. Subcommittee on Research, Development, and Radiation. *Fallout, Radiation Standards, and Countermeasures.* 2 vols. Washington, DC: Government Printing Office, 1963.
U.S. Congress, Senate Committee on Armed Services. Preparedness Investigating Subcommittee. *Military Aspects and Implications of Nuclear Test Ban Proposals and Related Matters.* Washington, DC: Government Printing Office, 1963.
U.S. Congress, Senate Committee on Foreign Relations. *Nuclear Test Ban Treaty.* Washington, DC: Government Printing Office, 1963.
U.S. Department of State. *Documents on Disarmament, 1945–1959.* 2 vols. Washington, DC: Government Printing Office, 1960.
Uno, Kathleen S. "The Death of 'Good Wife, Wise Mother'?" In *Postwar Japan as History,* edited by Andrew Gordon, 293–322. Berkeley: University of California Press, 1993.
UNSCEAR 2000 Report. 2 vols. New York: United Nations, 2000.
Vasil'ev, A. P., ed. *Rozhdennaia atomnym vekom: Sbornik istoricheskikh ocherkov, dokumentov i vospominanii veteranov k 40–letiiu sozdaniia v SSSR sluzhby spetsial'nogo kontrolia Ministerstva oborony* [Birth of the Atomic Age: Collection of Historical Assessments, Documents, and Reminiscences of Veterans toward 40 Years of the Establishment of the USSR Service of Special Control of the Ministry of Defense]. Moscow: Sluzhba spetsial'nogo kontrolia, 1998.
———. "Sistema dal'nego obnaruzheniia iadernykh vzryvov i Sovetskii atomnyi proekt" [The System of the Distant Observation of Nuclear Explosions and the Soviet Atomic Project]. In *Istoriia Sovetskogo atomnogo proekta: Dokumenty, vospominaniia, issledovaniia* [History of the Soviet Atomic Project: Documents, Reminiscences, and Investigations], edited by V. P. Vizgin, vol. 2, 237–278. St. Petersburg: Izdatel'stvo Russkogo Khristianskogo Gumanitarnogo Instituta, 2002.
Voronin, S. N., et al. *To vremia ukhodit v istoriiu* [That Time Becomes a History]. Sarov, Russia: RFIaTs-VNIIEF, 2008.

Walker, J. Samuel. *Permissible Dose: A History of Radiation Protection in the Twentieth Century*. Berkeley: University of California Press, 2000.

———. *Three Mile Island: A Nuclear Crisis in Historical Perspective*. Berkeley: University of California Press, 2004.

———, and George T. Mazuzan. *Containing the Atom: Nuclear Regulation in a Changing Environment, 1963–1971*. Berkeley: University of California Press, 1992.

Walker, John R. *British Nuclear Weapons and the Test Ban, 1954–73: Britain, the United States, Weapons Policies and Nuclear Testing: Tensions and Contradictions*. Farnham, UK; Burlington, VT: Ashgate, 2010.

Wang, Jessica. *American Science in an Age of Anxiety: Scientists, Anticommunism, and the Cold War*. Chapel Hill: University of North Carolina Press, 1999.

Warren, Stafford L. Interview by Adelaide Tusler for the Center for Oral History Research, University of California, Los Angeles, September 15, 1966, vol. 2. Accessed October 1, 2019. http://oralhistory.library.ucla.edu/viewItem.do?ark=21198/zz0008zcbp&title=Interview%20of%20Stafford%20L.%20Warren%20(1983).

Weart, Spencer R. *Nuclear Fear: A History of Images*. Cambridge, MA: Harvard University Press, 1988.

Wellock, Thomas R. "Engineering Uncertainty and Bureaucratic Crisis at the Atomic Energy Commission, 1964–1973." *Technology and Culture* 53, no. 4 (2012): 846–884.

Werner, Cynthia, and Kathleen Purvis-Roberts. "After the Cold War: International Politics, Domestic Policy and the Nuclear Legacy in Kazakhstan." *Central Asian Survey* 25, no. 4 (2006): 461–480.

Whittemore, Gilbert F., Jr. "The National Committee on Radiation Protection, 1928–1960: From Professional Guidelines to Government Regulation." PhD diss., Harvard University, 1986.

Widner, Thomas E., and Susan M. Flack. "Characterization of the World's First Nuclear Explosion, the Trinity Test, as a Source of Public Radiation Exposure." *Health Physics* 98, no. 3 (2010): 480–497.

Williams, Philip M. *Hugh Gaitskell: A Political Biography*. London: Jonathan Cape, 1979.

Winchester, Simon. *Krakatoa: The Day the World Exploded, August 27, 1883*. New York: Harper-Collins, 2003.

Windholz, George. "The 1950 Joint Scientific Session: Pavlovians as the Accusers and the Accused." *Journal of the History of the Behavioral Sciences* 33, no. 1 (1997): 61–81.

Wittner, Lawrence S. *The Struggle Against the Bomb, vol. 1: One World or None: A History of the World Nuclear Disarmament Movement Through 1953*. Stanford, CA: Stanford University Press, 1993.

———. *The Struggle Against the Bomb, vol. 2: Resisting the Bomb: A History* of the World Nuclear Disarmament Movement, 1954–1970. Stanford, CA: Stanford University Press, 1997.

Wolfe, Audra J. *Freedom's Laboratory: The Cold War Struggle for the Soul of Science.* Baltimore: Johns Hopkins University Press, 2018.

———. "What Does It Mean to Go Public? The American Response to Lysenkoism, Reconsidered." *Historical Studies in the Natural Sciences* 40, no. 1 (2010): 48–78.

Yamamoto, Mari. *Grassroots Pacifism in Post-War Japan: The Rebirth of a Nation.* London: RoutledgeCurzon, 2004.

Yamazaki, Masakatsu, and Kenzō Okuda. "Bikini jiken go no genshiro dōnyūron no taitō" [Rise of the Call for the Introduction of Atomic Power Reactors After the Bikini Incident]. *Kagakusi kenkyū* no. 230 (2004): 83–93.

Zachmann, Karin. "Risk in Historical Perspective: Concepts, Contexts, and Conjunctions." In *Risk: A Multidisciplinary Introduction*, edited by Claudia Klüppelberg, Daniel Straub, and Isabell M. Welpe, 3–35. Cham, Switzerland: Springer International Publishing, 2014.

Zakharov, I. K., and V. K. Shumnyi. "K 50-letiiu 'pis'ma trekhsot'" [Toward the 50th Year of the "Letter of 300"]. *Vestnik VOGuS* 9, no. 1 (2005): 12–33.

Zalasiewicz, Jan, Will Steffen, Reinhold Leinfelder, Mark Williams, and Colin Waters. "Petrifying Earth Process: The Stratigraphic Imprint of Key Earth System Parameters in the Anthropocene." *Theory, Culture & Society* 34, no. 2–3 (2017): 83–104.

Zalasiewicz, Jan, et al. "When Did the Anthropocene Begin? A Mid-Twentieth Century Boundary Level Is Stratigraphically Optimal." *Quaternary International* 383 (2015): 196–203.

Ziegler, Charles A., and David Jacobson. *Spying Without Spies: Origins of America's Secret Nuclear Surveillance System.* Westport, CT: Praeger, 1995.

Zierler, David. *The Invention of Ecocide: Agent Orange, Vietnam, and the Scientists Who Changed the Way We Think About the Environment.* Athens: University of Georgia Press, 2011.

Zimmerman, Jonathan. *Innocents Abroad: American Teachers in the American Century.* Cambridge, MA: Harvard University Press, 2006.

Zubok, Vladislav M. *Zhivago's Children: The Last Russian Intelligentsia.* Cambridge, MA: Harvard University Press, 2009.

Zwigenberg, Ran. "'The Coming of a Second Sun': The 1956 Atoms for Peace Exhibit in Hiroshima and Japan's Embrace of Nuclear Power." *Asia-Pacific Journal: Japan Focus* 10, no. 6 (February 4, 2012). https://apjjf.org/2012/10/6/Ran-Zwigenberg/3685/article.html.

INDEX

acceptable risk, 9–10, 12, 15, 76, 137, 171–80, 176, 180
Adamskii, Viktor B., 182–83
Advisory Committee on X-ray and Radium Protection (ACXRP), 9, 20, 25, 64, 65
Alexander, Lyle E., 29
Allison, John M., 55
All-Union Scientific Research Institute of Experimental Physics (VNIIEF), 33–34
Alsop, Joseph, 74
Alsop, Stewart, 74
American Association for the Advancement of Science (AAAS), 61–62, 68
Anderson, Clinton P., 151, 153
Anthropocene, the: Anthropocene Working Group of the Subcommission on Quaternary Stratigraphy, 199; and the Cold War, 4–5, 6, 12, 15, 190, 196–97; Crutzen on, 4–5; as nuclear, 6, 10, 13, 16, 43, 59, 60, 62–63, 104, 111, 134, 145, 150, 194; and Trinity test, 16
Anthropocene Working Group of the Subcommission on Quaternary Stratigraphy, 199
atmospheric tests: by China, 3, 188, 196; by France, 3, 131, 162, 164, 196; long-range detection of, 14, 17–18, 21–22, 23, 33–34, 38–40, 111, 113, 157, 181, 188, 191; number conducted, 3–4, 16–17, 163; by Soviet Union, 2, 15, 17, 21, 23, 31, 32–33, 34–35, 52, 61, 75, 83, 103, 113, 122, 130–31, 133, 136, 157–58, 161, 162–63, 166–70, 174–75, 182, 184, 190–91, 194; test sites, 2–3, 4, 18, 19–20, 22–23, 24–25, 26, 31, 32, 34, 36–37, 37, 38, 39, 40, 43, 44–45, 62, 85, 167, 168, 175, 191, 196, 198; by United Kingdom, 2, 17, 36, 38, 39, 83, 90, 102, 131, 134–35, 144, 156–57, 158–59, 161, 168, 171, 190–91, 194; by United

States, 2, 12, 14, 17, 22–23, 24–25, 34, 35, 37, 41–42, 45, 61, 73, 75, 83, 90, 92, 103, 106, 107, 117, 118–19, 131, 132, 134–35, 138, 154, 156–57, 158–61, 163, 168, 169, 170–71, 174, 175, 181, 182, 185, 187, 190–91, 193, 194; yield of, 3–4, 5–6, 16–17, 163. *See also* radioactive fallout; strontium-90

Atomic Energy Detection System, 22

Atomic Scientists' Association (ASA), 102

Attlee, Clement, 36, 81

Australia, 36–39, 85, 180

average risk, 9, 13, 121, 134, 137, 148–49, 151

Baby Tooth Survey, 138, 177, 248n68

Bacq, Zénon, 85

Bainbridge, Kenneth T., 19, 20

Beadle, George W., 95–96

Beale, Oliver Howard, 39

BEAR. *See* Biological Effects of Atomic Radiation (BEAR), Committees on

Belgium, 85, 130

Beria, Lavrentii, 32, 33, 72

Bevan, Aneurin, 81, 145

Bevin, Ernest, 36

Bhabha, Homi J., 85

Bikini Atoll tests, 18, 43, 45, 48, 50, 51, 118. *See also* Castle Bravo test

Biological Effects of Atomic Radiation (BEAR), Committees on (National Academy of Sciences, US), 89; cancer, 96–98; genetics, 89, 90, 91–93, 95–96; relationship to nuclear test ban debate, 90, 91, 101–108; vs. MRC, 12–13, 90–91, 94, 95, 98, 99–100, 101–2, 103, 107–8, 112, 192; report of, 89–90, 101–2, 103, 105, 106, 107–8, 113–14, 125–26

bone cancer, 89–90, 97–99, 100

Borst, Lyle B., 26

Boss, Willis R., 46, 56

Bradbury, Norris E., 26

Brazil, 85, 181

Brecon, Lord, 147

Brezhnev, Leonid, 168

Bronk, Detlev W., 82, 88

Brooke, Henry, 147

Brown, Harold, 164

Brown, Harrison, 114–15

Brues, Austin M., 98, 112, 129–30

Bugher, John C., 26, 45–46, 47, 56, 73

Bundy, McGeorge, 171

Burnazian, Avetik I., 33

Burney, Leroy E., 151

Callaghan, James, 81

Campaign for Nuclear Disarmament (CND), 145, 146, 177, 180

Canada, 7, 24, 36, 67, 84, 85, 86, 103–4, 107, 114, 170; milk contamination in, 144, 160, 169, 248n68; and nuclear test ban debate, 103–4, 107; snow contamination in, 25; wheat contamination in, 150, 169

carbon-14 (C-14), 128, 170. *See also* radiocarbon dating

Carling, Ernest Rock, 94

Carson, Rachel, *Silent Spring*, 4

Carter, T. C., 66

Caster, William O., 140

Castle Bravo test, 60, 61, 63, 81, 82, 121, 191; and *Lucky Dragon*, 41–43, 52–53, 54, 58, 59, 61, 73, 118, 125, 176; significance of, 12, 14, 40, 59–60, 64, 73, 78, 80, 86; and tuna contamination, 14, 41–43, 46–50, 51, 52–54, 55–59, 119, 120, 191

Catcheside, David G., 66, 67

cerium-141, 16
Chetverikov, Sergei S., 69
children: baby teeth from, 138, 177, 248n68; genetic damage to 13, 81, 96, 126; iodine-131 in, 175; and milk, 13, 142, 180; strontium-90 in, 96–97, 105, 144; as symbol of protest, 13, 163, 177, 178, 179; and motherhood, 13, 54, 177, 180; and risk, 7, 13, 26, 100, 126, 149, 178, 179, 180, 186, 194
China, 33, 34, 85, 86, 164, 184; atmospheric nuclear testing by, 3, 188, 196; refusal to sign PTBT, 2, 188, 196; relations with India, 86, 183
Churchill, Winston, 17, 81–82, 102
Citizens' Committee for a Nuclear Test Ban, 187
Clark, Herbert M., 25, 26
clean bomb, 104–5, 107, 108, 128–29, 134, 155, 167, 170, 195
Cockcroft, John, 100
Cold War, 13, 19, 22, 23, 26–27, 63–64, 70, 73, 75, 78, 86, 125, 127, 134, 139, 143, 159, 160, 163, 166, 171–72, 176, 185, 191, 193, 198, 199; and the Anthropocene, 4–5, 6, 12, 15, 190, 196–97; Cuban Missile Crisis, 1, 183, 184; and Japan, 43, 59–60, 119; nuclear arms race, 2, 4, 5–6, 14, 17, 19, 28, 31–32, 34, 35–36, 51, 80, 81, 102, 137, 155, 156, 164, 170, 178, 181, 183, 184, 186–87, 188–89; and science, 10–11, 71, 76, 79, 88, 122–23, 125, 129–30, 141–42
Cole, David, 147
Commoner, Barry, 138
consumers: in Japan, 14, 42–43, 46–48, 51, 53–54, 55, 60, 118, 191; in the United Kingdom, 148, 180;
in the United States, 48, 56, 137, 142–44, 151–52, 160, 176, 180
Consumers Union (CU) milk survey, 137, 142–44, 148, 151–52
Court-Brown, Michael, 100
Cousins, Norman, 126, 178, 184, 187
Crick, Francis, 124
Crutzen, Paul, on the Anthropocene, 4–5
Cuban Missile Crisis, 1, 183, 184

Daigo Fukuryū Maru (Lucky Dragon). See *Lucky Dragon* incident
Dean, Gordon, 26
Delbrück, Max, "On the Nature of Gene Mutation and Gene Structure," 63
Denmark, 170
Diefenbaker, John G., 150
DNA, 10–11, 63, 124
Dobzhansky, Theodosius, 75
Dodd, Thomas J., 184, 185
Doll, Richard, 100
Donaldson, Lauren R., 45, 46, 56–57
Douglas-Home, Alec, 1
Drosophila research, 62–63, 69, 71, 72, 78
Dubinin, Nikolai P., 69, 77–78, 125, 127, 128, 129; opposition to Lysenko, 70–71, 77, 78–79, 87, 191
Dulles, John Foster, 81, 83–84, 106, 131, 132–33, 155, 156, 194
Dunham, Charles L., 103, 141
Dunning, Gordon M., 140, 141

East Germany, 166, 169, 185
Eastman Kodak, 16, 23
Eden, Anthony, 102, 194
Eighteen-Nation Committee on Disarmament (ENCD), 181
Einstein, Albert, 19, 84

Eisenbud, Merril, 23–24, 30–31, 46, 47, 112, 138–39
Eisenhower, Dwight D., 61, 122, 150, 154, 164; Atoms for Peace proposal, 77, 83; attitudes regarding nuclear warfare, 151, 156; relationship with Macmillan, 107, 131, 132, 133, 156–57, 161; test-ban policies, 81–82, 105, 106, 107, 131, 132, 133–34, 155, 156, 157, 158, 159–61, 194
Emel'ianov, Vasilii S., 157
Evans, Robley D., 98–99

Failla, Gioacchino, 65, 66, 68, 92, 93–94
Federation of American Scientists (FAS), 80–81, 82, 83, 106, 192
Fermi, Enrico, 28
France, 29, 52, 85–86, 107, 156; atmospheric nuclear testing by, 3, 131, 162, 164, 196; refusal to sign PTBT, 2, 188, 196
Frank-Kamenetskii, David A., 33
Freeman, Orville L., 139, 148, 149
Fremlin, John H., 146–47
Fuchs, Klaus, 36, 66–67

Gaitskell, Hugh, 81, 156
Gascoyne-Cecil, Robert, 82
Geiger counters, 21–22, 23, 27, 32, 41, 51, 136, 137
genetics and geneticists: DNA, 10–11, 63, 124; genetic damage from radiation, 9, 12, 13, 14–15, 25, 56, 58, 60, 62–64, 65, 66, 67–68, 69, 72, 73–76, 77–79, 89–90, 91–96, 99, 108, 122–23, 125, 126, 127, 128, 191, 198; linear non-threshold (LNT) hypothesis, 9, 25, 56, 60, 62–63, 67, 69, 91–92, 127, 128, 198; in Japan, 57–58, 59; in Soviet Union, 10–11, 60, 64, 68, 69–73, 77–80, 124–25, 128, 129, 166, 191, 192; target theory, 63; in United Kingdom, 10, 60, 64, 65–67, 78–79, 87, 89–90, 94–96, 123; in United States, 10, 60, 61–63, 64–66, 67–68, 69, 73–77, 78–79, 86–87, 89, 90, 91–94, 95–96, 122–23, 191
Genetics Society of America (GSA), 76
Geneva Conference on the Peaceful Uses of Atomic Energy, 79, 83, 85, 119
Glass, H. Bentley, 65
Glushchenko, Ivan E., 72
Gray, Louis H., 99
Greater St. Louis Citizens' Committee for Nuclear Information, 138, 177
Green, Earl L., 75
Green, Howard, 150
Gregg, David T., 155
Gromyko, Andrei, 1, 127, 130

Haddow, Alexander, 100, 102
Hailsham, Lord, 185
Hammerskjöld, Dag, 109, 110
Harley, John H., 140, 141
Harriman, W. Averell, 185, 187
Hempelmann, Louis H., 20, 21
Herter, Christian, 156, 158
High Altitude Sample Program, 115, 152–53
Himsworth, Harold P., 94, 100
Hiroshima, 30, 31–32, 42, 55, 66, 80, 97, 118, 197
Hirschfelder, Joseph O., 18–19
Hiyama, Yoshio, 56–57, 58–59, 60, 118, 119–21
Hollister, Hal, 117
humanism, 13, 30, 64, 126–27, 129, 134, 141, 178, 192, 194. *See also* innocent bystanders
Humphrey, Hubert, 149

Iceland, 170
ICRP. *See* International Commission on Radiological Protection
India, 30: and nuclear test ban debate, 2, 78, 81–82, 83, 85, 121, 130, 132; relations with China, 86, 183; underground testing by, 196, 202n4
innocent bystanders, 13, 15, 29, 111, 134, 180, 192; Kennedy's views on, 163, 164, 171, 186–87, 194; Sakharov's views on, 128–29, 163, 167, 194; Schweitzer's views on, 126, 178, 194
International Commission on Radiological Protection (ICRP), 68, 92, 93–94, 95, 124; on permissible dose, 9, 32, 76, 93–94, 96, 98, 103, 119, 120–21, 172. *See also* International X-ray and Radium Protection Committee
International Congress of Radiology, second, 9
International Co-operative Alliance, 53
International Council of Scientific Unions, 83–84
International X-ray and Radium Protection Committee (IXRPC), 9
iodine-131 (I-131), 172–73, 174, 175, 198
Isotopes, Inc., 144
Israel, 202n4
Ivy Mike test, 45, 46

Jack, Homer, 178
Jackson, Henry M., 184
Japan, 7, 19, 30, 37, 104, 107, 132, 170, 196; Atomic Bomb Effect Research Commission (ABERC), 51, 56, 57–58, 59; consumerism in, 53–55, 60; contaminated rain in, 50–52, 55; contaminated rice in, 13, 51, 55, 110, 120–21, 134, 149; contaminated tuna in, 14, 41–43, 45, 46–50, 51, 52–54, 55–59, 119, 191; *Lucky Dragon* incident, 41–43, 52–53, 54, 58, 59, 61, 73, 118, 125, 176; Ministry of Agriculture and Forestry (MAF), 46; Ministry of Foreign Affairs, 119, 120; Ministry of Health and Welfare (MHW), 46, 47–48, 51, 56, 58, 59, 118–19; and nuclear test ban debate, 42, 52, 53, 54–55, 104, 107, 119–20, 121, 132, 134, 192, 196; pacifism in, 54, 118, 176; permissible dose in, 14, 47, 50, 51–52, 55–56, 57–59, 60, 118, 120–21, 191, 192; radiation protection in, 12, 14, 41, 42, 43, 46–50, 51–52, 55–60, 117–19, 120–21; relations with Soviet Union, 52; relations with United States, 12, 14, 15, 19, 43, 44, 46–47, 48, 52–53, 55–56, 57–58, 59–60, 118, 120, 121, 134, 191; South Pacific Mandate, 44; and UNSCEAR, 15, 85, 109, 110, 111, 117–21, 134, 192. *See also* Hiroshima; Kishi, Nobusuke; Nagasaki
Japan Council against Atomic and Hydrogen Bombs, 55
Japan Federation of Tuna Fishers, 53
Japanese Consumers' Co-operative Union, 53
Johnson, Lyndon B., 188
Joint Committee on Atomic Energy (JCAE), 115, 151, 152, 157, 172

Kapitsa, Pyotr L., 124
Keely, E. P., 145–46
Kennedy, John F., 169, 171–72, 174; relationship with Khrushchev, 165, 166, 168, 181–82, 183–84,

185; relationship with Macmillan, 166, 171, 181, 184; test-ban policies, 164–65, 166, 168, 170–71, 181, 183, 184, 185, 186–88; views on innocent bystanders, 163, 164, 171, 186–87, 194
Khariton, Yulii B., 33–34, 167–68
Khrushchev, Nikita: de-Stalinization policy, 125; nuclear testing policies, 166, 167, 168, 170, 171–72, 194; relationship with Kennedy, 165, 166, 168, 181–82, 183–84, 185; relationship with Lysenko, 71; relationship with Sakharov, 129, 167, 171, 182–83; test-ban policies, 130–31, 132, 156–57, 158, 165, 181–86, 193
Kikoin, Isaak K., 33
Killian, James R., 157
Kishi, Nobusuke, 120
Kistiakowsky, George B., 158, 159
Kol'tsov, Nikolai K., 69
Korean demilitarized zone, 5
Korean War, 23
Kraevskii, Nikolai, 129
Krakatoa, 21
Krotkov, Fedor G., 79
Kuboyama, Aikichi, 42, 55
Kulp, J. Laurence, 30, 115, 116–17
Kurchatov, Igor V., 32, 33, 128, 129
Kusumoto, Masahiro, 58
Kuzin, Aleksandr M., 78, 127, 129
Kuznetsov, Vasilii V., 85

Laboratory of Measuring Instruments of the USSR Academy of Sciences (LIPAN), 33
Lanier, Raymond R., 76
Lapp, Ralph E., 104
Lea, Douglas E.: on target theory, 63
Lebedinskii, Andrei V., 123–24, 129–30, 131
Leipunskii, Ovsei I., 128

"Letter of 300," 79
leukemia, 89–90, 97, 100, 110, 129–30
Lewis, Edward B., 62, 174
Libby, Willard F., 27, 74, 95, 96–97, 100–101, 105, 128, 133, 138, 150–51, 152, 153, 195; and Project Sunshine, 27–29, 30, 97, 114–15, 116
Lindbergh-Holm, Karl, 136
linear non-threshold (LNT) hypothesis, 9, 25, 56, 60, 62–63, 64, 67, 69, 91–92, 97, 99, 100, 127, 198
Livanov, Mikhail N., 123–24
Livingston, M. Stanley, 80
Lloyd, Hilda, 177
Lloyd, Selwyn, 131, 159
LNT. *See* linear non-threshold (LNT) hypothesis
Lodge, Henry Cabot, Jr., 83, 84
Lonsdale, Kathleen, 177
Loper, Herbert B., 153
Loutit, John F., 94
Lucky Dragon incident, 41–43, 52–53, 54, 58, 59, 61, 73, 118, 125, 176. *See also* Castle Bravo test
Lysenko, Trofim, 10–11, 69–70, 75, 76, 78–79, 121, 122–23, 127; and agriculture, 71–72, 77; and ban on genetics, 12, 64, 70–72, 122, 128; opposition to, 70–71, 77–79, 87, 124–25, 128, 134, 191, 192; relationship with Khrushchev, 71

MacArthur Line, 44
Machta, Lester, 152, 153
Macmillan, Harold, 121, 168, 171–72, 174; and fallout contamination in Wales, 144, 147, 156, 161; relationship with Eisenhower, 107, 132, 157, 159, 161; relationship with Kennedy, 166, 171, 181, 184; test-ban policies, 107,

131, 132, 133, 156–57, 158, 159, 164–65, 181, 184, 185
Magee, John L., 18–19, 21
Manhattan Project, 18–22, 25, 35, 80
Marley, W. G., 35, 94, 100, 112, 174
Marx, Karl, on relations of production, 8
maximum permissible body burden (MPB), 96–99, 100, 105
Mayneord, William V., 94, 99, 112, 113
McCarthy, Eugene, 149
McCone, John A., 132, 133, 153, 157, 159. *See also* U.S. Atomic Energy Commission
Medical Research Council (MRC) (UK), 64, 65–67, 82, 84, 87, 96–97, 112, 145, 146, 172, 173, 174; vs. BEAR, 88–91, 94–95, 98, 99–100, 101–2, 107–8, 192; relationship to nuclear test ban debate, 90, 91, 101–108; report of, 12–13, 14–15, 88–91, 101–2, 103, 105, 107–8, 113–14, 125–26, 144, 194
Menon, Krishna, 86
Menzies, Robert, 37
Messel, Harry, 38
Mexico, 85, 181
Michelson, Irvin, 151
Michurin, Ivan V., 69–70
milk contamination, 13, 29, 110; in Canada, 144, 150, 169; in United Kingdom, 146, 147, 163, 173, 174, 180; in United States, 137, 139, 140, 142–44, 148, 151–52, 160, 163, 173, 175, 176, 180; vs. rice, 120, 121, 134
Minnesota: Atomic Development Problems Committee (ADPC), 139, 140–41; fallout surveys in, 15, 137, 139–42, 148, 160, 193; political conditions in, 139, 193; Rural Cooperation Power Association/Elk River Station, 140; vs. Wales, 145, 156, 193; wheat contamination in, 148–51, 156
Mitchell, Joseph S., 99
Miyake, Yasuo, 51, 52
Morgan, Thomas Hunt, 62
MPB. *See* maximum permissible body burden
MRC. *See* Medical Research Council
Muller, Hermann J., 71, 75–76, 122–23; *Drosophila* research by, 63, 69; on radiation protection, 64–65, 68, 93
Murrow, Edward R., 189
Mutscheller, Arthur, 9

Nagasaki, 30, 31–32, 42, 55, 66, 80, 97, 118, 197
Nakasone, Yasuhiro, 58
Nasser, Gamal, 85
National Advisory Committee on Radiation (NACOR), 154
National Committee for a Sane Nuclear Policy (SANE), 176, 178, 180
National Committee on Radiation Protection (NCRP), 9, 66, 94, 124; on permissible dose, 9, 25, 51–52, 55, 56, 58, 59, 64, 67–68, 76, 91, 93, 96, 98, 103, 136, 140, 172. *See also* Advisory Committee on X-ray and Radium Protection
National Radiation Surveillance Network, 24, 40, 138, 141, 142
Naval Research Laboratory, 29
Neel, James V., 92
Nehru, Jawaharlal, 2, 81–82
Newell, Robert R., 65
New Zealand, 180, 196
Nixon, Richard, 164
Noble, Allan, 106

294 INDEX

Nolan, James F., 20
North Dakota, 140
North Korea, 202n4; underground testing by, 196
Norway: and nuclear test ban debate, 103–4, 107; snow contamination in, 103; vs. Soviet Union, 170
nuclear arms control and disarmament, 1, 2, 6–7, 83, 87, 131, 132, 133, 134, 155, 156, 158, 165, 167, 178, 180, 181, 183, 184, 186, 188, 190, 192; as comprehensive, 1, 78, 81, 101, 121–22, 158, 165; movements, 2, 52–55, 60, 80–81, 102–3, 111, 126–28, 139, 141–42, 143, 145, 146, 176–80, 187; in the United Nations, 101, 103, 105, 107, 122, 128, 181; verification of, 2, 81, 131, 132, 133, 138, 155–56, 157, 158, 159, 162, 165, 181, 183–84, 185, 186, 188, 195
nuclear proliferation, 2, 131, 164, 171, 181, 183, 186, 188, 196
nuclear test ban: as arms control, 1–2, 78, 81–82, 101, 121, 134, 156, 165, 181, 183, 184, 185, 186, 188, 190; atmospheric vs. comprehensive, 2, 15, 132, 138, 158, 159, 160–61, 163, 164–65, 168, 180, 181–82, 183–84, 185, 188; environmental dimension of, 2–4, 6, 14, 15, 90, 91, 102, 108, 133–34, 138, 156–57, 158–61, 163, 168, 185–87, 188–89, 190, 193–95; debate in India, 2, 78, 81–82, 85, 121, 130, 132; debate in Japan, 42, 52, 53, 54–55, 119–20, 121, 132, 134, 192, 196; debate in Soviet Union, 12, 78, 85, 101, 121–22, 123, 125, 127, 128, 129, 130, 131, 132, 155, 157–58, 165, 166–67, 181–86, 191, 192, 193–94; debate in United Kingdom, 81–82, 102, 107, 131, 132, 146, 156–57, 158–59, 161, 164–65, 168, 171, 180, 181, 183, 184, 185, 194; debate in United States, 81–82, 105, 106, 107, 114, 128, 131–34, 149, 152, 155–56, 157, 158–61, 162, 163, 164–65, 168, 180, 181, 183, 184, 185, 186–88, 194, 195; movements for, 126–28, 141–42, 143, 146, 149, 152, 163, 176–80, 187, 192, 193, 194; negotiations, 2, 4, 6, 15, 101, 102, 138, 150, 155, 156–57, 158–59, 160, 161, 162, 164–65, 178, 180, 181, 185, 194, 195; opponents of, 81–82, 105, 108, 114–15, 133, 134, 138, 141, 155–56, 160, 187, 188, 195; proposed by Stevenson, 101, 105–6, 141; relationship to the BEAR/MRC reports, 90, 91, 101–108; relationship to diet surveys, 137, 138, 149, 155, 156–57, 159, 160–61; relationship to UNSCEAR, 85, 87, 103, 107, 110, 111, 114, 121, 130–34, 194. *See also* Partial Test Ban Treaty
nuclear test limitation, 80–81, 104, 108, 132; and Canada, 103–4, 107; and Japan, 104, 107; and Norway, 103–4, 107; and United Kingdom, 102, 106–7, 131, 133, 194; and United States, 105, 106–7, 195
nuclear test moratorium, 6, 15, 83, 85, 110, 111, 130–31, 132–34, 137, 148, 155, 158, 160, 161, 165, 176, 181–82, 185, 193, 194; end of, 148, 158, 161, 162–63,

165–68, 169, 170–71, 172, 174, 178, 180, 182
nuclear warfare, 1, 6, 18, 26–27, 28, 29, 30–31, 40, 80, 129, 168, 183, 191; Eisenhower's attitudes regarding, 151, 152, 156
Nuzhdin, Nikolai I., 78, 79

Oparin, Aleksandr I., 71, 77–78, 127
Operation Castle, 34, 42, 48. *See also* Castle Bravo test
Operation Crossroads, 44–45
Operation Hurricane, 36, 37, 38
Operation Redwing, 104
Operation Sandstone, 22
Operation Totem, 36, 37–38, 39
Operation Troll, 50
Oppenheimer, J. Robert, 141
Orbeli, Leon A., 71, 123
Ormsby-Gore, David, 168

Pakistan: underground testing by, 196, 202n4
Paris Summit of 1960, 159
Partial Test Ban Treaty (PTBT): as arms control, 1–2, 186, 188, 190; environmental dimension of, 2–4, 6, 14, 15, 163, 185, 188–89, 190, 194, 196; refusal to sign, 2, 188, 196; relationship to underground testing, 2, 4, 186, 188, 195; Senate ratification, 186–88; signing of, 1–2, 186
Pauling, Ava Helen, 126
Pauling, Linus, 126, 127, 128, 141–42, 144, 194
Pearson, Paul B., 57–58
Pendleton, Robert, 175
Penny, William, 35, 36, 37, 38
permissible dose: and ICRP, 9, 32, 76, 93–94, 96, 98, 103, 119, 120–21, 172; in Japan, 14, 47, 50, 51–52, 55–56, 57–59, 60, 118, 120–21, 191, 192; and NCRP, 9, 25, 51–52, 55, 56, 58, 59, 64, 67–68, 76, 91, 93, 96, 98, 103, 136, 140, 172; in Soviet Union, 9–10, 32; in United Kingdom, 9–10, 37–38, 65–66, 67, 90, 94–95, 98–99, 100–101, 105, 173, 174, 192; in United States, 9–10, 14, 20, 22–23, 25–26, 40, 46–47, 55, 57, 60, 64, 90, 91, 92–93, 94, 95–97, 98, 99, 105, 136, 138–39, 140, 172–73, 174–76, 191, 192; UNSCEAR on, 110, 197–98. *See also* maximum permissible body burden; radiation protection; tolerance dose
Phillips, T. Alun, 147–48
Plowden, Edwin, 82
Pochin, Edward E., 112, 114
politics of risk, 6–14, 15, 86, 110, 127, 134–35, 163, 168, 180, 190; defined, 6–8; legacies of, 195–99; local turn in, 135, 137–38, 148, 160–61, 193; and nuclear test ban debate, 13–14, 193–95; and relations of definition, 8, 10, 12, 13, 14, 191–92; and reports of MRC and NAS/BEAR, 107–8
Potsdam Conference, 31
Powell, Richard, 82
Project Gabriel, 27, 28–29
Project Sunshine, 26–31, 40, 97, 111, 112–13, 114–17, 142, 144, 149
Pugwash conference, 127

Quirk, R. N., 145

race: in Asia, 13, 121, 134, 192; and nuclear paternalism, 29
radiation protection, 8–9, 11, 60, 123, 172–74; Advisory Committee on X-ray and Radium Protection (ACXRP), 9, 20, 25, 64, 65; in

Canada, 67, 114, 150, 168–69; International Commission on Radiological Protection (ICRP), 9, 32, 68, 76, 92, 93–94, 95, 96, 98, 103, 119, 120–21, 124, 172; International X-ray and Radium Protection Committee (IXRPC), 9; in Japan, 12, 14, 41, 42, 43, 46–50, 51–52, 55–60, 117–19, 120–21; and Lysenkoism, 77, 79, 122; Muller on, 64–65, 68, 93; National Committee on Radiation Protection (NCRP), 9, 25, 51–52, 55, 56, 58, 59, 64, 65, 66, 67–68, 76, 91, 93, 94, 96, 98, 103, 124, 136, 140, 172; in Norway, 103; in Soviet Union, 9–10, 14, 17–18, 31–35, 39, 40, 62–63, 72, 77–78, 79, 122–24, 129–30, 167, 191; in Sweden, 114; in United Kingdom, 4, 9–10, 14, 17–18, 35–39, 40, 62–63, 64, 65–67, 68, 81, 82, 84, 86, 87, 89–90, 94–95, 96, 98–101, 103, 105, 108, 111–12, 113, 114, 123, 135, 136–37, 144, 145–48, 160, 163, 172–74, 191, 192; in United States, 4, 9–10, 14, 17–21, 22–23, 24–26, 31, 40, 43, 46–47, 48, 55–58, 59, 60, 61, 62–63, 64–66, 67–68, 73–77, 82, 83–84, 86, 87, 89–90, 91–93, 94, 95–99, 100–101, 103, 105, 106, 107, 108, 111–13, 114–17, 118, 123, 128, 129–30, 135, 136–37, 138–39, 140, 141, 142–44, 148–49, 150–54, 159, 160, 163, 172–76, 191, 192. *See also* maximum permissible body burden; permissible dose; tolerance dose

radioactive fallout: acute radiation syndrome, 4, 14, 20–21, 23, 32–33, 34–35, 38, 39, 42, 51, 63, 167; genetic damage from, 9, 12, 13, 14–15, 25, 56, 58, 60, 62–64, 68, 73–77, 78, 80, 87, 89–90, 91–96, 99, 108, 122–23, 125, 126, 127, 128, 191, 198; global dispersion of, 2–3, 5–6, 7, 10, 11, 14, 18, 21–22, 28–31, 40, 43, 61–63, 74, 75, 85, 86, 95, 96–97, 102, 108, 109–17, 126, 127, 128–29, 130, 131, 133–34, 137, 152–53, 169–70, 186, 190–91, 193, 197, 198; "hot spots", 7, 13, 15, 110, 114, 137, 138–39, 144–48, 149, 160, 172, 173, 193; intelligence value of, 17–18, 21–22, 33–34, 38–40, 111, 191, 208n16; neurological effects of, 123–24; risks of cancer from, 14, 27, 89–90, 96–99, 100, 101, 102, 108, 110, 123, 127, 129–30, 175, 192; from Trinity test, 16, 18–22. *See also* atmospheric tests; strontium-90

radiocarbon dating, 27, 28
rainwater contamination, 50–52, 55
RAND Corporation, 27, 28
RDS-1 test, 32–33
RDS-6 test, 34–35
relations of definition, 8–9, 10, 12, 13, 14, 40, 64, 86, 191–92
rice contamination, 13, 51, 55, 110, 120–21, 134, 149
Roberts, Frank, 183
Rockefeller Foundation, 30, 82
Roosevelt, Franklin D., 19, 139–40
Rotblat, Joseph, 177
Rural Electrification Administration, 139–40
Rusk, Dean, 1, 165–66, 168, 171, 181, 185
Russell, Bertrand, 80, 84, 127, 180
Russell, Liane, 66
Russell, Richard, 184, 187
Russell, William L., 66, 92–93
Russell-Einstein Manifesto, 84, 127

Sakharov, Andrei D., 33, 34–35, 168; relationship with Khrushchev, 129, 167, 171, 182–83; views on innocent bystanders, 128–29, 163, 167, 194
SANE. *See* National Committee for a Sane Nuclear Policy
San Francisco Peace Treaty, 44
Schweitzer, Albert, 76, 126, 127, 128, 178, 194
Science Council of Japan (SCJ), 56, 57
Seaborg, Glenn, 187
Selove, Walter, 106
Shipman, Thomas L., 23
Shot Easy test, 25
Shot Harry test, 25
Shot Sedan test, 175
Shunkotsu Maru, 48–49
Sievert, Rolf, 114
Sisakian, Norair M., 127
Slavskii, Efim P., 167, 182, 183
Smith, Nicholas M., Jr., 27
snow contamination, 25, 103
Snyder, Laurence H., 106
South Africa, 202n4
Soviet Union: Academy of Medical Sciences, 79; Academy of Sciences, 33, 70, 71, 72, 77, 78, 125, 127–28, 129; Agricultural Ministry, 33; Defense Ministry/Main Intelligence Directorate, 33; genetics and geneticists in, 10–11, 60, 64, 68, 69–73, 77–80, 124–25, 128, 129, 166, 191, 192; Institute of Genetics, 69, 72, 78; relations with Japan, 52; Lenin All-Union Academy of Agricultural Sciences, 69, 70–71, 72; Novaya Zemlya test site, 18, 167, 170, 175; and nuclear test ban debate, 12, 78, 85, 101, 121–22, 123, 125, 127, 128, 129, 130, 131, 132, 155, 157–58, 165, 166–67, 181–86, 191, 192, 193–94; nuclear testing in the atmosphere by, 2, 15, 17, 21, 23, 31, 32–33, 34–35, 52, 61, 75, 83, 103, 113, 122, 130–31, 133, 136, 157–58, 161, 162–63, 166–70, 174–75, 182, 184, 190–91, 194; permissible dose in, 9–10, 32; radiation protection in, 9–10, 14, 17–18, 31–35, 39, 40, 62–63, 72, 77–78, 79, 122–24, 129–30, 167, 191; Radium Institute, 33; Semipalatinsk test site, 18, 31, 32, 34, 167, 168; State Meteorological Service, 33; vs. UK, 10–11, 12, 14, 15, 36, 37, 78–79, 84–85, 113, 122, 123, 131, 132, 168, 181–82, 183–86, 188–89; underground nuclear testing by, 2, 155–56, 157–58, 167–68, 182, 188, 195; and UNSCEAR, 15, 110, 111, 113, 122–24, 129–31, 134, 192–93; vs. US, 10–11, 12, 14, 15, 19, 21, 22, 23, 26–27, 28, 31–34, 73, 77, 78–79, 84–85, 113, 121–23, 128, 129–30, 131, 132, 133, 157–58, 159, 162–63, 164, 165, 166, 168, 169–70, 181–82, 183–86, 188–89. *See also* Cold War; Cuban Missile Crisis; Gromyko, Andrei; Khrushchev, Nikita; Stalin, Joseph
Spock, Benjamin, 178, 179, 187
Sputnik, 130, 132, 164
Sr-90. *See* strontium-90
Stalin, Joseph, 31–32, 33, 70, 122, 125
Starners, Richard, 176
Stassen, Harold A., 105, 106
Stern, Curt, 65, 68, 76–77, 93
Stevenson, Adlai, 101, 105–6, 126, 141
St. George, Utah, fallout in, 25
Stimson, Henry L., 19
Strauss, Lewis L., 61, 62, 73–74, 83,

104, 107, 132, 195. *See also* U.S. Atomic Energy Commission
strontium-89, 172–73
strontium-90, 26, 28, 56, 104, 105, 111, 113, 114, 115, 116–17, 128, 138, 140, 150, 156, 172–73, 198; as carcinogenic, 14–15, 27, 55, 96–99, 101, 102, 129–30, 192; maximum permissible body burden (MPB), 89–90, 96–99, 100, 105, 121, 138–39; in milk, 110, 120–21, 139, 140, 142–43, 147, 151–52; in rice, 110, 120–21, 149; in soil, 100–101, 110, 114, 116, 138; tuna, 58–59; in Wales, 101, 144, 146–47; in wheat, 148–49, 150–51, 159. *See also* clean bomb; Project Sunshine
Sturtevant, Alfred H., 61–63, 68, 73–75, 96
Suez Crisis of 1956, 107
Sugawara, Tomiko, 54
Summerskill, Edith, 81
Sweden, 82–83, 85, 86, 114, 130, 170, 181
Symington, Stuart, 184

Teller, Edward, 104, 131, 133, 187, 188, 195; and the clean bomb, 104–5, 167, 195; and underground testing, 155–56, 158
Thompson, Llewellyn E., 182
Timofeev-Ressovsky, Nikolai V., 69, 72, 124; "On the Nature of Gene Mutation and Gene Structure," 63
tolerance dose, 9, 66; and ACXRP, 9, 20, 25, 65. *See also* maximum permissible body burden; permissible dose
Tompkins, Paul, 175–76
Tremblay, Paul, 169
Trinity test, 16, 18–22, 23, 35
Truman, Harry S., 19, 22, 23, 31, 33

"Tsar Bomba" test, 167, 170, 174–75
Tsuzuki, Masao, 118, 119, 121
tuna contamination, 14, 41–43, 45, 46–50, 51, 52–54, 55–59, 119, 191
Turkevich, Anthony L., 21
thyroid cancer, 175

Undén, Östen, 83
underground tests: detection of, 155–56, 158, 160, 162, 165, 183–84, 185, 195; by India, 196, 202n4; by North Korea, 196, 202n4; number of, 3; by Pakistan, 196, 202n4; relationship to PTBT, 2, 4, 186, 188, 195, 197; by Soviet Union, 2, 155, 157–58, 167–68, 182, 183–84, 188, 195; by United Kingdom, 165–66; by United States, 2, 133, 138, 155, 157–58, 160, 163, 164, 165–66, 170, 188, 195; yield of, 3
United Kingdom: Atomic Energy Authority (UKAEA), 145, 146, 147–48; Atomic Energy Research Establishment (AERE), 35, 38, 66, 67, 94, 99, 100, 111, 144; relations with Australia, 37, 38; Christmas Island test site, 18; Committee of 100, 180; Conservative Party, 81; Emu test site, 18, 36–37, 38, 39; genetics and geneticists in, 10, 64, 65–67, 87, 89–90, 94–96, 123; Harwell facilities, *see* United Kingdom: Atomic Energy Research Establishment; Institute of Animal Genetics, 66; House of Commons, 81, 88, 94, 102, 144, 145; Labour Party, 81, 144, 145, 156; London Metropolitan Water Board (MWB), 66–67; Maralinga test site, 18, 36–37; Ministry of

Agriculture, Fisheries and Food (MAFF), 38, 145, 174; Ministry of Health, 174; Monte Bello Islands test site, 17, 18, 36–37, 38; National Council for the Abolition of Nuclear Weapon Tests (NCANWT), 102–3; National Peace Council, 102; and nuclear test ban debate, 81–82, 102, 106–7, 131, 132, 133, 146, 156–57, 158–59, 161, 164–65, 168, 171, 180, 181, 183, 184, 185, 194; nuclear testing in the atmosphere by, 2, 17, 36, 38, 39, 83, 90, 102, 131, 134–35, 144, 156–57, 158–59, 161, 168, 171, 190–91, 194; permissible dose in, 9–10, 37–38, 65–66, 67, 90, 94–95, 98–99, 100–101, 105, 173, 174, 192; radiation protection in, 4, 9–10, 14, 17–18, 35–39, 40, 62–63, 64, 65–67, 68, 81, 82, 84, 86, 87, 89–90, 94–95, 96, 98–101, 103, 105, 108, 111–12, 113, 114, 123, 135, 136–37, 144, 145–48, 160, 163, 172–74, 191, 192; Royal Society, 82; vs. Soviet Union, 10–11, 12, 14, 15, 17, 36, 37, 39–40, 43, 62–63, 78–79, 84–85, 113, 122, 123, 131, 132, 133, 168, 181–82, 183–86, 188–89; underground tests by, 165–66; relations with United States, 12–13, 15, 35–36, 38, 89, 94–95, 103, 106–7, 110–11, 113, 131, 132, 133–34, 156–57, 158, 159, 161, 164–66, 168, 171, 181–82, 184; vs. United States, 14–15, 36, 37, 89–90, 94–95, 98, 99, 100–101, 102, 107, 108, 113–14, 145–46, 148, 156, 161, 192; and UNSCEAR, 15, 85–86, 89, 109, 111–17, 121, 122–23, 129, 130, 131–32, 133, 134, 137; Wales, 15,
38, 101, 114, 137, 144, 144–45, 145, 146–47, 148, 156, 160, 161, 193. *See also* Attlee, Clement; Churchill, Winston; Eden, Anthony; Macmillan, Harold; Medical Research Council
United Nations, 1, 44, 64, 80, 81, 82, 83, 84–86, 87, 103, 104, 107, 110, 113, 126, 131, 132, 150, 165, 169, 170, 181, 192; Conference on the Human Environment, 4, 196; Subcommittee on Disarmament, 101, 103, 105, 107, 122, 128; Treaty on the Prohibition of Nuclear Weapons, 199. *See also* Hammerskjöld, Dag; U Thant
United Nations Scientific Committee on the Effects of Atomic Radiation (UNSCEAR), 13, 14, 108; and Belgium, 85, 130; and Canada, 85, 104, 114, 150, 169; and China, 85–86; and India, 85–86, 121, 130; and Japan, 15, 85, 104, 109, 111, 117–21, 192; mission, 109; and Norway, 103, 107; relationship to nuclear test ban debate, 85, 87, 103, 107, 110, 111, 114, 121, 130–34, 194; report of, 109–10, 111, 129–130, 150, 192–93, 194, 195, 197; and Soviet Union, 15, 85, 109, 111, 113, 121–31, 133, 134, 137, 192; and Sweden, 85–86, 114; and United Kingdom, 15, 85–86, 89, 109, 111–17, 121, 122–23, 129, 130, 131–32, 133, 134, 137; and United States, 15, 85–86, 89, 109, 111–17, 118–19, 121, 122–23, 129–30, 131, 132, 133, 134, 137
United States: Atomic Energy Act, 35, 132, 140; Bikini Atoll test site, 18, 43, 45, 48, 50, 51, 80, 118;

Christmas Island test site, 18; Congress, 35, 36, 132, 149, 151, 157, 172, 183, 184, 186, 187–88; congressional hearing, 115, 138, 152, 153, 175; Enewetak test site, 18, 22, 43, 45–46; Federal Radiation Council (FRC), 154, 172–73, 175–76; Food and Drug Administration (FDA), 46–47, 48, 55; genetics and geneticists in, 10, 60, 61–63, 64–66, 67–68, 69, 73–77, 78–79, 86–87, 89, 90, 91–94, 95–96, 122–23, 191; relations with Japan, 12, 14, 15, 19, 43, 44, 46–47, 48, 52–53, 55–56, 57–58, 59–60, 118, 119, 120, 121, 134, 191; Johnston Island test site, 18; National Academy of Sciences (NAS), 12, 14–15, 64, 82, 84, 87, 88–89, 101, 102, 108, 111–12, 113–14, 139, 192; Nevada Test Site, 18, 22–23, 24–25, 26, 37, 40, 85, 175; New Deal, 139–40; and nuclear test ban debate, 81–82, 105, 106–7, 114, 128, 131–34, 149, 152, 155–56, 157, 158–61, 162, 163, 164–65, 168, 180, 181, 183, 184, 185, 186–88, 194, 195; nuclear testing in the atmosphere by, 2, 12, 14, 17, 22–23, 24–25, 34, 35, 37, 41–42, 45, 61, 73, 75, 83, 90, 92, 103, 106, 107, 117, 118–19, 131, 132, 134–35, 138, 154, 156–57, 158–61, 163, 168, 169, 170–71, 174, 175, 181, 182, 185, 187, 190–91, 193, 194; permissible dose in, 9–10, 14, 20, 22–23, 25–26, 40, 46–47, 55, 57, 60, 64, 90, 91, 92–93, 94, 95–97, 98, 99, 105, 136, 138–39, 140, 172–73, 174–76, 191, 192; Public Health Service (USPHS), 142–43, 151, 152, 154, 174–75; radiation protection in, 4, 9–10, 14, 17–21, 22–23, 24–26, 31, 40, 43, 46–47, 48, 55–58, 59, 60, 61, 62–63, 64–66, 67–68, 73–77, 82, 83–84, 86, 87, 89–90, 91–93, 94, 95–99, 100–101, 103, 105, 106, 107, 108, 111–13, 114–17, 118, 123, 128, 129–30, 135, 136–37, 138–39, 140, 141, 142–44, 148–49, 150–54, 159, 160, 163, 172–76, 191, 192; vs. Soviet Union, 10–11, 12, 14, 15, 19, 21, 22, 23, 26–27, 28, 31–34, 73, 77, 78–79, 84–85, 113, 121–23, 128, 129–30, 131, 132, 133, 157–58, 159, 162–63, 164, 165, 166, 168, 169–70, 181–82, 183–86, 188–89; Tennessee Valley Authority, 139–40; Trinity test site, 16, 18, 19, 20–21; underground tests by, 2, 133, 138, 155, 157–58, 160, 163, 164, 165–66, 170, 188, 195; relations with United Kingdom, 12–13, 15, 35–36, 38, 89, 94–95, 103, 106–7, 110–11, 113, 131, 132, 133–34, 156–57, 158, 159, 161, 164–66, 168, 171, 181–82, 184; vs. United Kingdom, 14–15, 36, 37, 89–90, 94–95, 98, 99, 100–101, 102, 107, 108, 113–14, 145–46, 148, 156, 161, 192; and UNSCEAR, 15, 85–86, 89, 109, 111–17, 118–19, 121, 122–23, 129–30, 131, 132, 133, 134, 137; use of defoliants in Vietnam War, 5. *See also* Biological Effects of Atomic Radiation (BEAR), Committees on; Cold War; Cuban Missile Crisis; Eisenhower, Dwight D.; Kennedy, John F.; National Committee on Radiation Protection; Truman, Harry S.

UNSCEAR. *See* United Nations Scientific Committee on the Effects of Atomic Radiation
U.S. Atomic Energy Commission (USAEC), 22; General Advisory Committee (GAC), 27, 133; and genetics, 25, 61, 65–66, 67, 68, 73–74, 75, 91, 95; Health and Safety Laboratory (HASL), 23–24, 40, 46, 112, 138–39, 141, 142; permissible dose, 25–26, 46–47, 51–52, 55, 56, 57, 67; Project Gabriel, 27, 28–29; Project Sunshine, 26–31, 40, 97, 111, 112–13, 114–17, 142, 144, 149; as source of information, 25, 47, 50, 61, 73–74, 80, 82, 95, 97, 103, 104, 105, 107, 138–39, 140–41, 142, 144, 145–46, 148, 149, 150–51, 152, 153–54, 159, 160, 195; and tuna contamination, 45–47, 55, 56, 57–58, 59. *See also* Bugher, John C.; Eisenbud, Merril; Libby, Willard F.; McCone, John A.; Seaborg, Glen; Strauss, Lewis L.; Teller, Edward; Warren, Shields
U.S.-Japan Radiobiological Conference, 57–58
U Thant, 1

Vasilevskii, Aleksandr, 34
Vavilov, Nikolai I., 69, 70
verification, 2, 81, 131, 132, 133, 138, 155–56, 157, 158, 159, 162, 165, 181, 183–84, 185, 186, 188, 195
Vienna Summit of 1961, 165, 166
Vietnam War defoliants, 5
Vinogradov, Aleksandr P., 125
Visscher, Maurice B., 140–42, 148–49, 151

Wadsworth, James J., 155
Wales: fallout surveys in, 15, 38, 101, 114, 137, 144, 146–47, 148, 156, 160, 161; political conditions in, 145, 193
Warren, Shields, 67, 112
Warren, Stafford, 19
Watkinson, Ernest, 114
Watson, James, 124
Weaver, Warren, 91, 92, 95–96
Welander, Arthur D., 57
wheat contamination, 148–51, 156, 159, 160, 169
Wiesner, Jerome, 165
Wilson, Dagmar, 176
Winterbottom, Ian, 88
Wisconsin, 140
women, 13, 99, 194: and bone cancer, 98–99; in Canada, 248n68; in Japan, 42, 53–54, 191; in Sweden, 83; in United Kingdom, 81, 103, 163, 177, 180; in United States, 26, 163, 176–177, 180, 187
Women's Association for Radiation Information (WARI), 177, 180
Women Strike for Peace (WSP), 176–77, 180
World Conference against Atomic and Hydrogen Bombs, 55

X-ray film, 16, 23, 40

Yasui, Kaoru, 54

Zavenyagin, Avraamii P., 72
Zhdanov, Andrei A., 70
Zhukov, Georgy, 130
Zimmer, Karl G., "On the Nature of Gene Mutation and Gene Structure," 63
Zinc-65, 58
Zorin, Valerian A., 122, 128

The authorized representative in the EU for product safety and compliance is:
Mare Nostrum Group
B.V Doelen 72
4831 GR Breda
The Netherlands

www.ingramcontent.com/pod-product-compliance
Lightning Source LLC
Chambersburg PA
CBHW031901220426
43663CB00006B/711